HVAC Simplified

This publication was prepared in cooperation with TC 7.1, Integrated Building Design.

About the Author

Stephen P. Kavanaugh, PhD, Fellow ASHRAE, has been a professor of mechanical engineering at The University of Alabama since 1985, where he teaches HVAC and is faculty advisor for the ASHRAE Student Chapter as well as a Habitat for Humanity Student Affiliate.

Kavanaugh is co-author of *Ground-Source Heat Pumps—Design of Geothermal Systems for Commercial and Institutional Buildings*, published by ASHRAE in 1997. He has presented over 100 engineering seminars for more than 2,500 designers on the topics of energy efficiency, ground-source heat pumps, and HVAC. He maintains the Web site www.geokiss.com, where there is more information about HVAC and ground-source heat pump design tools.

He is past chair and current program chair of ASHRAE Technical Committee 6.8, Geothermal Energy, as well as past chair of ASHRAE Technical Committee 9.4, Applied Heat Pumps and Heat Recovery.

Kavanaugh is also a Fellow of the American Society of Mechanical Engineers and a board member and past president of Habitat for Humanity—Tuscaloosa.

Updates and errata to this publication will be posted on the ASHRAE website at www.ashrae.org/publicationupdates.

Errata noted in the list dated March 8, 2016, have been corrected.

HVAC Simplified

Stephen P. Kavanaugh

Atlanta

ISBN 10: 1-931862-97-4
ISBN 13: 978-1-931862-97-4

1791 Tullie Circle, NE
Atlanta, GA 30329
www.ashrae.org

Printed in the United States of America

Library of Congress Cataloging-in-Publication Data

Kavanaugh, Stephen P.
HVAC simplified / Stephen P. Kavanaugh.
p. cm.
ISBN 1-931862-97-4 (softcover)
1. Heating. 2. Ventilation. 3. Air conditioning. I. Title.
TH7011.K38 2005
697--dc22

2005037135

ASHRAE Staff

Contents

Preface

The text contains twelve chapters, including three short chapters that review the fundamentals of refrigeration, heat transfer, and psychrometrics. Information from the *ASHRAE Handbook—Fundamentals* is summarized and supplemented with items from industry sources. The remaining chapters assemble information from ASHRAE handbooks, standards, manufacturer data, and design procedures commonly used by professional engineers. Topics include equipment selection and specification, comfort and IAQ, building assemblies, heating and cooling loads, air distribution system design, water distribution system design, electrical and control systems, design for energy efficiency, and design for economic value.

A suite of complementary spreadsheet programs that incorporate design and computation procedures from the text are provided on the CD that accompanies this book. These programs include psychrometric analysis, equipment selection, heating and cooling load calculation, an electronic "ductulator," piping system design, a ductwork cost calculator, and programs to evaluate building system demand and energy efficiency. Future updates to these programs can be found at www.ashrae.org/publicationupdates.

TARGET AUDIENCES

- Novice HVAC engineers that did not complete an HVAC course during their undergraduate program
- Engineers moving into the HVAC field from other areas
- Undergraduate engineering students in design-oriented curriculums
- Non-engineering professionals with strong building-related experience

DISTINGUISHING FEATURES

The text places an emphasis on design and the energy and economic evaluation of results.

A large amount of information is provided so that instructors and students do not have to frequently consult supplemental documents in order to complete HVAC system design. For example, one section provides an overview of the revised 2004 ASHRAE Standards 62.1 and 62.2 ventilation air system design mandates, a feature not even included in the ASHRAE Handbooks. Also, a large amount of manufacturer performance data is provided with correction methods when conditions vary from rated values. Performance data are provided for packaged ACs, RTUs, air and water heat pumps, furnaces, chillers (six product lines), compressors, cooling towers, hydronic coils, pumps, fans, electric motors, lighting products, and other equipment.

The text concludes with a chapter on design evaluation methods. Sections in this chapter include HVAC system demand computations, a discussion of simplified energy analysis, operating costs (including tables of maintenance, repair, and replacement costs), a presentation of methods to compute system installation costs (with references to more detailed sources), and economic analysis methods.

Preface

Author's Note

This text is an attempt to provide an understanding of fundamental HVAC concepts and to extend these principles to the explanation of simple design tools used to create building systems that are efficient and provide comfortable and healthy environments. My decision to include the term "Simplified" in the title is rooted in my own preference for simple and creative systems. However, this philosophy was perhaps better expressed by ASHRAE Presidential member Bill Coad when he quoted Albert Einstein: "Everything should be as simple as possible, but no simpler."[1]

Certainly, there are procedures and tools that provide higher levels of accuracy and detail than those I have included in this text. There are also some that are more straightforward and less complicated. However, I have attempted to provide a middle ground with regard to the level of difficulty by avoiding oversimplifications that are not suited to design and more complex methods that are more suited to analytical research.

More importantly, I encourage readers to extend this philosophy of simplification to building designs and HVAC systems. In some applications, complexity cannot be avoided. However, options also exist to extend complex technologies to modest buildings in which the added benefit is not economically justifiable for a client with limited financial and staffing resources. The engineer must apply the pearl of wisdom of Einstein to best meet the needs of 21st Century.

Steve Kavanaugh
December 2005

1. Coad, W.J. 1996. Indoor Air Quality: A Design Parameter. *ASHRAE Journal* 38(6):39–47.

Acknowledgments

I wish to thank the cadre of volunteer reviewers that were assembled by the cognizant ASHRAE technical committee, Integrated Building Design (TC 7.1), which provided a wealth of corrections and suggestions on a very compacted time schedule. Core committee members include Charles E. Gulledge III (Chair), Bill Williams, Dennis Knight, Peter Gryc, Bert Phillips, Robert McDowell, Jesse Sandhu, Stirling Walkes, and Chris Ott.

Additionally, the members of the ASHRAE Publications Committee and staff have provided encouragement and support for this effort. I also wish to thank my employer, The University of Alabama, for providing a sabbatical for the Spring 2004 semester, during which I completed the first five chapters of this work. I also wish to thank the many students; my mentors Drs. Jerald Parker, Harry Mei, and Gene Martinez; and my ASHRAE colleagues, who have encouraged me to do a better job of presenting the ideas and concepts that have helped me learn. Finally, I wish to express my special gratitude to Barbara Hattemer McCrary, my former student who became my teacher in the use of spreadsheets as an engineering tool, and Kevin Rafferty, who reviewed the entire book and has provided a wealth of technical and philosophical insight throughout our many years of collaboration.

Acknowledgments

1 Introduction to HVAC

The challenges presented to engineers in the heating, ventilating, and air-conditioning (HVAC) industry have much in common with those in other technical fields. HVAC engineers face constraints imposed by rapidly changing technologies, increased liability, a heightened emphasis on short-term profitability, complexity in codes and standards, and an overabundance of information and misinformation. However, the public nature of the HVAC industry also presents many opportunities for the engineer to be creative and influential.

The successful HVAC engineer must be technically competent and conscientiously diligent in pursuit of optimum designs. However, in the 21st century, engineers must also be persuasive, to ensure their expertise is respected, passionate in their concern for the public health and environment, and open to all perspectives in pursuit of high quality and economically viable solutions. HVAC engineers encounter a wide variety of individuals ranging from owners of modest homes to employees of large industrial complexes. They must communicate and appreciate the needs of the broadest range of people because their clients include everyone who inhabits buildings.

In the real world there are many ethical conflicts, and although ethical lapses in legal, political, medical, and other professions are often more publicized, the HVAC industry is not without the potential for malpractice, conflict of interest, and dishonesty. Examples of inferior workmanship and unnecessarily inflated costs are evident in the industry. Rules, codes, standards, and guidelines evolve with input from those having self-serving agendas as well as those seeking the common good. Organizations such as the American Society of Heating, Refrigerating and Air-Conditioning Engineers (ASHRAE) and the National Society of Professional Engineers (NSPE) offer guidance to members for ethical behavior. A summary is offered here as a starting point for HVAC design.

ENGINEERS' CREED AND ASHRAE CODE OF ETHICS

An excerpt from the Engineers' Creed of the National Society of Professional Engineers (NSPE) calls on the engineer to:

> ... give the utmost performance;
>
> ... participate in none but honest enterprise;
>
> ... live and work according to the laws of man and the highest standards of professional conduct;
>
> ...place service before profit, the honor and standing of the profession before professional advantage, and the public welfare above all other considerations....

The ASHRAE Code of Ethics (ASHRAE 2004a) states:

> As members of a Society,
>
> "organized and operated for the exclusive purpose of advancing the arts and sciences of heating, refrigeration, air conditioning, and ventilation, the allied arts and sciences, and related human factors for the benefit of the general public" [ASHRAE 2004b],
>
> We recognize that honesty, fairness, courtesy, competence and integrity must characterize our conduct. With the foregoing in mind
>
> - Our efforts shall be directed at all times to the enhancement of the public health, safety and welfare.
> - Our services shall be offered only in areas of our competence.
> - Our products shall be offered only in areas of their suitability.
> - Our public statements shall be issued only in an objective and truthful manner.
> - Our endeavors shall carefully avoid conflicts of interest and the appearance of conflicts of interest.
> - The confidentiality of clients' and employers' business affairs, proprietary information, and procedures shall be respected.
>
> Approved by the Board of Directors
> June 22, 1986

THE BUILDING INDUSTRY ENVIRONMENT

The HVAC engineer operates in a world in which technology is rapidly changing and conventional roles, social behavior, working relationships, and partnerships are being challenged and reinvented. Figure 1.1 is an organizational chart of conventional working relationships for a large building project. The building owner's principal point of contact is with an architect. Architects are the lead designers and project managers. They will hire the engineers to design the building components in their areas of expertise. A mechanical design firm is typically responsible for the HVAC systems and interrelated plumbing systems (water, hot water, gas or oil piping, steam distribution, fire sprinklers, waste water, etc.). Design and construction teams often include many additional specialists not noted in Figure 1.1.

Design consists of

- communicating with the client(s) to determine needs and requirements,
- assessing the situation and performing computations to determine the most appropriate systems and options,
- taking steps to comply with building codes and regulations, and
- communicating design results and intent with drawings and specifications.

An important component in this process is to attempt to design the HVAC system within the preliminary budget allotment. Tools exist to pre-estimate building costs from historical data that can be adjusted for building type, location, and accessories (Means 2004). Although the architects are responsible for designing a building that is within budget, this may not be possible if the owners desire more than they can afford.

Two Design Processes

In the plan-bid process, the owner or developer hires the architect and the architect hires engineers to design the building and provide drawings and specifications for contractor and subcontractor bidding. Steps can be taken to prequalify suitable contractors and subcontractors, or the invitation to provide bids can be open to any firm. Qualified general contractors, in cooperation with subcontractors, review drawings and submit competitive bids. If the low bid is acceptable, building proceeds.

A building has a good chance of being a long-lasting, low-maintenance, quality structure, which is completed on time and within budget, *if*

1. the architect is a good project manager who designs buildings to meet the customer's needs (rather than making an architectural statement),

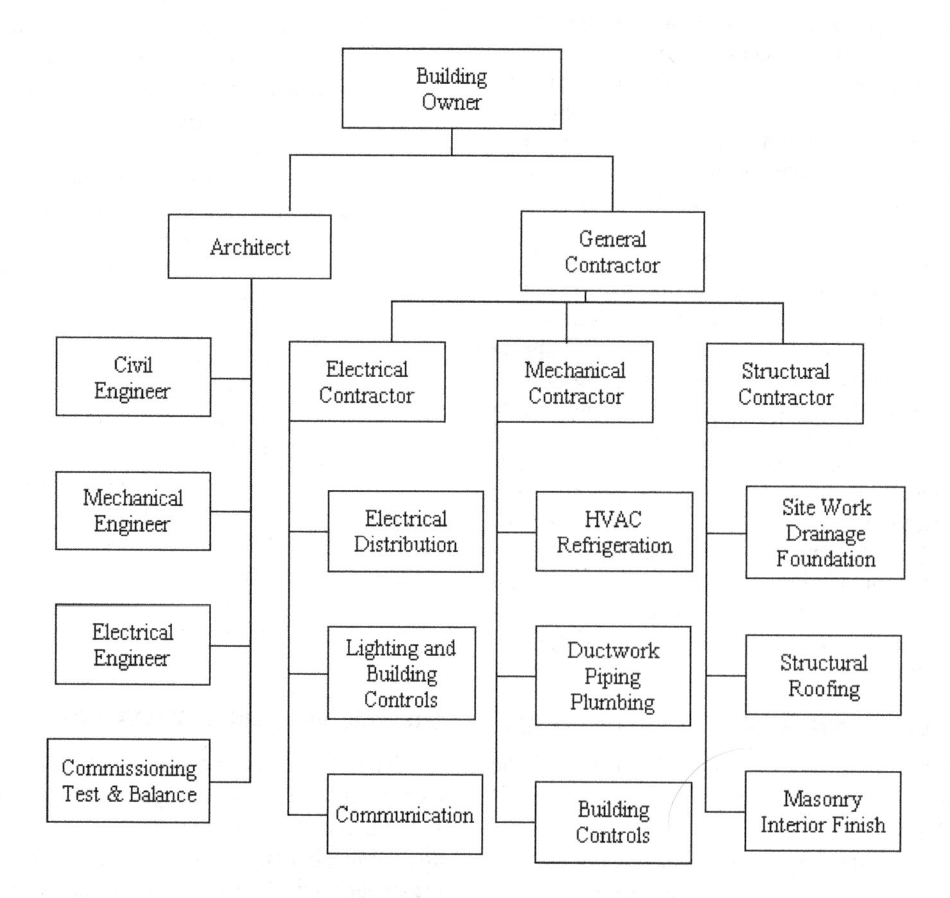

Figure 1.1 Plan-bid (a.k.a. owner-architect) building team structure.

2. who has hired competent engineers that provide thorough, code-compliant drawings and specifications in a timely manner,
3. so that experienced and qualified contractors can provide sound bids.

If the low bid is too high, the building is "value-engineered," or downsized, to meet the available budget. Great care and open communication are necessary to make adjustments that do not alter the building's intended functions and operating characteristics. Although the time schedule is often abbreviated, design drawings and specifications must be modified with thoughtful consideration being given to the impacts of lower-cost substitutions. The project is then re-bid and initiated. However, costs may skyrocket due to many noncompetitive bid "change orders" that are required because of the lack of time available to thoroughly redesign the building to meet the lower budget.

Figure 1.2 is the design-build alternative to the plan-bid method of designing and constructing buildings. In this situation, the owner hires a design-build contractor who employs either in-house architects and engineers or hires external firms. The project manager is the design-build general contractor and all the subcontractors work for the design-build firm. Subcontractors can also be outside firms.

This arrangement fosters better cooperation among participants, and the cost of change orders is the responsibility of the design-build firm. However, the level of checks and balances among architects, engineers, and contractors is reduced, which may leave the owner-customer more at risk. Savvy owners will likely hire a building professional to oversee the work of the design-build firm.

THE ROLE OF AN HVAC ENGINEER

Successful engineering firms consistently obtain projects, which ensures steady income during normal business cycles. These firms develop a positive reputation by establishing trust with architects, construction managers, and owners. Their reputation is enhanced by a record of promptness, dependability, and completing projects on time and within budget. HVAC engineers in such a firm do the following:

1. *Establish relationships with other design team members* to maximize cooperation and coordination of system integration and construction.

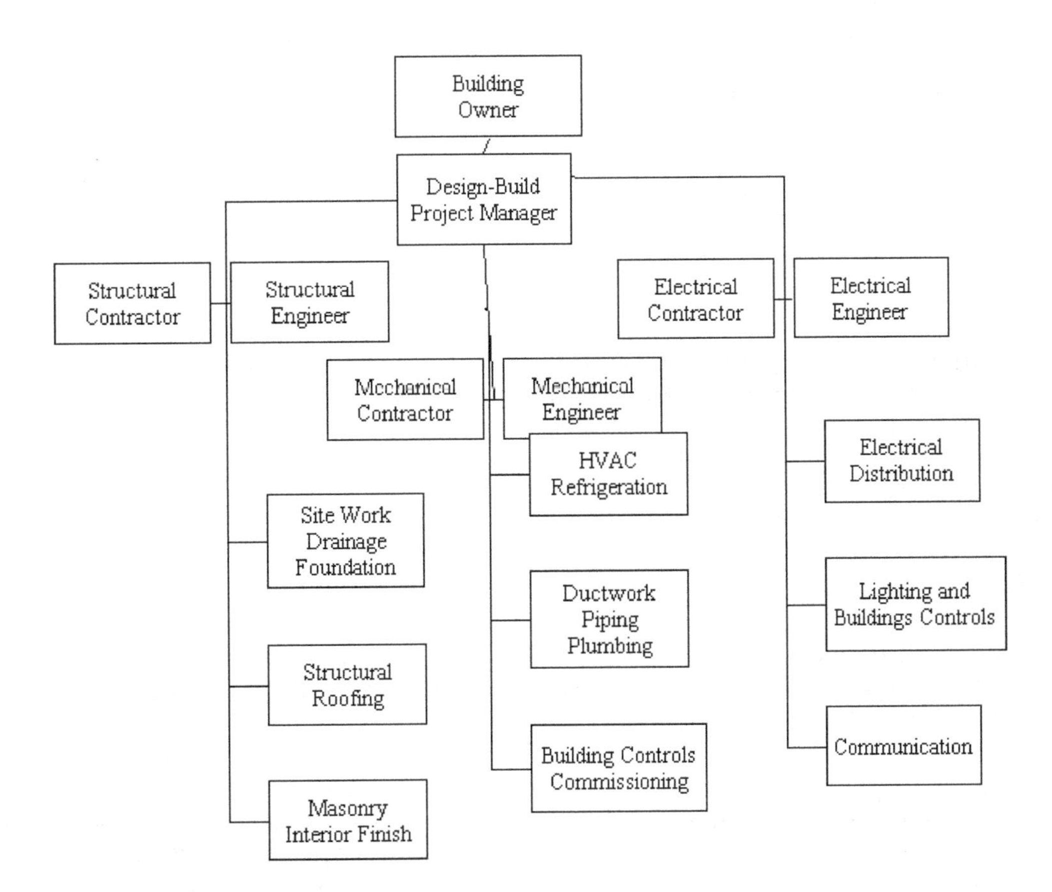

Figure 1.2 Design-build organization chart.

2. *Review the plans and divide the building into zones.* Large zones are typically less expensive (the thermostat is located in a central room; all other areas/rooms receive air, but comfort is dictated by conditions where the thermostat is located). Having many smaller zones is more expensive because more individual units (or terminals) and thermostats are required. Balancing is more difficult, but comfort in most areas is better since there is more control. How much can the client afford? (Simple example: Should you have one or two units for a two-story house?)
3. *Calculate required heat loss and gain for each zone.* A lot of information is necessary regarding thermal and moisture transmission properties of the structure, regional design criteria (temperature, humidity levels, latitude, elevations, wind speed, etc.), the number of people occupying the building and their schedules, desired indoor conditions to maximize comfort and productivity, the lighting loads, the type of heat/moisture-generating equipment in the building, the required amount of fresh air or infiltration, and the proposed location of the ductwork. In some cases this includes communicating with the architect or construction manager.
4. *Select the systems and related equipment to heat, ventilate, air-condition, and humidify or dehumidify each zone.* This often involves decisions to either (1) select higher-cost equipment that may use less energy and provide higher levels of comfort or (2) opt for less expensive equipment that is more economically and environmentally costly and may not provide good comfort or long service life.
5. *Select and locate the registers and diffusers to supply and return air to each zone.* A large number of low-velocity diffusers are quieter and generally provide better comfort, but they cost more. Having fewer registers requires high outlet velocities, which create more noise with less even air distribution. The fan energy requirement is also increased. Large return air registers are more expensive, but they are quieter and have low friction losses.
6. *Route and size ductwork to connect equipment to diffusers and registers.* Large ducts cost more and are more difficult to route and install, but they generate less noise (lower velocity) and require less fan energy with resulting heat generation.
7. *Select and size fans.*
8. *Route and size piping for water-based systems.* Large pipes cost more and are more difficult to route and install, but they generate less noise (lower velocity) and require less pump energy.
9. *Select and size pumps.*
10. *Select related accessories.* This includes, but is not limited to, humidification equipment, supplemental dehumidification equipment, ventilation air precooling/preheating equipment, vibration and noise mitigation systems, and seismic control precautions.
11. *Select control systems and specify operating sequences and control strategies.*
12. *Redo steps 2 through 11 if bids are too high.*
13. *Review testing and balancing (TAB) and/or commissioning reports* to ensure systems are installed as specified and operate as designed to best meet the owner's needs, resulting in a safe, healthy, and comfortable working environment that operates in an efficient and environmentally sustainable manner.

The Challenge

Building systems, code requirements, and design procedures are becoming increasingly complex. Some engineers may be tempted to default design responsibilities to equipment vendors (equipment selection, control specification, etc.) and contractors (ductwork quality, piping material, etc.). This can place the customer at risk if the vendor chooses equipment that is more expensive than necessary, with proprietary components that cannot be replaced or added with competitively priced alternatives. The engineer should do the design work rather than allowing contractors to dictate specifications and procedures that can compromise the long-term quality of the building.

Good engineers must consider input from owners, vendors, and contractors, but ultimately they must take the time and effort to design the systems, prepare specifications, and verify that construction and operation practices follow design intent so that the building provides long-term occupant safety and health with the lowest possible maintenance requirements, operating costs, and environmental impact.

REFERENCES

1. NSPE. Engineers' Creed. National Society of Professional Engineers, Alexandria, VA.
2. ASHRAE. 2004a. Code of Ethics. American Society of Heating, Refrigerating and Air-Conditioning Engineers, Atlanta. http://www.ashrae.org/template/AssetDetail/assetid/25318.
3. ASHRAE. 2004b. ASHRAE Bylaws. American Society of Heating, Refrigerating and Air-Conditioning Engineers, Atlanta. http://www.ashrae.org/template/AssetDetail/assetid/25318.
4. Means. 2004. *Mechanical Cost Data*. Kingston, MA: R.S. Means, Reed Construction Data.

2 HVAC Fundamentals: Refrigeration

REFRIGERATION CYCLES

The purpose of a refrigeration cycle is to transfer heat from a region of low temperature to a region of higher temperature. There is a variety of thermodynamic cycles for this purpose, but discussion in this section is limited to the single-stage vapor compression cycle, which is the dominant method used in air-conditioning (cooling-only) units and heat pumps (heating and cooling). Introductory discussions of additional cycles can be found in the *ASHRAE Handbook—Fundamentals* (ASHRAE 2005, chapter 1). The heart of the cycle is the compressor, which elevates the pressure and temperature of a refrigerant gas. At the elevated temperature, heat is rejected to the environment (cooling mode) or to the indoor space (heating mode). After the refrigerant has released heat, it is a liquid at high pressure and flows to an expansion device. This device reduces the pressure and temperature of the refrigerant, which initiates evaporation of the liquid at low temperature. At this temperature, heat is removed from the indoor space (cooling mode) or the environment (heating mode). The refrigerant gas then enters the compressor to repeat the cycle.

The Carnot Cycle

Figure 2.1 is a component diagram of a simple vapor compression refrigeration cycle operating in the cooling mode. Refrigerant enters the compressor as a low-temperature and low-pressure gas (point 1). Work (W) is required to drive the compressor so it can elevate the refrigerant to a higher pressure and temperature. The refrigerant exits the compressor as a superheated vapor (point 2) and enters the condenser. At the higher temperature, heat is rejected to the environment (air, water, or ground) through a condenser (heat exchanger), which lowers the temperature of the refrigerant vapor and then converts it to a liquid. The warm liquid refrigerant exits the condenser (point 3) and enters the expansion device. As the refrigerant leaves the expansion device (point 4), its pressure is now lower than the saturation pressure of the surrounding fluid so it begins to flash in the evaporator. The cooling effect is accomplished in the evaporator as the fluid (typically air or water) is passed through the opposite side of this heat exchanger and its temperature is reduced by the evaporating refrigerant. The refrigerant exits the evaporator as low-temperature and low-pressure gas to complete the cycle at point 1.

This process is referred to as the Carnot refrigeration cycle. It is an attractive thermodynamic process because heat is being "pumped" from one level (temperature) to another. Therefore, it is possible to deliver heat at a rate greater than the required input power. This provides an advantage over heating processes that convert heat from chemical or electrical energy that cannot exceed conversion efficiencies of 100%.

The efficiency, or dimensionless coefficient of performance (COP), of a Carnot refrigeration cycle is limited by the temperature of the heat source (evaporating temperature, t_e) and the heat sink (condensing tempera-

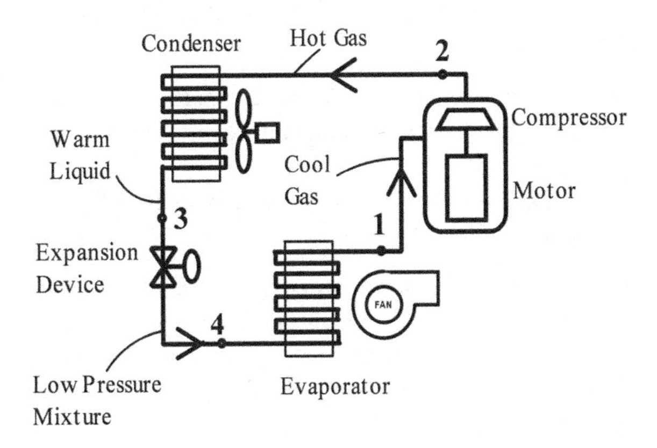

Figure 2.1 Simple vapor compression refrigeration cycle.

ture, t_c). If cooling is the intended effect of the cycle, the theoretical COP is

$$COP_c = \frac{t_e \text{ (in absolute temperature)}}{t_c - t_e} \quad . \tag{2.1}$$

If the intended effect is heating (heat pump in heating mode), the theoretical COP is

$$COP_h = \frac{t_c \text{ (in absolute temperature)}}{t_c - t_e} \quad . \tag{2.2}$$

The COP values generated with Equations 2.1 and 2.2 assume the process is completely reversible. However, the second law of thermodynamics is used to describe the irreversibilities in real processes due to pressure drops, mechanical friction, heat transfer, etc., in terms of entropy (Btu/lb·°F) flow.

Thermodynamics Definitions

adiabatic compression: compression of a gas during which no heat is exchanged with the surroundings.

adiabatic efficiency: (1) efficiency with which work is done with respect to heat gains or losses. (2) (indicated efficiency) ratio of the work absorbed in compressing a unit mass of refrigerant in a compressor to the work absorbed in compressing the same mass in an ideal compressor. See *isentropic process*.

adiabatic expansion: expansion of a fluid during which no heat is exchanged with the surroundings.

adiabatic process: thermodynamic process during which no heat is extracted from or added to the system.

enthalpy (heat content): thermodynamic quantity equal to the sum of the internal energy of a system plus the product of the pressure-volume work done on the system; $h = u + pv$, where h = enthalpy or total heat content, u = internal energy of the system, p = pressure, and v = volume.

entropy (ratio): ratio of the heat absorbed by a substance to the absolute temperature at which it was added.

> ***specific entropy***: entropy per unit mass of a substance.

irreversible process: a process in which available energy is lost due to the creation of entropy. The energy loss in a system or component is determined by multiplying the absolute temperature of the surroundings by the entropy increase.

isentrope: line of equal or constant entropy.

isentropic process (reversible adiabatic): thermodynamic change at constant entropy.

The Ideal Vapor Compression Cycle with Refrigerants

An important concept to remember in these systems is the relationship between the pressure of the refrigerant and the saturation temperature at that point. The high-pressure side of the system (compressor discharge to expansion device inlet) operates at a relatively constant pressure. The corresponding saturation temperature (temperature at which a pure fluid changes state at a given pressure) is referred to as the condensing temperature (t_c). The refrigerant enters the condenser above this temperature, but once it has been desuperheated and begins to condense, the temperature will be essentially constant until it is 100% liquid. In an ideal system the refrigerant will exit the condenser at this state, but in a real system it is typically subcooled several degrees before entering the expansion device. On the low-pressure side of the system (expansion device outlet to compressor inlet), the saturation temperature is referred to as the evaporating temperature (t_e). The refrigerant typically enters the evaporator at this temperature and remains relatively constant until it is 100% vapor. In an ideal system the refrigerant will exit the evaporator at this condition, but in a real system it is typically superheated several degrees before entering the compressor inlet (suction).

An ideal system would include an isentropic compression ($s_1 = s_2$) from point 1 to point 2, constant pressure desuperheating and condensing ($p_2 = p_3$) to a saturated liquid state at point 3, adiabatic expansion ($h_3 = h_4$) from a saturated liquid at p_3 to a mixture of liquid and gas at p_4, and a constant pressure evaporation ($p_4 = p_1$) through the evaporator to point 1, a saturated vapor at t_e.

Both ideal and "real" vapor compression cycles can be analyzed with pressure-enthalpy (p-h) diagrams. In this case, enthalpy (internal energy + pressure × volume) is used to quantify the energy of the refrigerant per unit mass at various points in the cycle. Figure 2.2 is a p-h diagram that is used to visualize the process thermodynamically (see note). More accurate and complete p-h diagrams are available in Appendix B, "Refrigerant Pressure-Enthalpy Charts and Sea Level Psychrometric Chart," Figures B.1 through B.3. Tables 2.1 and 2.2 will also be used in this discussion to provide improved accuracy for the properties of R-134a. Table 2.1 lists the saturation temperatures, enthalpy, and entropy of saturated liquid and gaseous R-134a ($C_2H_2F_4$) for pressures of 25 psia through 225 psia. Saturated liquid values correspond to points on the left side of the dome-shaped curve in Figure 2.2. Saturated vapor lines correspond to points on the right side of the dome-shaped line. Table 2.2 is a set of properties for super-

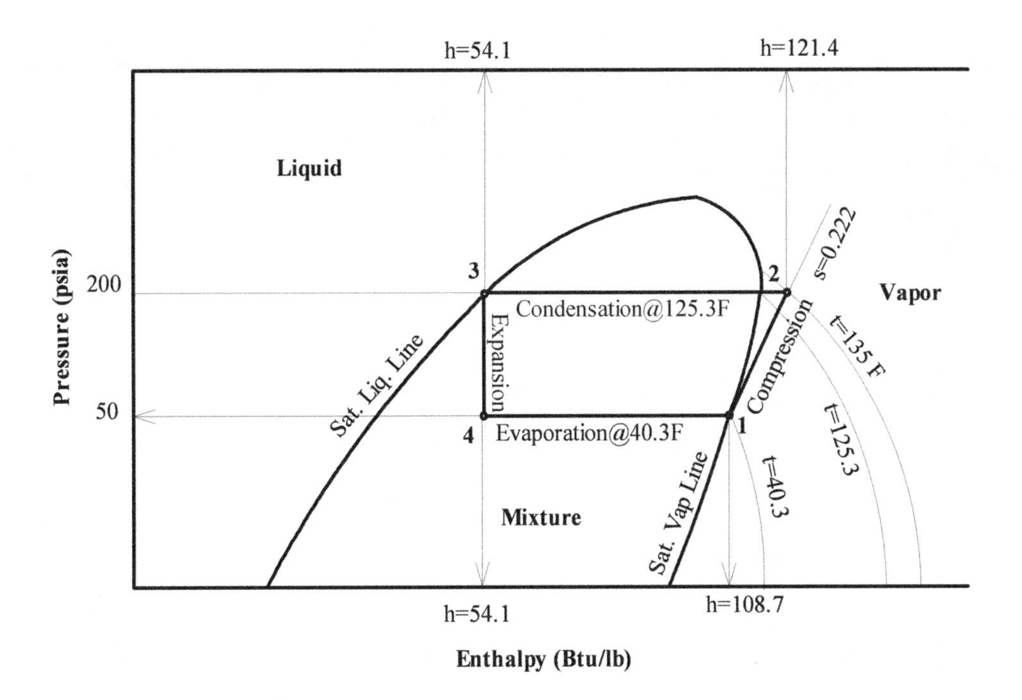

Figure 2.2 R-134a pressure-enthalpy diagram for ideal vapor compression cycle.

Table 2.1 Properties of Saturated R-134a Liquid and Vapor

p (psia)	25	50	75	100	125	150	175	200	225
t_{sat} (°F)	7.2	40.3	62.2	79.2	93.2	105.2	115.8	125.3	133.9
h_{liq} (Btu/lb)	14.3	24.8	32.0	37.7	42.5	46.8	50.6	54.1	57.4
h_{vap} (Btu/lb)	104.1	108.7	111.7	113.8	115.4	116.7	117.8	118.6	119.3
s-liq (Btu/lb·°F)	0.0322	0.0538	0.0678	0.0784	0.0872	0.0946	0.1013	0.1072	0.1127
s-vap (Btu/lb·°F)	0.2245	0.2217	0.2204	0.2196	0.2190	0.2184	0.2179	0.2174	0.2169

heated R-134a, which correspond to values to the right of the dome in Figure 2.2.

Note: The reader is alerted to the use of *h* as the symbol for *heat transfer coefficient* (in heat transfer), *head* (in fluid mechanics), and *enthalpy* (in thermodynamics).

Consider a vapor compression system operating at a compressor inlet (suction) pressure of 50 lb/in.^{2} absolute (psia) (= 35.3 lb/in.^{2} gauge [psig]) and an outlet (discharge) pressure of 200 psia (185.3 psig). The evaporating temperature (t_e) at 50 psia pressure is 40.3°F and point 1 is located on the saturated vapor line as shown in Figure 2.2. Table 2.1 is consulted for more exact values at this point.

h_1 = 108.7 Btu/lb, p_1 = 50 psia, t_1 = 40.3°F
(t_{sat}@50 psia),
and
s_1 = 0.2217 Btu/lb·°F

The process continues from point 1 via isentropic compression to the condensing pressure of 200 psia along a line of constant entropy ($s_1 = s_2 =$ 0.2217 Btu/lb·°F) to point 2. Since this process moves the refrigerant into the superheated vapor region, Table 2.2 applies. To find point 2, it is necessary to locate the point where p_2 = 200 psia and s_2 = 0.2217 Btu/lb·°F. Values in Table 2.2 indicate this occurs when t_2 = 135°F. Thus,

h_2 = 121.4 Btu/lb, p_2 = 200 psia, $t_2 \approx$ 135°F
(t_{sat}@50 psia),
and
s_2 = 0.222 Btu/lb·°F

From point 2 the process continues along a line of p = 200 psia from a superheated vapor to a point on the saturated vapor line. At this point the temperature of the refrigerant is 125.3°F, the condensing temperature at 200 psia. The temperature and pressure remain constant as the condensation process continues until the refrigerant is 100% liquid at point 3 on the saturated liquid line.

Table 2.2 Properties of Superheated R-134a Vapor

P-abs.	t-sat.	P-gauge	P-abs	t-sat.	P-gauge	P-abs.	t-sat.	P-gauge
25 psia	**7.2°F**	**10.3 psig**	**50 psia**	**40.3°F**	**35.3 psig**	**75 psia**	**62.2°F**	**60.3 psig**
***t* (°F)**	***h* (Btu/lb)**	***s* (Btu/lb·°F)**	***t* (°F)**	***h* (Btu/lb)**	***s* (Btu/lb·°F)**	***t* (°F)**	***h* (Btu/lb)**	***s* (Btu/lb·°F)**
10	105.1	0.226	45	109.8	0.224	65	112.3	0.222
15	105.9	0.228	50	110.8	0.226	70	113.5	0.224
20	106.6	0.230	55	111.9	0.228	75	114.6	0.226
25	107.6	0.232	60	113.0	0.230	80	115.7	0.228
30	108.6	0.234	65	114.1	0.232	85	116.9	0.230
35	109.6	0.236	70	115.2	0.234	90	118.0	0.232
100 psia	**79.2°F**	**85.3 psig**	**125 psia**	**93.2°F**	**35.3 psig**	**150 psia**	**105.2°F**	**135.3 psig**
***t* (°F)**	***h* (Btu/lb)**	***s* (Btu/lb·°F)**	***t* (°F)**	***h* (Btu/lb)**	***s* (Btu/lb·°F)**	***t* (°F)**	***h* (Btu/lb)**	***s* (Btu/lb·°F)**
80	113.8	0.220	95	113.8	0.220	110	113.8	0.221
85	115.2	0.222	100	117.2	0.222	115	119.3	0.223
90	116.4	0.224	105	118.4	0.224	120	120.6	0.225
95	117.6	0.227	110	119.7	0.227	125	121.9	0.227
100	118.8	0.229	115	120.9	0.229	130	123.2	0.230
105	120.0	0.231	120	122.2	0.231	135	124.5	0.232
175 psia	**115.8°F**	**160.3 psig**	**200 psia**	**125.3°F**	**185.3 psig**	**225 psia**	**133.9°F**	**210.3 psig**
***t* (°F)**	***h* (Btu/lb)**	***s* (Btu/lb·°F)**	***t* (°F)**	***h* (Btu/lb)**	***s* (Btu/lb·°F)**	***t* (°F)**	***h* (Btu/lb)**	***s* (Btu/lb·°F)**
120	119.0	0.220	130	120.0	0.200	135	113.8	0.217
125	120.4	0.222	135	121.4	0.222	140	121.2	0.220
130	121.7	0.225	140	122.9	0.225	145	122.6	0.222
135	123.1	0.227	155	124.2	0.227	150	124.0	0.225
140	124.4	0.229	150	125.5	0.229	155	125.5	0.227
145	125.8	0.231	155	127.0	0.232	160	127.0	0.230

Note that the constant temperature line is horizontal within the dome-shaped curve created by the saturated liquid and vapor lines. Table 2.1 can be consulted for the saturated liquid properties at 200 psia.

$h_3 = 54.1$ Btu/lb, $p_3 = 200$ psia, $t_3 = 125.3$°F (t_{sat}@200 psia),
and
$s_3 = 0.1072$ Btu/lb·°F

From point 3 the process continues down a line of constant enthalpy ($h_3 = h_4$) until it intersects a horizontal line representing the evaporating pressure of 50 psia, which corresponds to a temperature of 40.3°F. At this point (4) the refrigerant is a mixture of liquid and vapor (mostly liquid).

$h_4 = 54.1$ Btu/lb, $p_4 = 50$ psia, and $t_4 = 40.3$°F (t_{sat}@50 psia)

The final leg of the process is constant pressure evaporation at $p = 50$ psia and $t = 40.3$°F from point 4 to point 1, which provides the desired refrigeration effect. The values generated in this process can be used to find the maximum possible cooling effect (q_c) for R-134a at the selected evaporating and condensing temperatures. Additionally, the minimum amount of power (w_{Comp}) required to compress the refrigerant and the maximum theoretical refrigerant cooling (COP_c) can be found. For any refrigerant flow rate (m_r):

$$q_c = m_r(h_1 - h_4) \quad (2.3a)$$

$$w_{Comp} = m_r(h_2 - h_1) \quad (2.3b)$$

$$COP_c = \frac{q_c}{w_{Comp}} = \frac{(h_1 - h_4)}{(h_2 - h_1)} \quad (2.3c)$$

In cases where heating is the desired effect (i.e., heat pump heating mode, heat pump water heater), the desired effect is desuperheating and condensation (h_2h_3). Equation 2.3 is modified to find q_h, w_{Comp}, and COP_h for heating.

$$q_h = m_r(h_2 - h_3) \quad (2.4a)$$

$$w_{Comp} = m_r(h_2 - h_1) \quad (2.4b)$$

$$COP_h = \frac{q_3}{w_{Comp}} = \frac{(h_2 - h_3)}{(h_2 - h_1)} \quad (2.4c)$$

It should be noted that the values computed for COP_c and COP_h in Equations 2.3c and 2.4c are idealized theoretical values for R-134a and should be less than the corresponding theoretical values of Equations 2.1 and 2.2, which assume both an "ideal" refrigerant and reversible processes. Thus, the values computed using Equations 2.1 and 2.2 should be greater than those found using Equation 2.3c and 2.4c for the same evaporating and condensing temperatures.

Operating Characteristics of Real Vapor Compression Cycles

A familiarity with this process and the values generated in the preceding discussion is valuable even though the process is idealized. Limits are set and the system designer can determine the relative importance of operating conditions and component selection that will have the greatest impact on total system optimization. However, there are a number of constraints that limit the capacity and efficiency of actual vapor compression cycles. Several are listed here:

- Refrigerant enters the compressor as a superheated vapor to protect it from damage by liquid refrigerant (it is essentially incompressible and may remove oil from bearing surfaces). Therefore, the inlet gas density will be slightly lower compared to saturated vapor, thereby reducing mass flow rate and compressor capacity.
- There are pressure drops (across ports, valves, and mufflers), friction in the compressor and fluid, and heat generated by the compressor drive. Thus, the pressure difference from inlet to outlet will be greater than ideal, and entropy will be generated since the processes are irreversible.
- There is pressure drop and internal heat generation through the condenser with a corresponding change in saturation temperature as the pressure changes.
- Condensers are heat exchangers of finite size and will operate at temperatures elevated above the idealized condensing temperatures.
- The refrigerant must be subcooled several degrees below the saturation temperature to ensure the refrigerant is 100% liquid since the performance of many expansion devices is compromised by the presence of vapor.
- Heat transfer and entropy generation are encountered when refrigerant passes through real expansion devices.
- Evaporators are heat exchangers of finite size and will operate at temperatures below the idealized evaporation temperatures.
- The evaporation temperature in many cases must be suppressed to low values to meet the needs of the application (ice making, dehumidification, etc.).
- Many alternative refrigerants are zeotropic mixtures that do not behave as a pure substance. Thus, condensation and evaporation saturation temperatures change (glide) with phase concentration. This typically results in lower system efficiency and/or larger heat exchanger size.

Figure 2.3 is a p-h diagram that demonstrates the impact of a number of these constraints.

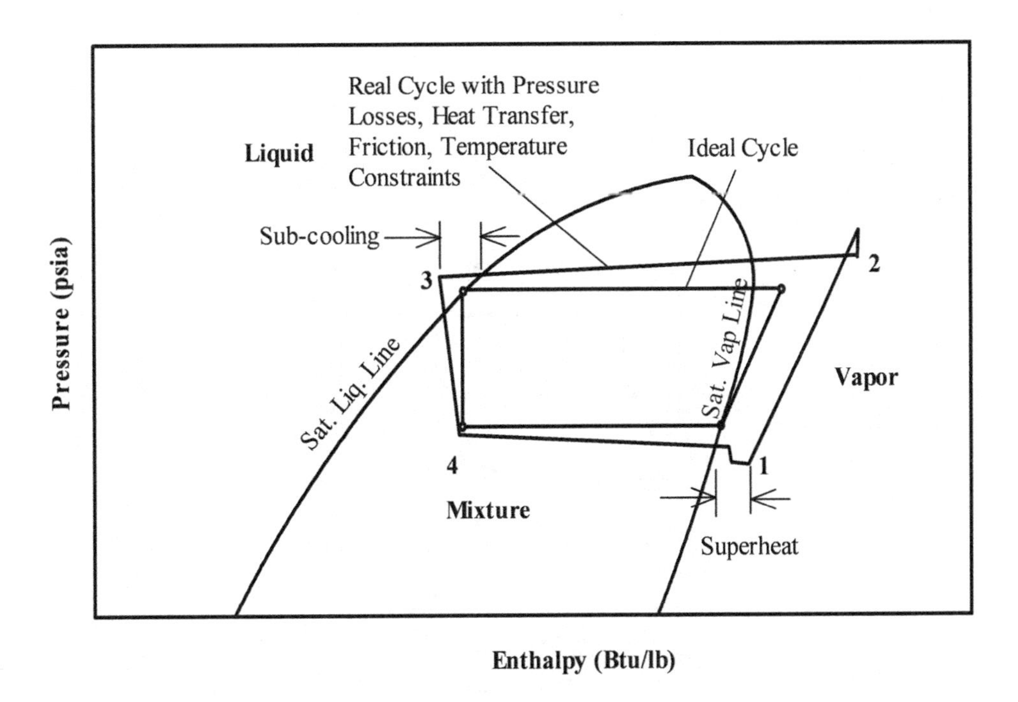

Figure 2.3 P-h process diagram of a real vapor compression cycle (ASHRAE 2005).

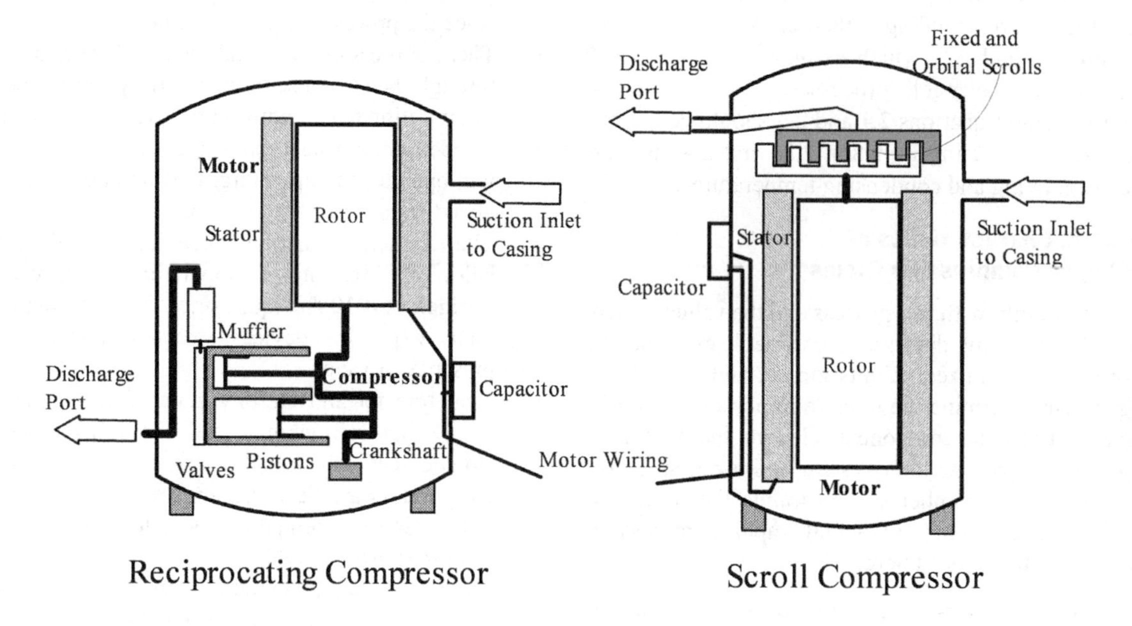

Figure 2.4 Hermetic compressors.

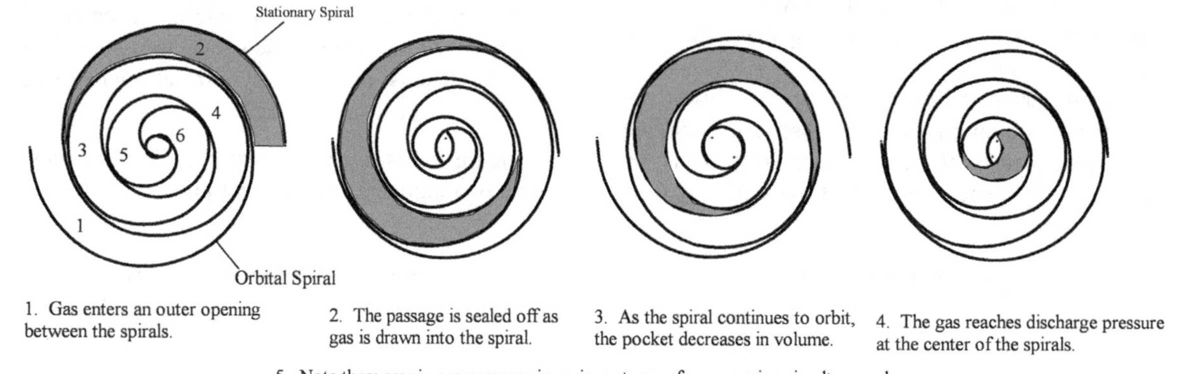

Figure 2.5 Scroll compressor operation.

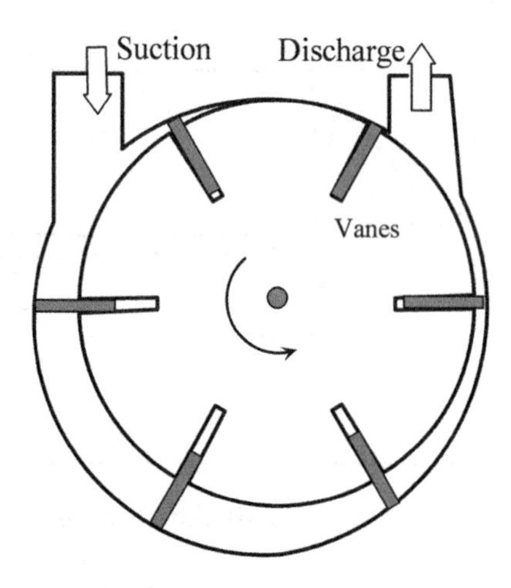

Figure 2.6 Rotary vane compressor.

COMPRESSORS

A variety of compressor types are available to meet the many refrigeration and air-conditioning applications. Several types are discussed in this section, but more detail can be found in the *ASHRAE Handbook—HVAC Systems and Equipment* (ASHRAE 2004a, chapter 34). Figure 2.4 provides diagrams of two common hermetic designs. The term *hermetic* indicates the motor and compressor are encapsulated into a pressure vessel to eliminate the compressor drive shaft leakage that occurs with open compressor designs. The suction port opens directly into the casing, while the discharge port and electrical leads enter through pressure-sealed passages. Motor cooling and lubricating oil circulation are accomplished with strategic injection of the return gas. The reciprocating piston design shown in Figure 2.4 places the motor above the compressor. The orbiting scroll design has become popular because of the reduction in moving components. Advances in precision manufacturing have resulted in scroll compressors with slightly higher efficiency compared to reciprocating models. Figure 2.5 shows the scroll compression cycle.

Another type of compressor often used in a variety of applications is the rotary vane. One design is shown in Figure 2.6. Larger compressors are typically "open" drives in that they are connected through a shaft to a motor (or engine). Figure 2.7 is a diagram of an open drive twin-screw compressor used for larger applications. Figure 2.8 is a complete chiller package with a centrifugal compressor that is typically used in large commercial and industrial applications. Performance data for these larger machines will be provided in chapter 5.

The impact of operating characteristics on the actual performance of a compressor can be summarized in curves or tables. Table 2.3 is a typical presentation that includes the gross compressor refrigeration capacity, the required input power, and the refrigerant mass flow rate for a hermetic scroll compressor with R-134a refrigerant.

Table 2.3 provides the performance information for various combinations of evaporating and condensing temperatures at conditions that are typically encountered in air-conditioning applications. The temperature of the vapor entering the compressor is 20°F (superheated) above the evaporating temperatures in the table. Com-

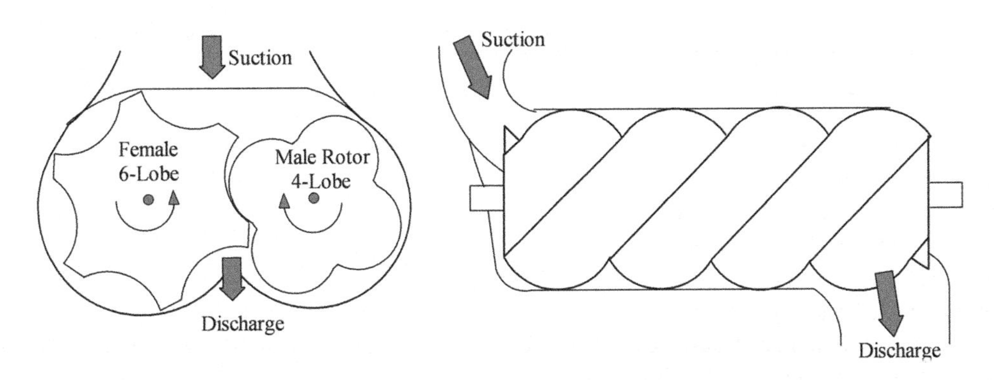

Figure 2.7 Rotary twin-screw compressor.

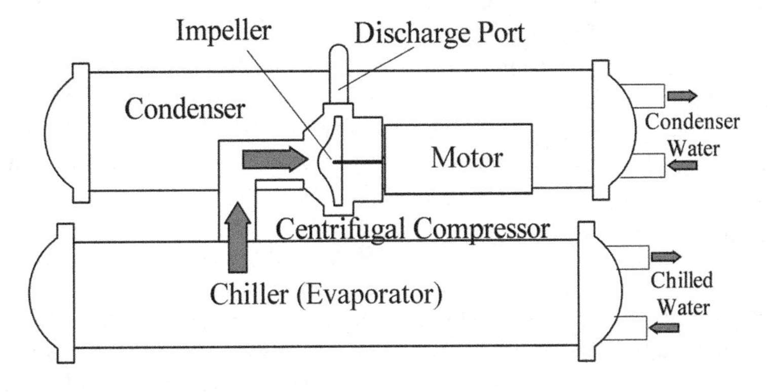

Figure 2.8 Chiller with centrifugal compressor.

Table 2.3 Performance of an R-134a Scroll Compressor (Copeland 2004a)

Conditions: 20°F suction superheat, 15°F liquid subcooling, 95°F ambient air.

Cond. Temp. (°F)	Capacity (q_r)—Power Input (kW)—Mass Flow Rate (m_r)—Evaporation Temperature (°F)	10	20	30	40	45	50	55
80	q_r - MBtu/h	16.0	20.3	25.6	31.9	35.5	39.4	43.7
	kW	1.45	1.43	1.42	1.41	1.40	1.39	1.38
	m_r - lb/h	212	264	326	399	440	484	530
90	q_r - MBtu/h	15.1	19.3	24.4	30.4	33.9	37.7	41.8
	kW	1.63	1.62	1.60	1.58	1.57	1.57	1.56
	m_r - lb/h	210	263	325	398	439	483	530
100	q_r - MBtu/h	14.3	18.3	23.1	29.0	32.3	35.9	39.8
	kW	1.83	1.82	1.80	1.78	1.77	1.76	1.75
	m_r - lb/h	208	261	323	396	437	481	530
110	q_r - MBtu/h	13.4	17.2	21.9	27.4	30.6	34.1	37.9
	kW	2.07	2.05	2.03	2.01	2.00	1.98	1.97
	m_r - lb/h	206	259	322	395	436	480	525
120	q_r - MBtu/h	12.5	16.1	20.6	25.9	28.9	32.2	35.8
	kW	2.33	2.32	2.29	2.27	2.25	2.24	2.23
	m_r - lb/h	203	256	319	392	433	477	525
130	q_r - MBtu/h	---	15.0	19.2	24.2	**27.1***	30.3	33.7
	kW	---	2.62	2.60	2.57	**2.55***	2.54	2.52
	m_r - lb/h	---	253	316	389	**430***	474	520
140	q_r - MBtu/h	---	---	17.8	22.5	25.3	28.2	31.5
	kW	---	---	2.94	2.91	2.89	2.88	2.86
	m_r - lb/h	---	---	311	385	426	470	515

* ARI-520 rating conditions: **EER = 27.1 ÷ 2.55 = 10.6**
COP = EER (MBtu/kWh) ÷ 3.412 MBtu/kWh.

pressor manufacturers suggest this level of superheat to protect the compressor from any non-evaporated liquid entering the compressor. However, air-conditioner and heat pump manufacturers often attempt to reduce the superheat in order to increase the density (or lower the specific volume) of the vapor entering the compressor. Higher density will result in higher mass flow rates and, therefore, greater cooling capacity with little change in the power required.

The performance rating conditions also consider that the liquid refrigerant leaving the condenser is subcooled 15°F below the condensing temperature. This practice will add to the capacity of the system since the enthalpy of the liquid is reduced and will result in a lower enthalpy mixture entering the evaporator. A standard rating point (ARI 1997) for air-conditioning applications is shown in bold in Table 2.3 with the figures of merit being refrigeration capacity (q_r) in Btu/h and energy efficiency ratio (EER = q_r/W Btu/Wh MBtu/kWh) at t_e = 45°F and t_c = 130°F.

The refrigeration capacity (q_r) and refrigerant mass flow rate (m_r) values in Table 2.3 can be corrected for different suction superheat temperatures by assuming a direct correlation with the ratio of the rated to actual specific volume (v_{Act}) to the specific volume for 20°F superheated vapor ($v_{SH = 20°F}$).

$$q_r \approx q_{SH = 20°F}\left(\frac{\rho_{Act}}{\rho_{SH = 20°F}}\right) \approx q_{SH = 20°F}\left(\frac{\upsilon_{SH = 20°F}}{\upsilon_{Act}}\right) \quad (2.5a)$$

$$m_r \approx m_{SH = 20°F}\left(\frac{\rho_{Act}}{\rho_{SH = 20°F}}\right) \approx m_{SH = 20°F}\left(\frac{\upsilon_{SH = 20°F}}{\upsilon_{Act}}\right) \quad (2.5b)$$

Refrigeration capacity values in Table 2.3 can be corrected for other subcooling temperatures (t_{SC}) by

$$q_r \approx q_{SC = 15°F} - m_r c_p(15°F - t_{SC}) \,. \quad (2.6)$$

The impact of alternative refrigerants is demonstrated by comparing the performance data shown in Table 2.4, which result when the traditional air-conditioning refrigerant (R-22) is substituted for R-134a in the same compressor. Efficiency with R-22 improves by approximately 5% and capacity increases by almost 50% at the standard rating point. This reduction in capacity is almost directly proportional to the mass density of the refrigerant at the compressor suction. For this reason, it is not likely that R-134a will be a popular R-22 substitute in heat pump and air-conditioning applications. More likely substitutes are R-407c and R-410a,

Table 2.4 Performance of an R-22 Scroll Compressor (Copeland 2004b)

Conditions: 20°F suction superheat, 15°F liquid subcooling, 95°F ambient air.

Cond. Temp. (°F)	Capacity (q_r)—Power Input (kW)—Mass Flow Rate (m_r)—Evaporation Temperature (°F)	10	20	30	40	45	50	55
	q_r - MBtu/h	25.3	31.7	39.0	47.5	52.0	57.0	62.5
80	kW	1.99	1.98	1.97	1.95	1.94	1.94	1.93
	m_r - lb/h	315	391	475	570	625	680	740
	q_r - MBtu/h	24.2	30.3	37.4	45.5	50.0	55.0	60.0
90	kW	2.25	2.24	2.23	2.21	2.20	2.19	2.17
	m_r - lb/h	313	388	473	570	620	675	735
	q_r - MBtu/h	22.9	28.9	35.7	43.5	47.8	52.5	57.5
100	kW	2.55	2.54	2.53	2.50	2.49	2.48	2.46
	m_r - lb/h	309	384	468	565	615	670	730
	q_r - MBtu/h	21.6	27.3	33.8	41.3	45.5	50.0	54.5
110	kW	2.89	2.88	2.86	2.84	2.82	2.81	2.79
	m_r - lb/h	303	378	462	560	610	665	725
	q_r - MBtu/h	20.1	25.6	31.9	39.1	43.1	47.4	52.0
120	kW	3.27	3.26	3.25	3.22	3.20	3.18	3.16
	m_r - lb/h	295	370	455	550	605	660	720
	q_r - MBtu/h	---	23.8	29.2	36.8	40.6*	44.7	49.1
130	kW	---	3.69	3.68	3.65	3.63*	3.61	3.59
	m_r - lb/h	---	361	447	545	595*	650	710
	q_r - MBtu/h	---	---	27.8	34.4	38.1	42.0	46.2
140	kW	---	---	4.16	4.13	4.12	4.09	4.07
	m_r - lb/h	---	---	437	535	585	640	700

* ARI-520 rating conditions: **EER = 40.6 ÷ 3.63 = 11.2.**

which provide improved capacities and in some cases higher efficiencies for compressors of equivalent size. However, they will operate at higher pressures.

In the US the COP of compressors traditionally has not been expressed in terms of the dimensionless parameter COP. Manufacturers have used the energy efficiency ratio (EER) with the units of capacity in Btu/h and power input in watts. Coincidentally, the value for EER in Btu/Wh is identical to the value if the units of capacity are expressed in the often used units of MBtu/h (1000 Btu/h) and power in kW. (Note the English unit symbol for 1000 is M while the international unit symbol for 1000 is k.)

To convert EER to COP,

$$COP = \frac{EER(Btu/W{\cdot}h)}{3.412(Btu/W{\cdot}h)} = \frac{EER(MBtu/kWh)}{3.412(MBtu/kWh)} \quad (2.7)$$

Manufacturers of larger compressors used to produce chilled water (chillers) have traditionally expressed performance in terms of an inverse efficiency: kW of input power per ton of refrigeration (kW/ton), where a ton is defined as the rate (12,000 Btu/h) required to convert 2000 lb (1 ton) of liquid water at 32°F to ice at 32°F during a 24-hour period. The heat of fusion for water-to-ice is 144 Btu/lb; thus,

$$q_r\left(\frac{Btu/h}{ton}\right) = \frac{2000(lb/ton) \times 144(Btu/lb)}{24\ h} = 12,000\ Btu/ton{\cdot}h \quad (2.8)$$

To convert kW/ton to EER and COP,

$$EER(Btu/W{\cdot}h) = \frac{12000(Btu/ton{\cdot}h)}{1000(w/kW) \times (kW/ton)} = \frac{12}{(kW/ton)} \quad (2.9)$$

$$COP = \frac{12000(Btu/ton\text{-}h)}{1000(W/kW) \times 3.412(Btu/W{\cdot}h) \times (kW/ton)} = \frac{3.52}{(kW/ton)} \quad (2.10)$$

REFRIGERANT CONTROLS

A variety of control devices are used to optimize the performance and protect the components of refrigerant devices. A few simple devices are discussed here, but more detail and a list of references are presented in the *ASHRAE Handbook—Refrigeration* (ASHRAE 2002, chapter 45).

The metering of refrigerant flow is critical to ensure efficient performance without risking damage to the compressor under widely varying loads. A common metering device is the thermostatic expansion valve (TXV), shown in Figure 2.9. The valve is located between the condenser and the evaporator. It is opened and closed by a diaphragm, which is connected by a small capillary tube to a bulb filled with an ideal gas (typically some type of refrigerant). The bulb is attached with insulating wrap to the refrigerant line between the evaporator outlet and the compressor inlet. Consider the situation with the valve partially open. As the load increases, more heat is available, so the gas leaving the evaporator will have an increasing temperature (superheat). The temperature in the bulb will also increase. Since the gas is ideal, the pressure will also increase, which expands the diaphragm and opens the valve to increase refrigerant flow. The temperature of the gas leaving the evaporator will then decrease since the mass flow has increased. If the valve is correctly sized, the correct valve position will be established without a great deal of oscillation (or "hunting").

Note the valve has a screw to set the superheat temperature by adjusting the spring tension on the diaphragm. There are two options for the valve shown. One is a "bleed port" design that permits a small amount of refrigerant to flow at all times, which permits the high-pressure side to bleed to the low-pressure side of the system more quickly when the compressor turns off. This speeds the time that the compressor is able to restart without having to work against a high differential pressure. The second variation is a bi-flow valve, which permits flow to go in both directions. This is necessary for heat pump operation when flow is reversed.

The capillary tube shown in Figure 2.10 is another refrigerant flow metering device that is often used in refrigerators, freezers, small air-conditioning units, and air heat pumps operating in the heating mode. The length and diameter of the "cap tube" are sized to provide a relatively constant mass flow rate at all conditions. The tube is sized to pass primarily liquid. Higher flow rates are regulated since the refrigerant will tend to flash as back pressure is reduced, thus creating a higher volume flow rate and more resistance to increased flow.

Short-tube (3/8 to 1/2 in.) restrictors operate in the same fashion as capillary tubes, but they are more compact and can be replaced more easily. They can also be used to eliminate the check valves in heat pump systems when two different metering devices are needed because cooling and heating mode mass flow rates are substantially different. In the forward flow direction, the body of the restrictor opens like a check valve. In the reverse direction, flow is forced through the small restrictor port that serves as the expansion device, as shown in Figure 2.10. A second restrictor (not shown in Figure 2.10) is open in the reverse flow direction and is the expansion device in the forward direction.

Refrigeration control can also be accomplished with electric or electronically activated valves. One of the most common control valves is the four-way reversing valve, which is used to reverse the mode of a heat pump from heating to cooling operation. Most heat pump designs set the U-shaped port to connect the outdoor coil to the compressor suction, which also allows flow from the compressor discharge to the indoor (heat-

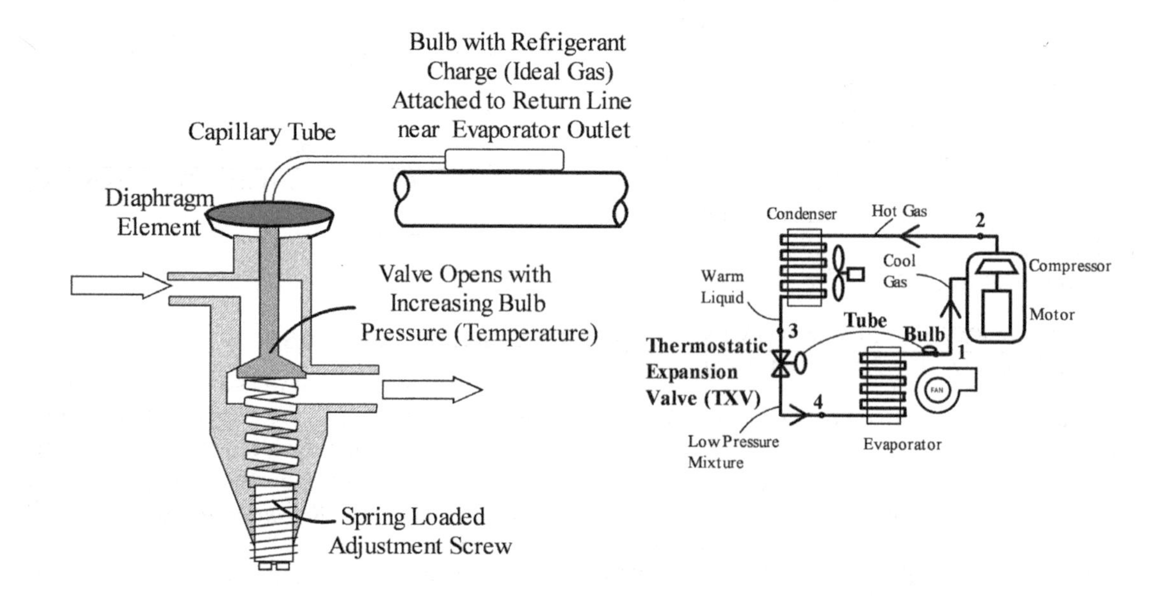

Figure 2.9 Thermostatic expansion valve (TXV).

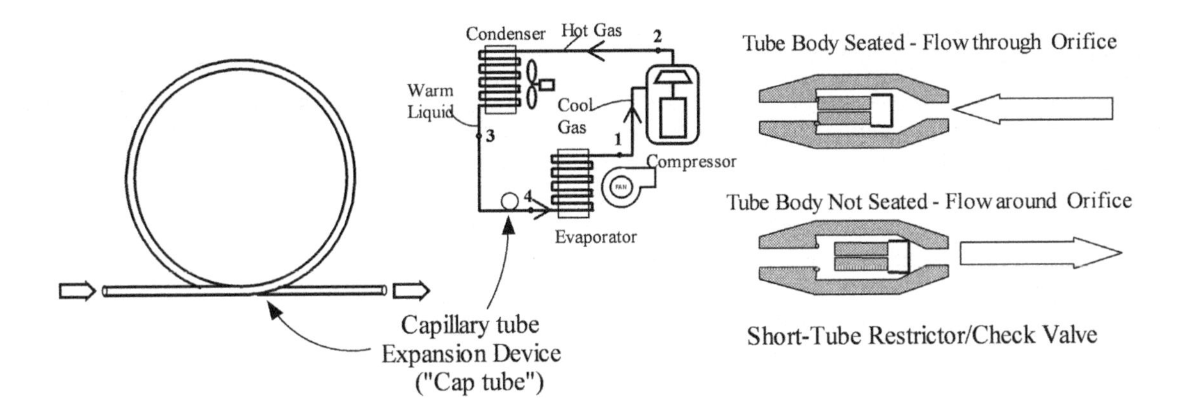

Figure 2.10 Capillary tube and short-tube restrictor expansion devices.

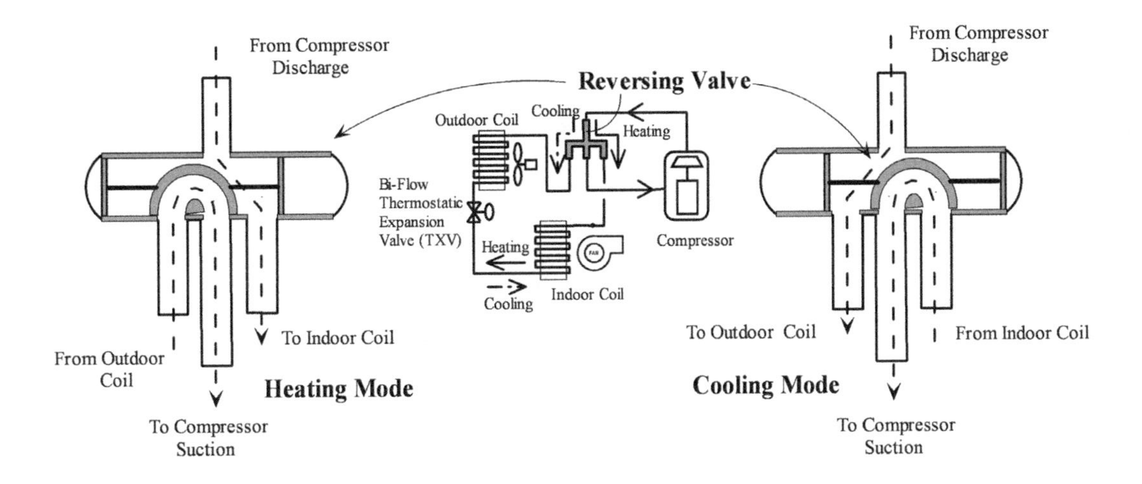

Figure 2.11 Heat pump four-way reversing valve.

ing) coil, as shown in the "Heating Mode" position of Figure 2.11. Thus, the hot gas from the compressor goes to the indoor coil, which is the condenser in this mode. The refrigerant flows from the indoor coil through the expansion device into the outdoor coil, which is the evaporator in this mode. When the coil is activated electrically, the U-shaped port is set so that gas flows from the indoor coil to the compressor suction. Hot compressor gas is routed to the outdoor coil, which is the condenser in the cooling mode. Refrigerant moves through the expansion device through the indoor coil, which is now the evaporator (cooling coil). Except for the reversing valve, heat pump operation is identical in this mode to a conventional air conditioner.

REFRIGERANTS

Refrigerants for vapor compression systems are a rapidly evolving topic due to a multitude of constraints including performance, application operating conditions, environment, safety, lubricant compatibility, and materials of construction compatibility. In the early development of refrigeration, flammable and hazardous fluids were common. More stable and inert refrigerants were developed by substituting fluorine and chlorine for some or all of the hydrogen in light hydrocarbons (methane, ethane, propane). Chlorofluorocarbons (CFCs) were widely used for low-temperature refrigeration, mobile air conditioning, and centrifugal chillers. Hydrochlorofluorocarbons (HCFCs) were commonly used in air-conditioning, heat pumps, and screw compressors. Azeotropic blends (mixtures that behave similarly to pure refrigerants) of CFCs and HCFCs were also common.

The chemical stability and low cost of CFCs that made them attractive refrigerants have resulted in unintended environmental consequences. For many years, CFCs from systems that needed repair were frequently vented to the atmosphere. Systems with minor leaks were recharged rather than repaired. CFCs were used for

Refrigerant Charging

The amount of refrigerant charge in a vapor compression system is critical to maintain efficient performance and minimize maintenance requirements. The system must also be free of noncondensable vapor (such as air) and water. Factory-charged systems and systems that have been repaired must be completely evacuated with a vacuum pump to remove all contaminants. The system is then charged with refrigerant, with great care being required to not re-inject air into the system. A highly recommended method of metering the correct charge into new, repaired, or poorly functioning systems is by "suction superheat." Manufacturers will provide a recommended suction superheat (typically 10°F to 20°F).

A refrigerant-charging manifold with two gauges is connected to the compressor suction and discharge. A small bulb thermometer is attached to the outside of the suction line, as shown in the figure below. Refrigerant is drawn into the suction operating compressor through the manifold. The low-pressure gauge records the suction pressure (in psig = psia –14.7) shown in the enlarged view on the left. Note the gauge has two scales that also provide the saturation temperature of R-134a and R-22. The gauge indicates the suction pressure is 40 psig, which corresponds to an evaporation temperature (t_e) of 45°F. However, the thermometer indicates the temperature is actually 60°F, which means the superheat is 15°F (60°F – 45°F). Note also the discharge pressure is 250 psig (t_c = 144°F), which is somewhat high for an R-134a system.

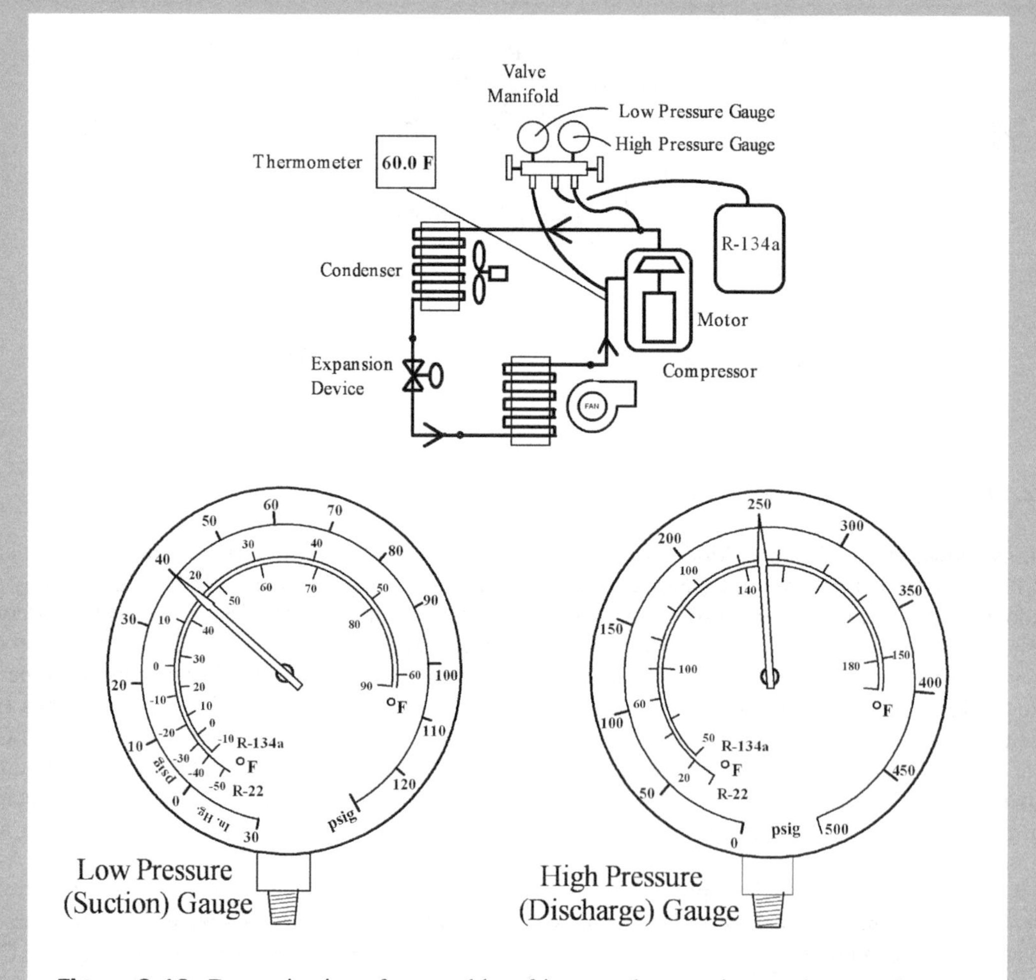

Figure 2.12 Determination of acceptable refrigerant charge using suction superheat.

cleaning and insulation manufacturing. These stable compounds diffused into the atmosphere where conditions allow them to break down. The released chlorine then reacts with and begins to deplete the atmosphere's protective ozone layer. HCFCs also contribute somewhat to ozone depletion, but they are less stable and tend to break down before reaching the stratosphere. The Montreal Protocol, originally signed in 1987 and continuously amended, is an international agreement to phase out the production of CFCs by 1996 and essentially phase out HCFCs by 2020 (UNEP 2003).

Currently a variety of hydrofluorocarbons (HFCs) and mixtures of HCFCs and hydrocarbons are being used as replacement refrigerants. There are advantages and disadvantages to all options that cannot be adequately addressed in this document. References are available for additional information (ASHRAE 2005, chapters 19, 20; ASHRAE 2002, chapter 5; McLindon et al. 2000). However, some general characteristics are listed in the following sections.

Refrigerant Numbering System

- For methane, ethane, and propane derivatives, the refrigerant number is related to the number of atoms in the compound structure by

  R- [carbon atoms –1][hydrogen atoms +1] [fluorine atoms] .

  For example,
  chlorodifluoromethane ($CHClF_2$) is
  R- [1–1][1+1][2] = R-022 or R-22.
  Propane (C_3H_8) is R-[3–1][8+1][0] = R-290.
  1,1,1,2-tetrafluoroethane (CH_2FCF_3) =
  R-[2–1][2+1][4]a = R134a with the "a" used to designate the molecule arrangement in the chain.
- Zeotropic blends of methane, ethane, and propane derivatives: R-4xx
- Azeotropic blends of methane, ethane, and propane derivatives: R-5xx
- Miscellaneous organic compounds: R-6xx
- Inorganic compounds: R-7xx

Thermophysical Properties

Higher theoretical vapor compression cycle efficiency is achieved by fluids with low vapor heat capacity. This is characteristic of fluids with simple molecular structure and low molecular weight.

High thermal conductivity is desirable to minimize heat exchanger size and low viscosity is desirable to minimize friction losses.

Many alternative refrigerants (i.e., R-410, a potential R-22 replacement) operate at higher pressure, which provides higher capacity but greater power consumption. This typically results in high compression ratios, greater loads on the compressor, shortened compressor life, and higher required rating pressure for heat exchangers and related components (Hughes 2003). There are a variety of potential replacements for R-22 applications that demonstrate a variety of benefits and limitations.

Azeotropic refrigerant mixtures exhibit characteristics similar to pure substances in that they evaporate and condense at a constant temperature when pressure is held constant. However, zeotropic mixtures do not exhibit constant temperature during evaporation and condensation. This undesirable trait (glide) results in a reduction in efficiency unless heat exchangers are enlarged to compensate. A more practical problem results from the fact that they also tend to have different component fractions in the gas phase at various temperatures. When leaks occur, the individual component losses are not in proportion to the original refrigerant components. Recharging is complicated because the makeup of the remaining refrigerant is not easily determined.

Safety and Environmental Issues

ASHRAE Standard 34 (ASHRAE 2004b) classifies refrigerants into six safety groups (A1, A2, A3, B1, B2, B3). Letter designations refer to toxicity, with class A being no identifiable toxicity and B being toxic. Numerical values indicate flammability, with 1 being no flame propagation in standard air, 2 having a low, lower flammability limit (LFL) and low heat of combustion (<8174 Btu/lb), and 3 having high LFL or a high heat of combustion.

Three factors are used to express the environmental impact of refrigerants. The ozone depletion potential (ODP) is a measure of a refrigerant's ability to destroy stratospheric ozone relative to R-11 (CCl3F). Other CFCs are less than 1.0 (ODP for R-12 is 0.82), HCFCs are much lower (ODP for R-22 = 0.034), and HFCs have 0 ODP. The second measure is global warming potential (GWP), which is a measure of the material's ability to trap radiant energy (greenhouse effect) relative to carbon dioxide. The third measure is also connected to global warming in that refrigerant thermal efficiency impacts the amount of energy and associated generation pollution and greenhouse gas emission required to drive vapor compression appliances. GWP and refrigerant efficiency are combined into a total equivalent warming impact (TEWI).

Material Compatibility

Refrigerants can react with tubing materials (i.e., ammonia with copper), sealant materials, and insulating materials used in the motors and wiring of hermetic compressors. Lubricants must be compatible with the selected refrigerant. Mineral oils worked well with CFCs (and to a lesser degree with HCFCs) since they are miscible and the oil is returned to the compressor with the refrigerant. However, alternative lubricants (polyolesters) are necessary with HFCs. Additional additives are sometimes required to combat the loss of the natural anti-wear characteristics of chlorine.

CHAPTER 2 PROBLEMS

2.1 Find the Carnot COP and the ideal COP for a system that uses R-134a refrigerant at an evaporating temperature of 45°F and a condensing temperature of 120°F. Find the suction pressure and discharge pressures in psia and psig and the temperature of the refrigerant leaving the compressor (assuming the ideal cycle conditions).

2.2 A scroll compressor (Table 2.3) with R-134a refrigerant operates with a 45°F evaporating temperature and a 120°F discharge saturation temperature. Find the cooling capacity (20°F suction superheat and 15°F liquid subcooling), compressor input power, EER, suction pressure, and discharge pressure (psig).

2.3 What increase in capacity and EER can be expected if the superheat is lowered to 10°F and the condensing temperature is lowered to 100°F? What is the disadvantage of doing this?

2.4 Sketch the atomic makeup of R-22, R-12, and R-123.

2.5 How can you determine if a refrigerant has chlorine in its structure from the R-xxx designation?

2.6 Compare the ideal COP of R-134a and R-22 at an evaporating temperature of 40°F with 20°F superheat and a condensing temperature of 120°F with 15°F subcooling with the actual compressor COPs calculated from the manufacturer's performance tables.

2.7 A set of pressure gauges on a manifold (see figure in "Refrigerant Charging" insert above) read 35 psig and a thermometer placed in close contact with the compressor inlet reads 67°F. The discharge pressure is 200 psig with an outdoor temperature of 95°F, and the refrigerant is R-134a. Is this system properly charged? If not, what range of temperature should be expected for these pressures?

2.8 A manufacturer recommends that their R-22 equipment operate with a suction pressure of 72 psig and a return gas temperature of 53°F with a specified air temperature (75°F) and flow rate (400 cfm/ton). What are the corresponding evaporating temperature and superheat?

2.9 With regard to the use of refrigerant mixtures as substitutes for CFCs, explain the difference between azeotropes and zeotropes. What is "glide"?

2.10 A refrigerant has an ASHRAE Standard 34 designation of A2 and B2. What does this mean? It also has an ODP of 0.75. Is this good, acceptable or unacceptable?

REFERENCES

ARI. 1997. *ARI Standard 520, Positive Displacement Condensing Units*. Arlington, VA: Air-Conditioning and Refrigeration Institute.

ASHRAE. 2005. *2005 ASHRAE Handbook—Fundamentals,* chapter 1, Thermodynamics and refrigeration cycles; chapter 19, Refrigerants; chapter 20, Thermophysical properties of refrigerants. Atlanta: American Society of Heating, Refrigerating and Air-Conditioning Engineers, Inc.

ASHRAE. 2004a. *2004 ASHRAE Handbook—HVAC Systems and Equipment*, chapter 34, Compressors. Atlanta: American Society of Heating, Refrigerating and Air-Conditioning Engineers, Inc.

ASHRAE. 2004b. *ANSI/ASHRAE Standard 34-2004, Designation and Safety Classification of Refrigerants*. Atlanta: American Society of Heating, Refrigerating and Air-Conditioning Engineers, Inc.

ASHRAE. 2002. *2002 ASHRAE Handbook—Refrigeration,* chapter 5, Refrigerant System Chemistry; chapter 45, Refrigerant Control Devices. Atlanta: American Society of Heating, Refrigerating and Air-Conditioning Engineers, Inc.

Copeland. 2004a. *Autogenerated Compressor Performance*. Bulletin 2.22AC60-40.0-PFV. Sidney, OH: Copeland Corp.

Copeland. 2004b. *Autogenerated Compressor Performance*. Bulletin 2.25AC60-40.0-PFV. Sidney, OH: Copeland Corp.

Hughes, H.M. 2003. Refrigerant Characteristics. *Air-Conditioning, Heating and Refrigeration News*, August 11, 2003. Troy, MI.

McLinden, M.O., S.A. Klein, E.W. Lemmon, and A.P. Peskin. 2000. *NIST Standard Reference Database 23, REFPROP*, Version 6.10. Gaithersberg, MD: National Institute of Standards and Technology.

UNEP. 2003. *Handbook for the International Treaties for the Protection of the Ozone Layer*. United Nations Environment Programme, Nairobi, Kenya.

3 HVAC Fundamentals: Heat Transfer

Design engineers must adopt effective methods and tools to deal with complex heat transfer phenomena that cannot be easily predicted from fundamental principles. The difficulty in applying fundamental principles arises from at least three circumstances: (1) heat transfer is a combination of the three complex and interconnected modes of conduction, convection, and radiation; (2) thermal properties and exact dimensions of many building materials and heat exchangers are not well established and change with exposure to the environment; and (3) climatic conditions and building occupancy levels vary dramatically with time and are difficult to predict. The best practices of heat transfer-related design are often a combination of simplified fundamental principles and historic field-measured data.

This chapter will present simplified heat transfer mechanisms and equations. Chapters 5 through 8 will provide procedures that combine these principles with measured data and simplified correlations, which permits computations that generate accurate results. Although more detailed and accurate computations are possible, a balance must be drawn given the constraints of available time, available funds, and the lack of accuracy for required input (weather, loads, occupancy patterns, building materials, construction quality, equipment performance, etc.).

HEAT TRANSFER MODES

Conductive Heat Transfer

Conductive heat transfer is the transport of energy via kinetic energy between particles in solids and in fluids near the boundary layers with solids. For a plane wall with constant properties throughout, the Fourier law is used to compute the heat transfer rate from the inside to the outside of the wall.

$$q = -(kA)\frac{t_1 - t_2}{\Delta x} = \frac{A\Delta t}{R} \tag{3.1}$$

where

q = heat transfer rate, Btu/h
k = thermal conductivity, Btu/h· °F· ft
A = cross-sectional area normal to the direction of heat flow, ft^2
t = temperature difference across wall, °F
x = thickness of wall, ft
R = (x/k) = unit area thermal resistance (R-value), h· °F· ft^2/Btu

This equation assumes that heat transfer occurs only in the direction indicated by the arrow in Figure 3.1. The negative sign in Equation 3.1 indicates a heat flow is positive in the direction of decreasing temperature.

"R-value" is a convenient method of expressing the effectiveness of insulating materials. In some cases, material thickness may vary from conventional values so that "R-value/unit thickness" is used in many cases. A common expression is "R per inch." Therefore, to obtain the R-value used in Equation 3.1,

$$R = (R/\text{in.}) \times \Delta x\ (\text{in.})\,. \tag{3.2}$$

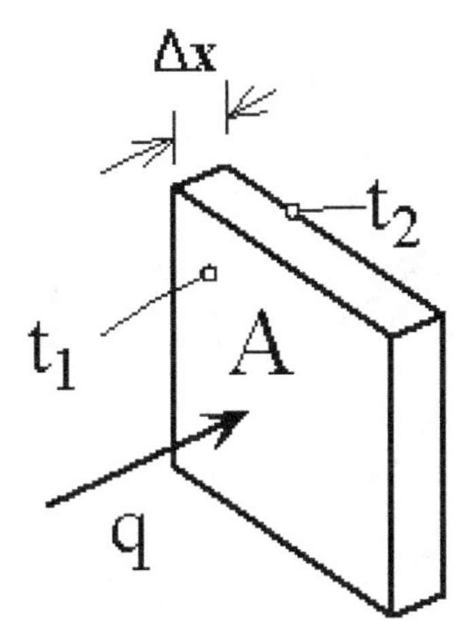

Figure 3.1 Wall conduction.

For a hollow cylinder (pipe) as shown in Figure 3.2, the conduction equation is

$$q = \frac{2\pi k L \Delta t}{\ln(r_o/r_i)} = \frac{L\Delta t}{R_L}, \tag{3.3}$$

where

L = cylinder length, ft;
r_o = outside diameter of cylinder, ft;
r_i = inside diameter of cylinder, ft; and
R_L = $[\ln(r_o/r_i)/2\pi k]$ = linear thermal resistance, h· °F· ft/Btu.

Many geometric arrangements are possible and can be solved using either derivations of three-dimensional partial differential equations or previously derived steady-state conditions presented in the form of "conduction shape factors" (S).

$$q = kS\Delta t \tag{3.4}$$

Tables of shape factors for various geometries are presented in introductory heat transfer texts (e.g., Holman [1990]) and heat transfer handbooks (Rohsenow et al. 1985).

Convective Heat Transfer

A boundary of molecules forms at the interface between solids and fluids, and heat is transferred primarily by conduction across the layer to the free stream fluids. The rate at which heat is moved across this boundary is influenced by the thermal properties of the fluid, the geometry of the surface, and the velocity of the fluid. A vast array of correlations has been developed to predict the convective heat transfer coefficient (h_c) used in the fundamental equation,

$$q = h_c A(t_s - t_\infty) = h_c A\Delta t = \frac{A\Delta t}{R_c}, \tag{3.5}$$

where

q = heat transfer rate, Btu/h;
h_c = heat transfer coefficient (a.k.a. film coefficient), Btu/h· °F· ft^2;
A = surface area of solid-fluid interface, ft^2;
t_s, t_∞ = surface temperature, free stream fluid temperature, °F;
Δt = temperature difference across boundary layer, °F; and
R_c = $(1/h_c)$ = unit area film resistance, h· °F· ft^2/Btu.

Note: The reader is alerted to the use of h as the symbol for both *heat transfer coefficient* (in heat transfer texts) and *enthalpy* (in thermodynamics texts).

In many heat exchanger applications, the film resistance is much higher than the thermal resistance of the solid (i.e., outside surface of a copper tube to air vs. conduction across a thin copper tube wall). Fins, such as the one shown in Figure 3.3, are commonly used to extend the solid-to-fluid surface area to enhance heat transfer. However, there are temperature gradients through the fin so that the entire fin surface is not at the same temperature as the base of the surface adjacent to the tube. Tables and equations for fin effectiveness (η_{fin}) as a function of geometry and fin thermal properties have been developed. The convection equation is modified.

$$\begin{aligned} q &= (\eta_{fin} h_c A_{fin} + h_c A_{base})(t_s - t_\infty) \\ &= (\eta_{fin} h_c A_{fin} + h_c A_{base})\Delta t \end{aligned} \tag{3.6}$$

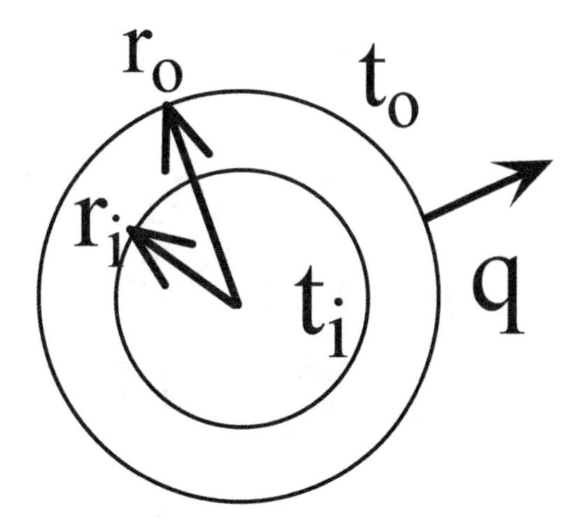

Figure 3.2 Conduction in tube.

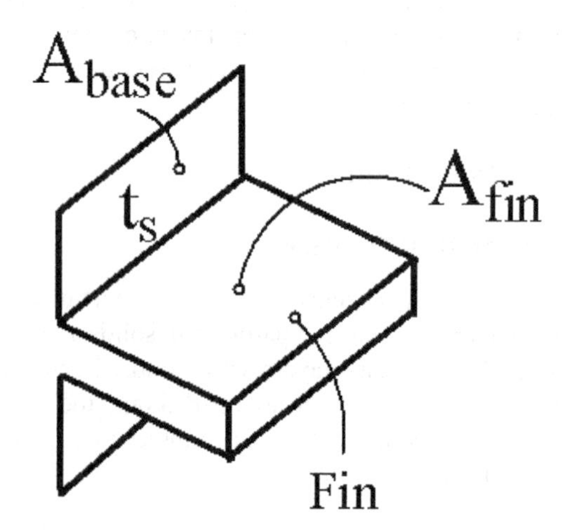

Figure 3.3 Rectangular fin (a.k.a. extended surface).

Natural Convection

As stated previously, the correlations for convective heat transfer are functions of the velocity of a fluid that is forced to flow (pumped, blown) across the solid boundary. However, velocity may also be induced by the temperature difference between a solid surface and the fluid. Consider the heating method of choice in many older classrooms—radiators (a misnomer since they are actually convector-conductor-radiators). Steam or hot water is circulated through the device, which results in an increase in surface temperature, a conduction of heat across the fluid boundary layer surrounding the surface, a decrease in local fluid density because of the temperature rise, and an induced velocity due to the buoyancy forces of the warm, lighter air. This rising air is replaced by colder air from below the device, and convection loops are established that circulate heated air throughout the classroom. This is referred to as natural, or "free," convection.

Often free convection occurs in combination with forced convection, and complex "free-force convection" equations apply. However, when free convection is the dominate mode, the convection coefficient (h_c) is based on correlations that involve not only thermal properties of the fluid but also fluid buoyancy, viscosity, and temperature difference across the boundary layer. Therefore, the computation is typically iterative since the heat transfer rate is computed by ($q = h_c A \Delta t$) and the convection coefficient (h_c) is dependent on boundary layer temperature difference (Δt).

Radiant Heat Transfer

Heat is also transferred via electromagnetic radiation between surfaces at different temperatures. The transfer of heat by radiation is a function of the absolute temperature (°R[K]) of a surface raised to the fourth power. The heat emitted from one surface is also a function of its emissivity (ε), with 1.0 being the value of a perfect blackbody. The ability of the receiving surface to transfer heat is a function of its absorptivity (α), reflectivity (ρ), and transmissivity (τ). All of these values are a function of radiation wavelength, temperature, and angle of incidence. The heat rate is also affected by the view factor (F_G), which is a fraction of the total output of the emitting surface that is incident on the receiving surface. A simplified radiant heat transfer relationship between two surfaces is (Holman 1990):

$$q = \varepsilon F_G \sigma A (T_1^4 - T_2^4) \tag{3.7}$$

where

ε = surface emissivity function (fraction of 1.0)

F_G = fraction of radiation incident upon receiving surface from emitting surface

σ = Stefan-Boltzman constant: 0.1714×10^{-8} Btu/h·ft²·°R⁴

A = area of receiving surface, ft²

T_1 = absolute temperature of surface 1, °R

T_2 = absolute temperature of surface 2, °R

Additionally, the surfaces may also be receiving heat from other surfaces so that a network of interconnected relationships among several surfaces might be necessary for accurate computations.

In some cases, it is possible to develop a radiation coefficient (h_r) or unit area radiation resistance (R_r) similar to the convective heat transfer coefficient. For special cases, this is an algebraic simplification of the above equation with the coefficient being a function of absolute temperature cubed so that the simplified equation takes the form of the convective heat transfer equation.

$$q = h_r A (t_s - t_\infty) = h_r A \Delta t = \frac{A \Delta t}{R_r} \tag{3.8}$$

The complexity of radiant heat transfer necessitates the use of simplifications. In the case of opaque surfaces, this takes the form of an equivalent surface temperature (or temperature difference) for building surfaces that are exposed to solar radiation. As will be demonstrated in later chapters, this results in a set of tables of equivalent temperature differences based on interior temperature, facing direction (south, east, etc.), and thermal properties of the building.

This simplification is also used for surfaces that transmit radiation, but another simplification is used to account for the transmitted energy through clear surfaces such as glass:

$$q = \tau A (q_{solar}/A)_{Incident} \tag{3.9}$$

where

τ = transmittance of surface (fraction)

A = area of surface normal to incident radiation, ft²

q_{solar}/A = solar insolation incident on the surface (clear day ≈ 320 Btu/h·ft²)

Combined Heat Transfer

Heat transfer typically occurs in all three modes simultaneously. In the case of a building wall, there is forced (fans blowing) or free (still air) convection from the interior space to the wall and heat is conducted through the solid wall (in some cases convection occurs in parallel if the material is semi-porous or if cavities exist). At the exterior surface, heat is transferred via forced convection and by radiation to the clear sky at night or from the sun during the day. In these cases, combined heat transfer can more easily be handled by using an overall heat transfer coefficient or thermal resistance.

$$q = U_{ov}A(t_o - t_i) = \frac{A(t_o - t_i)}{R_{ov}} \tag{3.10}$$

Computation can be simplified by adopting an electrical circuit analogy: thermal resistances to heat flow in series are additive and the inverse of thermal resistances is additive for parallel heat flow. Consider the simple wall in Figure 3.1 with convection coefficients (h_i) and (h_o) on the inside and outside walls. From Equations 3.1 and 3.5,

$$R_i = (1/h_i);\ R_w = (\Delta x/k);\ \text{and}\ R_o = 1/h_o \tag{3.11}$$

The electrical analogy is shown in Figure 3.4, and since the cross-sectional area (A) in the case of a plane wall is constant, Equation 3.12 applies.

$$(R_{ov} = R_i + R_w + R_0 = 1/h_i + (\Delta x)/k + 1/h_o) : \Rightarrow U_{ov} = 1/R_{ov} \tag{3.12a,b}$$

In many series heat flow cases, the overall thermal resistance can be obtained by adding the individual resistances if the cross-sectional area is constant. However, in cases where the cross section changes in the heat flow path, the (R/A) or (R_L/L) values are added. An overall coefficient must be based on a chosen area. Consider the case of heat flow through a tube that has parallel heat flow with varying cross-sectional area. Thermal resistances and areas are shown in Figure 3.5. The overall heat transfer coefficient is typically based on the outside area of the tube ($A_o = 2\pi r_o L$).

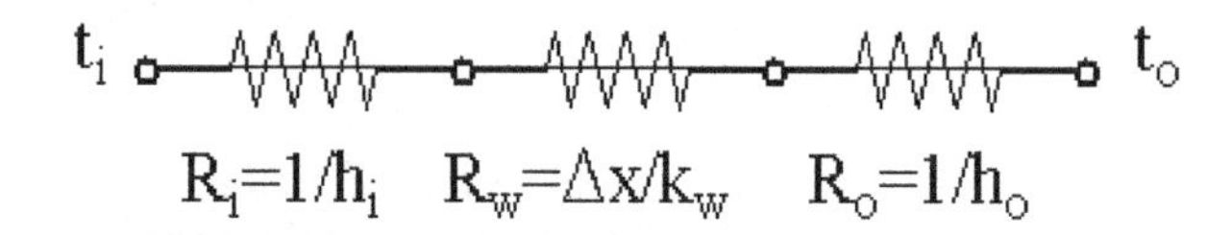

Figure 3.4 Wall thermal resistance.

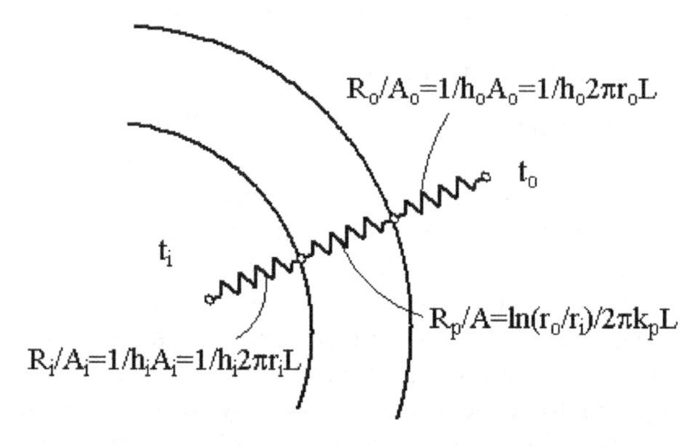

Figure 3.5 Thermal resistances in series heat flow through a tube.

$$q = U_{ov}A_o(t_o - t_i) = \frac{t_o - t_i}{R_i/A_i + \ln(r_o/r_i)/2\pi kL + R_o/A_o}$$

thus (3.13)

$$U_{ov} = \frac{1}{r_o/r_ih_i + \dfrac{r_o\ln(r_o/r_i)}{k} + 1/h_o} \tag{3.14}$$

In the cases of parallel heat flow, the cross-sectional areas are usually not equal. Thus, adding hA and $kA/\Delta x$ (inverse of R/A) is also recommended for parallel heat flow.

Consider the case shown in Figure 3.6, which represents a wall. Heat flow through sections B and C is in parallel if it is assumed that heat flow through the top and ends of the wall is negligible. To follow the electrical resistance analogy, the inverses of R/A must be added, inverted, and then added to the R/A values for section A and the internal and external convection components.

$$\frac{A_B}{R_B} + \frac{A_C}{R_C} = \frac{k_BA_B}{\Delta x_{B-C}} + \frac{k_CA_C}{\Delta x_{B-C}} : \quad \text{Inverse} \Rightarrow \frac{\Delta x_{B-C}}{k_BA_B + k_CA_C} \tag{3.15}$$

$$q = \frac{t_o - t_i}{R_i/A_A + R_A/A_A + \dfrac{\Delta x_{B-C}}{k_BA_B + k_CA_C} + 1/h_oA_A} = \frac{t_o - t_i}{1/h_iA_A + \Delta x_A/k_AA_A + \dfrac{\Delta x_{B-C}}{k_BA_B + k_CA_C} + 1/h_oA_A} \tag{3.16}$$

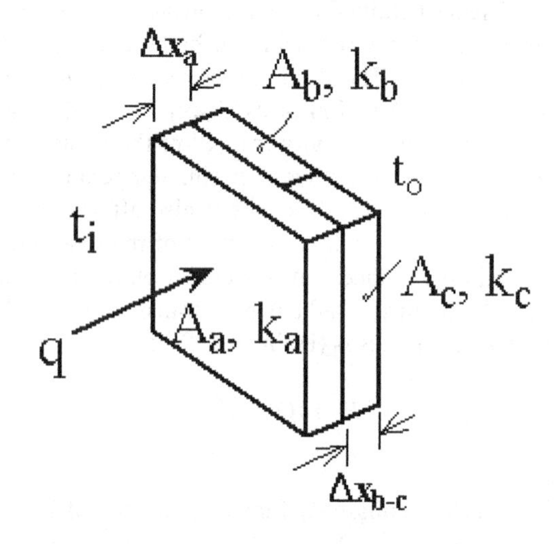

Figure 3.6 Wall with parallel heat flow.

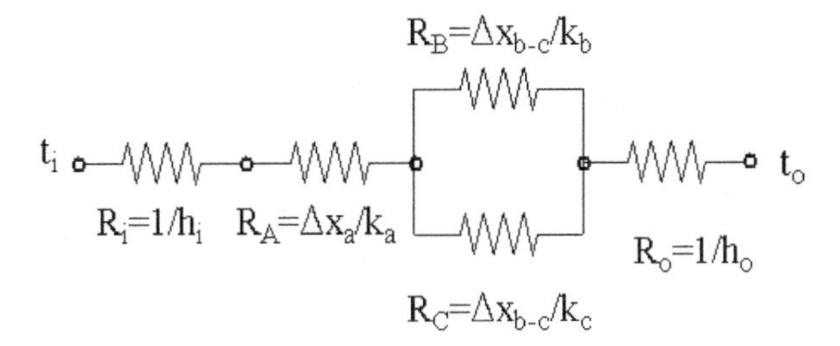

Figure 3.7 Heat resistance for Figure 3.6.

Even in simple walls, such as the one in Figure 3.6, the computation can be cumbersome. Chapter 7 presents a variation of the thermal resistance method that is useful for calculating the overall resistance of composite wall and roof sections. The structure of the calculation is consistent and well suited to spreadsheet formats.

HEAT EXCHANGER BASICS

A variety of heat exchangers is used in the HVAC industry. Figure 3.8 demonstrates one of the most simple variations, referred to as a coaxial tube-in-tube (or double-pipe) design. The flow pattern shown in the figure is counterflow, which is typically the preferred pattern to maximize heat transfer and minimize the required surface area interface between the hot and cold fluids. Fluid velocity is another important parameter in heat exchanger optimization. Higher velocities tend to improve heat transfer, which will enhance overall system performance. However, higher velocities also result in higher friction losses, which will drive up the required pump, fan, piping, and duct sizes needed to circulate the fluids.

Equation 3.13 is the starting point for the simple coaxial heat exchanger. The equation must be modified since the fluid temperatures between the hot (t_h) and cold (t_c) streams change throughout the heat exchanger. Therefore,

$$q = U_{ov}A_o(t_h - t_c)_{mean} = U_{ov}A_o(\Delta t_m). \quad (3.17)$$

It can be shown in the case of fluids with relatively constant specific heats, which do not experience a phase change in the heat exchangers, that the mean temperature difference (Δt_m) is the log-mean temperature difference (LMTD) (Holman 1990), which can be expressed in terms of the hot fluid inlet and outlet temperatures (t_{hi}, t_{ho}) and cold fluid inlet and outlet temperatures (t_{ci}, t_{co}).

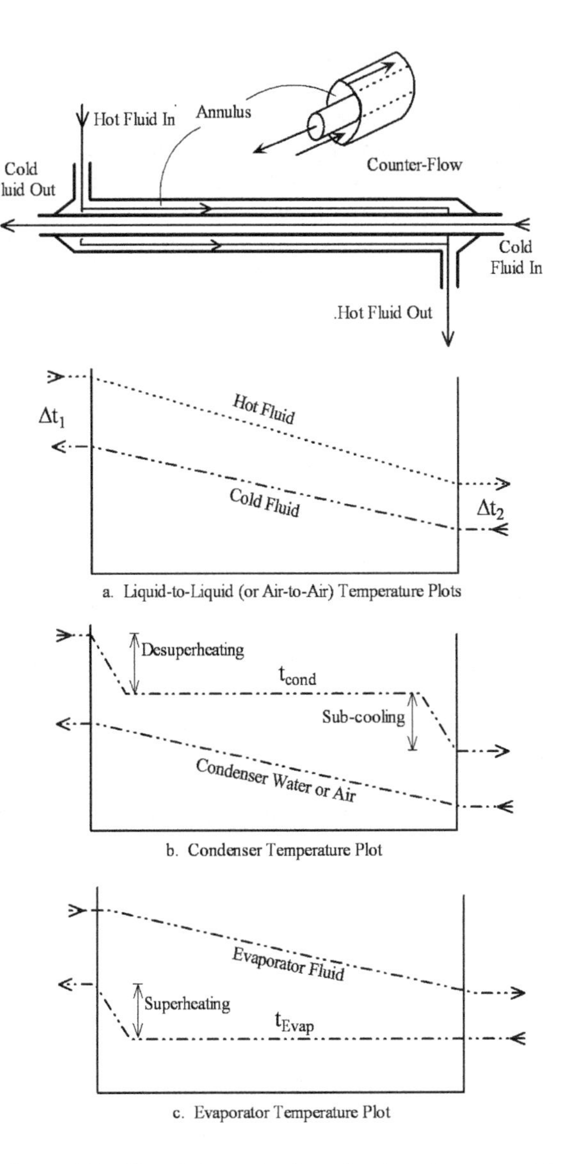

Figure 3.8 Coaxial counterflow heat exchanger with temperature plots.

$$q = U_{ov}A_o(\Delta t_m) = U_{ov}A_o\frac{(t_{hi}-t_{co})-(t_{ho}-t_{ci})}{\ln((t_{hi}-t_{co})/(t_{ho}-t_{ci}))}$$

$$= U_{ov}A_o\frac{\Delta t_1 - \Delta t_2}{\ln(\Delta t_1/\Delta t_2)} \quad (3.18)$$

$$= U_{ov}A_o \times \text{LMTD}, \text{ where LMTD} = \frac{\Delta t_1 - \Delta t_2}{\ln(\Delta t_1/\Delta t_2)}$$

The desired effect is to raise or lower a temperature in at least one of the heat exchanger fluids. For fluids with relatively constant specific heats that do not experience a phase change, the heat exchange rate (q) and fluid temperatures are related by

$$q = m_h c_{ph}(t_{ho} - t_{hi}) = |m_c c_{pc}(t_{co} - t_{ci})|. \quad (3.19)$$

The mass flow rate-specific heat capacity product is defined as the fluid capacity rate ($C \equiv mc_p$). Many heat exchanger performance correlations are based on the minimum fluid capacity rate or the ratio of the minimum to maximum. The minimum fluid capacity is noted by the stream that experiences the greater temperature change in the heat exchanger, which is the hot fluid ($C_{min} \equiv m_h c_{ph}$) in Figure 3.8, Plot a.

The coaxial heat exchanger can also be used as a condenser or evaporator. Temperature plots for these applications are also shown in Figure 3.8. If no subcooling, superheating, or desuperheating occurs, Equation 3.18 applies, except the LMTD is simply the average of Δt_1 and Δt_2. If subcooling, superheating, or desuperheating occurs, Equation 3.17 applies, but the value of Δt_m must be integrated over the entire length of the heat exchanger. An alternative is to subdivide the heat exchanger into two or three sections, apply Equation 3.18 to each, and find a weighted average of Δt_m.

In the case of a condenser, the heat rate is related to the refrigerant enthalpy ($h_{ho} - h_{hi}$) in the case of the condenser and temperature change in the non-phase-changing condensing fluid (water or air) by

$$q = m_h(h_{ho} - h_{hi}) = |m_c c_{pc}(t_{co} - t_{ci})| \,. \tag{3.20}$$

For an evaporator,

$$q = m_c(h_{co} - h_{ci}) = |m_h c_{ph} hi| \,. \tag{3.21}$$

A common design in the HVAC industry is a fin-tube coil, shown in Figure 3.9. This coil type is referred to as a direct expansion (DX) coil when used as an evaporator and as a hydronic coil when used with a liquid. Heat is transferred to or from air flowing around the fins and outer surface of the tubes to the liquid or refrigerant inside the tubes. Thin sheets of metal (typically aluminum) are used as fins to counteract low air-side heat transfer coefficients. Copper tubes are inserted perpendicularly through the fins, which are typically separated from adjacent fins by a 1/16 to 1/8 inch space. Tubes are usually 3/8, 1/2, or 5/8 inch in diameter, and tube-to-tube spacing is typically 2 to 4 diameters. There are several rows of tubes in the direction of airflow, with 3 or 4 being common in small unitary cooling equipment and up to 12 rows in large central air-handling unit coils. The ends of the tubes are connected to adjacent tubes with U-bends or to the inlet or outlet headers. Variations of this design are also used as condensers, with the number of rows typically being one or two since condenser fans are typically axial designs with low static pressure capability.

Design or selection of fin-tube coils requires the specification of the required face area (A_f), fin spacing (or the inverse fin pitch), number of tube rows, and face velocity (V_f) to optimize the heat transfer/head loss trade-off of the air side. In order to minimize air-side pressure losses, maximum face velocities should be limited to 500

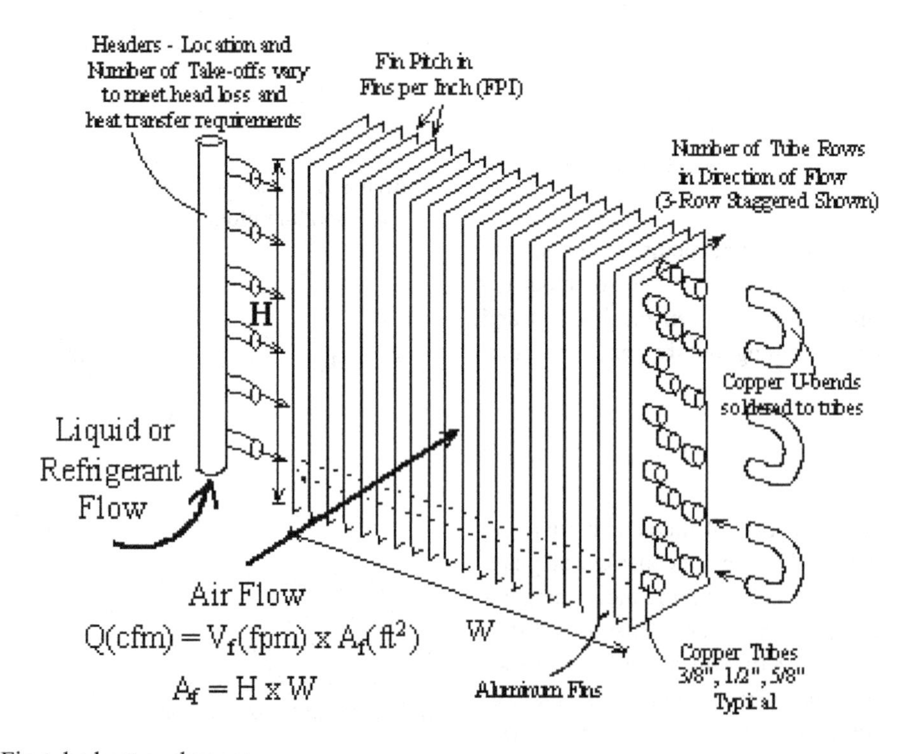

Figure 3.9 Fin-tube heat exchanger.

feet per minute (fpm). Fin pitch is typically 12 fins per inch (fpi) or less to minimize pressure loss and to allow proper cleaning. Recent revisions to ASHRAE Standard 62.1 have included a limit on air-side pressure loss (ASHRAE 2004). The tube-side design requires similar consideration including the optimum number of header take-offs and optimum tube diameter to adjust velocity in the tube for proper heat transfer without excessive friction losses. Coil sizing is typically performed by determining the capacity and friction losses at standard conditions and then correcting these values for a variety of inputs including air temperature and flow, liquid (or refrigerant) flow, and temperature, fin pitch, and number of rows. Chapter 5 contains a performance table and graphs for fin tube heat exchangers.

Most heat exchangers are more complex than the coaxial design. They are often a combination of counterflow or co-current flow (fluids traveling in same direction), cross flow, or a combination of counterflow and co-current and cross flow. Equation 3.17 must be modified to account for these designs with a correction factor (F_c).

$$q = F_c UovA_o(\Delta t_m) \quad (3.22)$$

Most heat transfer texts provide values for the correction factor for a variety of heat exchanger designs. The factors are determined from charts or equations that are functions of inlet temperature, outlet temperature, and capacity rate (Holman 1990). Two common HVAC heat exchangers are shown in Figure 3.10. Plate frame heat exchangers are compact, able to have low approach temperatures (t_2 in Figure 3.8), and relatively easy to disassemble for cleaning. They are commonly used in water-to-water applications to isolate interior loops that

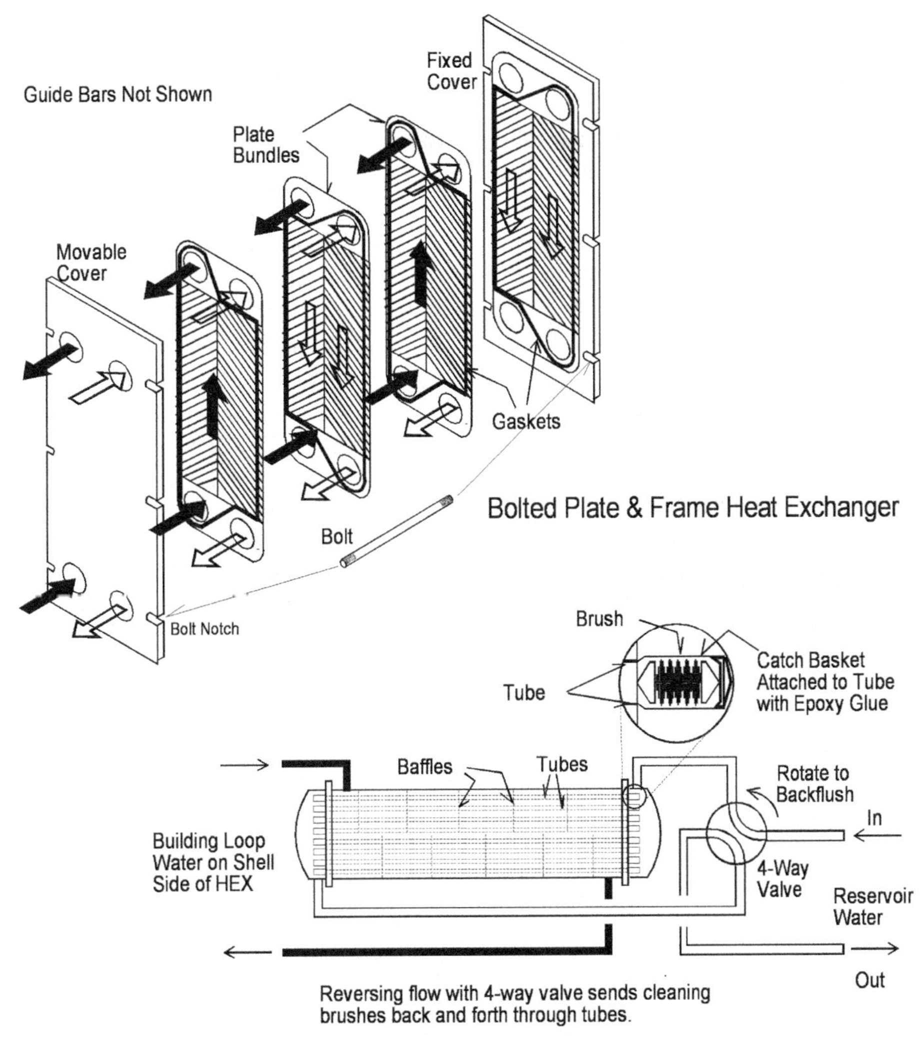

Figure 3.10 Common HVAC heat exchanger types (Kavanaugh and Rafferty 1997).

require relatively clean liquid from outdoor loops with poor water quality. Welded plate heat exchangers are used with refrigerant-to-water applications. Plate heat exchangers (PHEs) are also used as pressure breaks in high-rise buildings. Elevated pressure will result if a single closed loop connects the lower floor to an upper floor. Inserting one or more PHEs in intermediate floors creates multiple closed loops with reduced elevation head. Tube-in-shell heat exchangers are commonly used as condensers and evaporators for chillers. Water is typically circulated through the interior of the tubes (tube side) while the refrigerant is circulated around the outside of the tubes (shell side). Cleaning is typically accomplished by removing the end caps and "rodding" out the tube interior, or automated brush cleaning systems are incorporated as shown in Figure 3.10.

CHAPTER 3 PROBLEMS

3.1 A stream of water flowing at 25 gpm must be cooled from 80°F to 70°F with chilled water at 50°F flowing at 20 gpm in a coaxial counterflow heat exchanger with an overall U-factor of 450 Btu/h· ft^2· °F and 1.25 in. diameter inner tube. Calculate the required length of heat exchanger tubing.

3.2 Find the overall heat transfer coefficient for a schedule 40 steel pipe (d_o = 1.9 in., d_i = 1.61 in., k = 41 Btu/h· ft· °F) with an internal heat transfer coefficient of 48 Btu/h· ft^2· °F and an external coefficient of 20 Btu/h· ft^2· °F.

3.3 A wall is made of a 4 in. thick layer of masonry (0.9 Btu/h· ft· °F) and a 1 in. layer of insulation (k = 0.03 Btu/h· ft· °F). Find the overall thermal resistance if the inner and outer surfaces have heat transfer coefficients of 5.0 Btu/h· ft^2· °F.

3.4 Repeat problem 3.3 if an added layer of ½ in. plywood (0.2 Btu/h· ft· °F) covers 50% of the wall and the remaining 50% is covered by ½ in. thick additional insulation.

3.5 A condenser is to be fabricated from the heat exchanger tubing described in Problem 3.1 for a compressor that flows 950 lb/h of R-134a refrigerant. Find the total required heat transfer rate, the heat required to desuperheat the gas, and the required length of tubing if the overall U-factor is 500 Btu/h· ft^2· °F, the temperature leaving the compressor is 200°F, and the pressure is 185 psig. The condenser exit is saturated liquid at 185 psig and the water temperatures entering and leaving the condenser are 70°F and 80°F, respectively.

3.6 Hot waste water flowing at 20 gpm at 200° F is used to heat 15 gpm of incoming water at 85° F to 125° F in a coaxial-counterflow heat exchange. The copper (k = 220 Btu/h· ft· ° F) inside tube has an outer diameter of 1.125 in. and inside diameter of 1.00 in. Compute the required length of tube for an internal heat transfer coefficient of 750 Btu/h· ft^2· ° F and an outer heat transfer coefficient of 900 Btu/h· ft^2· ° F.

REFERENCES

ASHRAE. 2004. *ANSI/ASHRAE Standard 62.1-2004, Ventilation for Acceptable Indoor Air Quality.* Atlanta: American Society of Heating, Refrigerating and Air-Conditioning Engineers, Inc.

Holman, J.P. 1990. *Heat Transfer*, 7th ed. New York: McGraw-Hill.

Rohsenow, W.M., J.P. Hartnett, and E.N. Ganic. 1985. *Handbook of Heat Transfer Fundamentals*, 2d ed. New York: McGraw-Hill.

Kavanaugh, S.P., and K. Rafferty. 1997. *Ground-Source Heat Pumps: Design of Geothermal Systems for Commercial and institutional Buildings.* Atlanta: American Society of Heating, Refrigerating and Air-Conditioning Engineers, Inc.

4 HVAC Fundamentals: Psychrometrics

The term *moist air* is used to emphasize the importance of both dry air and water vapor in the practice of HVAC design. Psychrometrics deals with the thermodynamics of moist air. Although air is made up of a variety of components, the properties of moist air can be adequately addressed by considering only the two primary components, dry air and water vapor. The amount of moisture in air is very small, and properties are presented per unit mass of dry air.

FUNDAMENTALS OF MOIST AIR

An important step in the analysis and design of HVAC systems is to determine the properties of air in order to provide indoor air quality (IAQ) that promotes occupant comfort and health. Analysis must include consideration of the mixture of dry air and varying levels of moisture. This section reviews fundamental ideal gas concepts and introduces moist air terminology, equations, and tools from the *ASHRAE Handbook—Fundamentals* (ASHRAE 2005). This chapter summarizes the equations used to develop the fundamental graphical tool for moist air analysis and design, the psychrometric chart. The use of one or more of three options is allowed:

1. Analysis and design using the psychrometric chart.
2. Analysis and design using packaged psychrometric software available from vendors or manufacturers.
3. Analysis and design using the psychrometric spreadsheet (*PsychProcess.xls*) on the CD that accompanies this text. This program is in open code and can be modified to suit additional applications.

Dry air is primarily nitrogen (~78%), oxygen (~21%), argon (~1%), carbon dioxide (300–400 ppm),[1] and traces of other gases. It has a molecular weight (MW) of 28.96 lb/lb· mole, and the gas constant for air (R_a) can be determined from the universal gas constant (R).

$$R_a = \frac{R}{MW} = \frac{1545\ (\text{ft} \cdot \text{lb}_f/\text{lb}_{mole} \cdot {}^\circ\text{R})}{28.96\ (\text{lb}_m/\text{lb}_{mole})} = 53.34\ \text{ft} \cdot \text{lb}_f/\text{lb}_m \cdot {}^\circ\text{R} \tag{4.1}$$

The density (ρ) of dry air or its inverse specific volume ($\upsilon = 1/\rho$) can be calculated from ideal gas relationships. At the specified standard conditions of 60°F (520°R) and sea level atmospheric pressure (p = 14.696 psia), the specific volume of dry air (υ_a) is

$$\upsilon_a = \frac{1}{\rho} = \frac{R_a T}{p} = \frac{53.34\ (\text{ft} \cdot \text{lb}_f/\text{lb}_m \cdot {}^\circ\text{R}) \times 520^\circ\text{F}}{14.696\ (\text{lb}_f/\text{in.}^2) \times 144\ \dfrac{\text{in.}^2}{\text{ft}^2}} = 13.1\ \text{ft}^3/\text{lb}_m \tag{4.2}$$

The specific volume can be corrected by inserting an atmospheric pressure corrected for altitude (Z, in feet above sea level) into Equation 4.2 (ASHRAE 2005).

$$p\ (\text{psia}) = 14.696\ (1 - 6.8753 \times 10^{-6}\ Z)^{5.2559} \tag{4.3}$$

It is convenient to express the amount of moisture in air in terms of the humidity ratio (W), which is the mass of water vapor (M_w) per mass of dry air (M_a). Current practice in the US is to use the units of pound mass of water (lb_w) to pound mass of air (lb_a). Some documents continue the use of grains per pound mass of air, where 7000 grains = 1.0 pound mass.

$$W = \frac{M_w}{M_a} \equiv \frac{\text{lb}_w}{\text{lb}_a} \equiv 7000\ (\text{grains}/\text{lb}_w) \times \frac{\text{lb}_w}{\text{lb}_a} \tag{4.4}$$

The amount of water vapor required to saturate air increases with temperature. A widely used term is *relative humidity* (RH), which is the mole fraction (or percent) of water vapor (x_w) present in the air relative to the

1. The concentration of carbon dioxide (CO_2) in parts per million (ppm) in outdoor air is important since it has been widely used in combination with indoor CO_2 concentration as an indicator of adequate ventilation air.

Table 4.1 Thermodynamic Properties of Water at Saturation

Temp (°F)	35	40	45	50	55	60	65	70	75	80	85	90	95	100
p_s (psia)	0.010	0.122	0.0148	0.178	0.214	0.256	0.306	0.363	0.430	0.507	0.596	0.699	0.816	0.950
υ_f(ft³/lb)	0.0160	0.0160	0.0160	0.0160	0.0160	0.0160	0.0160	0.0161	0.0161	0.0161	0.0161	0.0161	0.0161	0.0161
υ_g (ft³/lb)	2946	2444	2036	1703	1431	1206	1021	867	739	633	543	468	404	350
h_f(Btu/lb)	3.0	8.03	13.1	18.1	23.1	28.1	33.1	38.1	43.1	48.1	53.1	58.0	63.0	68.0
h_{fg} (Btu/lb)	1073	1070	1068	1065	1062	1059	1057	1054	1051	1048	1045	1042	1040	1037
h_g (Btu/lb)	1076	1079	1081	1083	1085	1087	1090	1092	1094	1096	1098	1100	1103	1105

mole fraction of saturated air (x_{ws}) at the local temperature (t_a) and pressure (p). This ratio is also the partial pressure (p_w) of the vapor relative to the partial pressure of saturated vapor (p_{ws}). Table 4.1 provides properties of saturated water vapor, including the saturation pressures for temperatures encountered in HVAC applications.

$$\text{RH} = \frac{x_w}{x_{ws}@t_a} = \frac{p_w}{p_{ws}@t_a} \quad (4.5)$$

The saturation pressure (p_s) of water as a function of temperature is a very important correlation in psychrometric analysis. The following equation is recommended (ASHRAE 2005) for temperatures between 32°F (492°R) and 392°F (852°R). A similar equation is available for temperatures below 32°F (492°R) (ASHRAE 2005).

$$p_s = \exp(C_8/t + C_9 + C_{10}t + C_{11}t^2 + C_{12}t^3 + C_{13}\ln t) \quad (4.6)$$

where

p_{ws} ≡ psia
t ≡ °R
C_8 = -1.04403971×10^4
C_9 = -4.8932428
C_{10} = $-5.3765794 \times 10^{-3}$
C_{11} = 1.2890360×10^{-5}
C_{12} = $-2.4780681 \times 10^{-9}$
C_{13} = 6.5459673

MOIST AIR PROPERTY INDICATORS

Air temperature is easily measured, but determination of the amount of moisture in the air is more difficult. Relative humidity sensors are available and can be used with the local air temperature to determine moist air conditions. In addition to relative humidity, two other measurable indicators are the dew-point temperature (t_d) and the wet-bulb temperature (t_{wb} or WB).

The moist air dew-point temperature can be determined by measuring the temperature of a surface when moisture begins to condense. The dew-point temperature also corresponds to the saturation temperature, which is the temperature when the relative humidity of the air is 100%. A correlation for dew-point temperature (t_d ≡ °F) from 32°F to 200°F as a function of the partial pressure of water vapor (p_w ≡ psia) is (ASHRAE 2005):

$$t_d = 100.45 + 33.193(\ln p_w) + 2.319(\ln p_w)^2 + 0.17074(\ln p_w)^3 + 1.2063(p_w)^{0.1984} \quad (4.7)$$

The wet-bulb temperature of the air is determined by placing a thermometer bulb that is covered with a completely wetted wick in an airstream. Moist air near saturation will have a low evaporation rate, and the resulting modest cooling effect will only slightly depress the temperature of the wet bulb. Dryer air will result in a higher evaporation rate, increased cooling effect, and a greater depression of the wet-bulb temperature. Since the wet-bulb temperature is a measure of temperature depression of the air, the temperature of a dry-bulb thermometer (t_{wb} or WB) is also necessary to identify the air conditions.

As mentioned previously, the important indicators for moist air property determination are the humidity ratio (W) and the easily measured dry-bulb temperature. The humidity ratio is determined from the local atmospheric pressure (p) and the vapor pressure (p_w) using the relationships of molecular weight (MW), mole fractions (x), and the partial pressures of water (p_w) and air (p_a):

$$W = \frac{M_w}{M_a} = \frac{MW_w \times x_w}{MW_a \times x_a} = \frac{18.01528 \times x_w}{28.9645 \times x_a} = 0.62198\frac{p_w}{p_a} = 0.62198\frac{p_w}{p - p_w} \quad (4.8)$$

To determine humidity ratio when the dry-bulb temperature and the relative humidity are known, the value of p_w can be determined using Equation 4.5 while the value of p_{ws} is found using either Equation 4.6 or Table 4.1. When the dry-bulb temperature and the dew-point temperature are known, p_w in Equation 4.8 is the saturation pressure (p_{ws}) at the dew-point temperature, which can also be found using either Equation 4.6 or Table 4.1. When the dry-bulb temperature and wet-bulb temperature (t_{wb}) are known, the humidity ratio is found from

$$W\ (\mathrm{lb_w/lb_a}) = \frac{(1093 - 0.556t_{wb}\ (°\mathrm{F}))W_s\ (\mathrm{lb_w/lb_a}) - 0.24(t - t_{wb})\ (°\mathrm{F})}{1093 + 0.444t\ (°\mathrm{F}) - t_{wb}\ (°\mathrm{F})}, \quad (4.9)$$

where t and t_{wb} are in °F, and W_s is determined by inserting the saturation pressure of water vapor at the wet-bulb temperature into Equation 4.8.

In summary, to determine moist air properties, the dry-bulb temperature and the humidity ratio are needed. The humidity ratio can be determined from one of three indicators: wet-bulb temperature, relative humidity, or dew-point temperature.

Equations for Moist Air Properties

The thermodynamic properties of moist air can be determined from the dry-bulb temperature and humidity ratio. These include the enthalpy, specific volume (or its inverse, density), and specific heat. In the US, the current convention is to set base values at 0°F and compute the values at other temperatures. At 0°F, h_w = 1061 Btu/lb and h_a = 0 Btu/lb. The specific heat of air is 0.24 Btu/lb· °F and water vapor is 0.444 Btu/lb· °F. For moist air at dry-bulb temperature (t) and humidity ratio (W),

$$h\ (\mathrm{Btu/lb_a}) = 0.24t\ (°\mathrm{F}) + W\ (\mathrm{lb_w/lb_a})(1061 + 0.44t\ (°\mathrm{F}))\,. \quad (4.10)$$

The specific heat of moist air is

$$c_p\ (\mathrm{Btu/lb_a} \cdot °\mathrm{F}) = 0.24 + 0.444W\ (\mathrm{lb_w/lb_a})\,. \quad (4.11)$$

The specific volume of moist air is

$$\upsilon\ (\mathrm{ft^3/lb}) = 0.37059t\ (°\mathrm{F}) + 459.67[1 + 1.6078W\ (\mathrm{lb_w/lb_a})]/p\ (\mathrm{psia})\,. \quad (4.12)$$

THE PSYCHROMETRIC CHART AND COMPUTER CODE

Many of the properties of moist air discussed in the previous section can be estimated from a psychrometric chart, a plot of dry-bulb temperature (horizontal axis) and humidity ratio (vertical axis). A psychrometric chart for sea level is shown in Figure 4.1. Dry-bulb temperatures on the chart range from the freeze point of water (32°F) to 120°F. Humidity ratios range from dry air (W = 0) to 0.028 $\mathrm{lb_w/lb_a}$ (196 grains/$\mathrm{lb_a}$). Lines of constant humidity ratio are horizontal on the graph. Lines of constant temperature exhibit a slight tilt to the left of vertical, with the degree of tilt increasing with lower temperature. As shown in the figure, the left top portion of the plot is terminated at the saturation line, which represents both the 100% relative humidity curve and the plot for dew-point temperature. Lines of constant enthalpy appear as straight lines that slope down as the temperature increases. Lines of constant wet-bulb temperature are nearly parallel to lines of constant enthalpy. Lines of constant specific volume also slope down as temperature increases but at a much greater angle compared to enthalpy and wet-bulb lines. Lines of constant relative humidity curve up as temperature increases. This is a result of the fact that lines of constant RH are formed by plotting a set of points that are a fractional

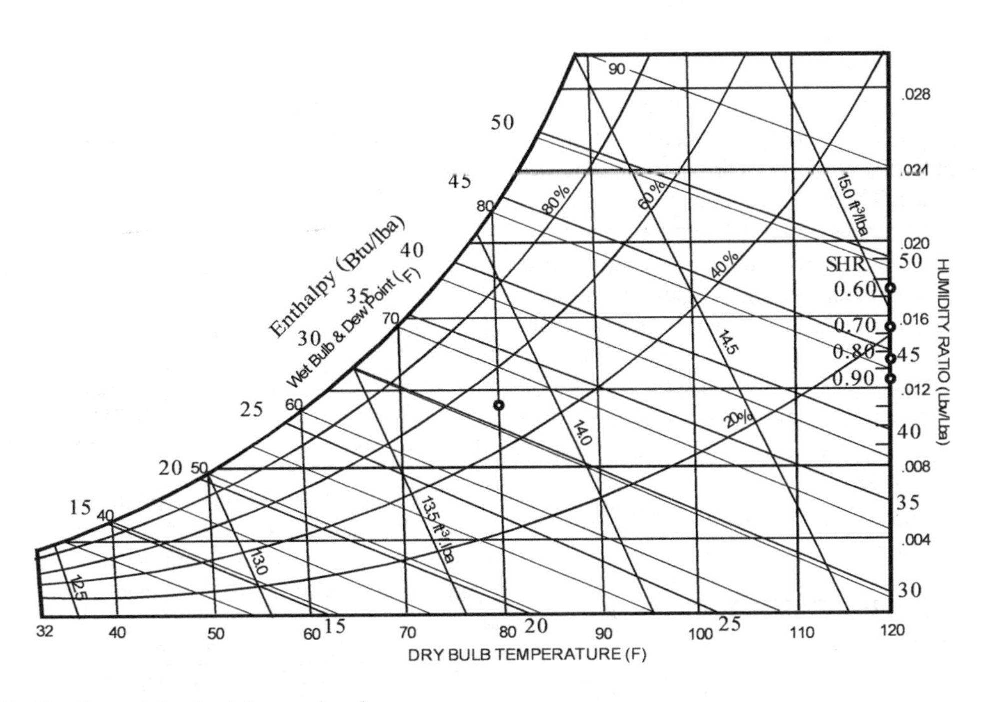

Figure 4.1 Psychometric chart for sea level.

distance between the saturation curve (RH = 100%) and the horizontal dry-bulb temperature line that represents zero relative humidity (RH = 0%).

Use of the psychrometric chart can be demonstrated at the standard indoor air rating point for cooling equipment, which is noted on the chart with a small round circle. The point has a dry-bulb temperature of 80°F and a wet-bulb temperature of 67°F. Using this condition as a starting point on the chart, other properties are found.

- Humidity ratio—Move right horizontally to the axis and read, $W \approx 0.0112$ lb_w/lb_a
- Relative humidity—Interpolate between the 40% and 60% lines, RH ≈ 52%
- Enthalpy—Interpolate between the 30 and 35 lines, $h \approx 32$ Btu/lb
- Dew point—Move left horizontally to the saturation curve and read, $t_d \approx 61$°F
- Specific volume—Interpolate between the 13.5 and 14.0 lines, $\upsilon \approx 13.8$ ft³/lb

Estimating values using reduced-size psychrometric charts, such as Figure 4.1, can lead to some error. For greater detail and improved accuracy, refer to Appendix B, Figure B.4, or use the psychrometric properties spreadsheet (*PsychProcess.xls*) on the accompanying CD.

Much greater accuracy can be realized and is often necessary with a larger psychrometric chart or computer-based computations. It is also important to recognize that most psychrometric charts are for sea level and significant error can result if they are used to find properties for higher elevations.

The computer program shown in Figure 4.2 incorporates the equations presented in this section. It can be used to compute the same psychrometric properties found on the chart. The required input values are dry-bulb and wet-bulb temperatures (°F) and elevation (in feet above sea level). It is not limited to a single elevation like psychrometric charts. The program listing does not include dew-point temperature, but a code for Equation 4.7 can be added.

The spreadsheet program *PsychProcess* (on the accompanying CD) has a "Properties" worksheet that computes psychrometric properties using the equations in the preceding sections. It uses Visual Basic® macros for specific humidity, relative humidity, and dew-point temperature modified from the code shown in Figure 4.2.

SIMPLIFIED PSYCHROMETRIC EQUATIONS OF HVAC PROCESSES

The psychrometric chart or equivalent equations and computer codes can be used to predict the results of HVAC processes, including heating, cooling, dehumidifying, humidifying, energy recovery, and mixing. Outlet conditions (temperature [t_2], enthalpy [h_2], humidity ratio [W_2], specific volume [υ_2]) can be calculated if inlet conditions (temperature [t_1], enthalpy [h_1], humidity ratio [W_1], specific volume [υ_1]), airflow rates (Q), total heat transfer (q), and either sensible (q_s) or latent heat transfer (q_L) are known. If inlet air conditions (t_1, h_1, W_1, υ_1), airflow rate (Q), and outlet conditions (t_2, h_2, W_2, υ_2) are known, the total, sensible, and latent heat transfer rates (q, q_s, q_L) can be found.

Sensible (Only) Air Heating or Cooling

Values for the sensible heating and cooling process shown in Figure 4.3 can be determined from *ASHRAE* (2005) (Point 1 = inlet, Point 2 = outlet):

$$q_{s(1\text{-}2)}\ (\text{Btu/h}) = m_a(h_2 - h_1)$$
$$= \frac{Q\ (\text{cfm}) \times 60\ \text{min/h}}{\upsilon\ (\text{ft}^3/\text{lb})} c_p\ (\text{Btu/lb}\cdot{}^\circ\text{F}) \times (t_2 - t_1)\ ({}^\circ\text{F}) \tag{4.13}$$

At standard air conditions of t = 70°F and RH = 50%, c_p = 0.243 Btu/lb· °F, υ = 13.5 ft³/lb,

$$q_s\ (\text{Btu/h}) \approx 1.08 \times Q\ (\text{cfm}) \times (t_2 - t_1)\ {}^\circ\text{F}. \tag{4.14}$$

(***Note:*** If standard conditions are specified as 60°F and 100% RH, replace 1.08 with 1.1.)

Mixing of Airstreams 1 and 2 (Sensible and Latent)

A psychrometric chart can be used to graphically find the mixed air conditions (point 3) when air at point 1 is mixed with air at point 2 on the chart. Points 1 and 2 are located on a psychrometric chart, and a line is drawn between them (as shown in Figure 4.4). Point 3 (which will be on line 1-2) is located by measuring the length of the line from 1 to 2 and finding the length of the line from 1 to 3 by applying Equation 4.15.

$$\overline{1-3} = \overline{1-2}\frac{m_2}{m_3} = \overline{1-2}\frac{m_2}{m_1 + m_2} = \overline{1-2}\frac{Q_2/\upsilon_2}{Q_1/\upsilon_1 + Q_2/\upsilon_2}$$
$$\text{and if } \upsilon_1 \approx \upsilon_2$$
$$\overline{1-3} \approx \overline{1-2}\frac{Q_2}{Q_1 + Q_2} \tag{4.15}$$

The other conditions at point 3 are read from the chart. The conditions of the mixed air can also be found from the results of total and sensible energy balances and air and moisture mass balances:

$$t_3 = \frac{t_2 + t_1 m_1/m_2}{1 + m_1/m_2};\ h_3 = \frac{h_2 + h_1 m_1/m_2}{1 + m_1/m_2};\ W_3 = \frac{W_2 + W_1 m_1/m_2}{1 + m_1/m_2} \tag{4.16a,b,c}$$

```
'Convert Wet Bulb Temperature to Degree Rankin
RT = wb + 459.67

'Compute Atmos. Pressure in psia from elevation in feet above sea level
pt = 14.696 * (1 – 0.0000068753 * ElevInFt) ^ 5.2559

' Use Equation 4.6 to find saturation pressure at the wet bulb temperature
c8 = -10440.4
c9 = -11.29465
c10 = -0.027022355
c11 = 0.00001289036
c12 = -0.000000002478068
c13 = 6.5459673
pws = Exp(c8 / RT + c9 + c10 * RT + c11 * RT ^ 2 + c12 * RT ^ 3 + c13 * Log(RT))

' Use Equation 4.8 to find W at saturation
wsat = (pws * 0.62198) / (pt – pws)

' Use Equation 4.9 to find W
wnom = (1093 – 0.556 * wb) * wsat – 0.24 * (db – wb)
wdenom = 1093 + 0.444 * db – wb
W = wnom / wdenom

'Rearrange Equation 4.8 to find water vapor pressure
pw = (W * pt) / (0.62198 + W)
RT = db + 459.67

' Use Equation 4.6 to find pressure at the dry bulb temperature
ps = Exp(c8 / RT + c9 + c10 * RT + c11 * RT ^ 2 + c12 * RT ^ 3 + c13 * Log(RT))
pa = pt – pw

'Combine Equations 4.5 and 4.8 to find relative humidity
RelHum = 100 * (W * pa) / (0.62198 * ps)

'Use Equation 4.10 to find h in Btu/lb
TDB=RT-459.67
Enthalpy = 0.24*TDB+W*(1061*0.444*TDB)

'Use Equation 4.11 to find specific heat in Btu/lb-°F
SpcHt = 0.24+0.444*W

' Use Equation 4.12 to find specific volume in cubic feet per lb
SpcVol = 0.3705*RT*(1+1.6078*W)/pt
```

Figure 4.2 Psychrometric computer program.

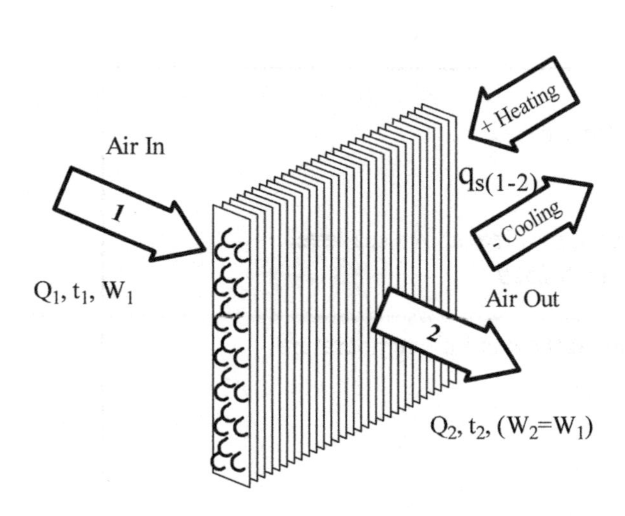

Figure 4.3 Air coil sensible cooling or heating (no dehumidification, no humidification).

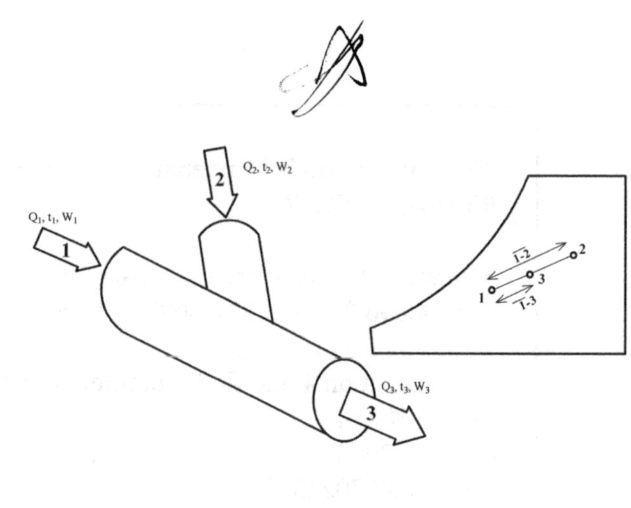

Figure 4.4 Location of mixed air condition on psychrometric chart.

And if $\upsilon_1 \approx \upsilon_2$,

$$t_3 = \frac{t_2 + t_1 Q_1/Q_2}{1 + Q_1/Q_2}; h_3 = \frac{h_2 + h_1 Q_1/Q_2}{1 + Q_1/Q_2}; W_3 = \frac{W_2 + W_1 Q_1/Q_2}{1 + Q_1/Q_2} \tag{4.17a,b,c}$$

Total Cooling
(Temperature Reduction and Dehumidification)

Total cooling indicates the temperature and the humidity ratio are reduced as the air moves through the cooling coil from point 1 to point 2, as shown in Figures 4.5 and 4.6. The importance of this process is becoming more critical as designers attempt to enhance the moisture removal capability of HVAC systems. Total cooling coils are typically the primary component to accomplish this task. So it is important to match the sensible and latent capacities of these coils to the corresponding loads on the building. Water is condensed on the coil surface, collected into a pan, and drains out of the system into a P-shaped trap (to prevent air from being drawn through the water outlet tube). From the *ASHRAE Handbook—Fundamentals* (ASHRAE 2005), neglecting the small amount of energy in the condensate,

$$q_{(1\text{-}2)}\ (\text{Btu/h}) = m_a(h_2 - h_1) = \frac{Q\ (\text{cfm}) \times 60\ \text{min/h}}{\upsilon\ (\text{ft}^3/\text{lb})}(h_2 - h_1)\ (\text{Btu/lb}) . \tag{4.18}$$

At the specified standard air conditions of t = 70°F and RH = 50%,

$$q\ (\text{Btu/h}) \approx 4.44 \times Q\ (\text{cfm}) \times (h_2 - h_1)\ \text{Btu/lb}_a . \tag{4.19}$$

(*Note:* If standard conditions are specified as 60°F and 100% RH, replace 4.44 with 4.5.)

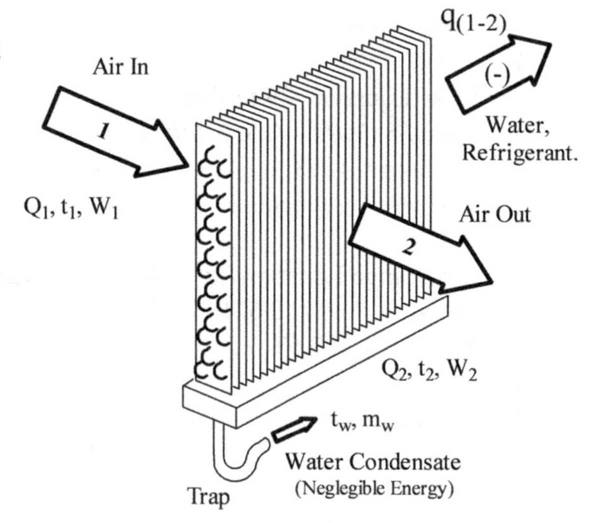

Figure 4.5 Air coil total cooling.

The latent cooling capacity (q_L) or dehumidification capacity of the cooling coil can be determined from

$$q_{L(1\text{-}2)} = m_w h_{fg} = m_a h_{fg}(W_2 - W_1)$$

$$q_{L(1\text{-}2)}\ (\text{Btu/h}) = \frac{Q\ (\text{cfm}) \times 60\ \text{min/h}}{\upsilon\ (\text{ft}^3/\text{lb})}[h_{fg}\ (\text{Btu/lb})\ (W_2 - W_1)\ (\text{lb}_w/\text{lb}_a)] . \tag{4.20}$$

At standard air conditions of t = 70°F and RH = 50%,

$$q_L\ (\text{Btu/h}) \approx 4680 \times Q\ (\text{cfm}) \times (W_2 - W_1)\ \text{lb}_w/\text{lb}_a . \tag{4.21}$$

The sensible heat ratio (SHR) is defined as

$$\mathrm{SHR} = \frac{q_s}{q} = \frac{q_s}{q_s + q_L}. \quad (4.22)$$

The SHR of the coil is an important concept since both temperature and humidity must be maintained to control indoor conditions. The SHR_{Unit} of the coil must be selected to match the SHR_{Bldg} of the building load in order to simultaneously keep both the temperature and the humidity within comfortable limits. Coils with higher SHRs may reduce temperature too quickly for adequate moisture removal to occur. Coils with low SHRs may dry out the air too much and the indoor air may become too dry. However, low SHR is somewhat self-correcting since the dry air will result in a coil with a higher SHR (which will have a lower dehumidification capability).

The SHR can also be determined graphically from the psychrometric chart by plotting a straight line from the inlet air condition (point 1) to the outlet air condition. A second line is drawn parallel to the first line through the 80°F/67°F line-up point to the right vertical axis of the chart. Points representing SHR = 0.60, 0.70, 0.80, and 0.90 are located on this axis. The SHR is estimated by interpolating between these points and the intersection of the second line and the vertical axis.

The psychrometric chart in Appendix B uses an SHR protractor in the upper-left portion of the chart rather than the line-up points shown on Figure 4.1. This reduces clutter on the chart but makes constructing a parallel line somewhat more difficult (Figure 4.6).

Bypass Factor/Apparatus Dew Point

Many manufacturers provide the bypass factor and an apparatus dew point so that outlet air conditions can be determined. The bypass factor (BF) is defined as the equivalent fraction of air that bypasses the cooling coil while the remaining portion of the air is completely cooled to the apparatus dew-point temperature (t_{adp}) on the saturation line of the psychrometric chart. The apparatus dew point is essentially the average surface temperature on the outside of the cooling coil. The location of the outlet conditions can be found by drawing a line from the inlet condition point (1) on the psychrometric chart to t_{adp}, as shown in Figure 4.7, and then calculating the outlet dry-bulb temperature by rearranging Equation 4.23a to solve for t_2 using Equation 4.23b. Point (2) will be located on the line connecting (1) and t_{adp}.

$$BF = \frac{t_2 - t_{adp}}{t_1 - t_{adp}} \rightarrow t_2 = BF(t_1 - t_{adp}) + t_{adp} \quad (4.23a,b)$$

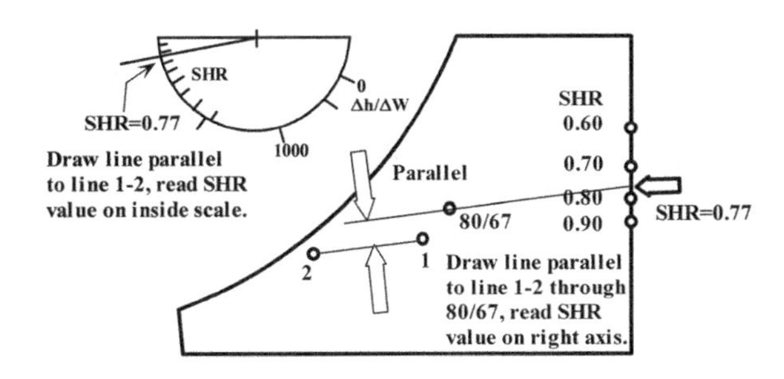

Figure 4.6 Two methods to estimate sensible heat ratio from psychrometric charts.

The sensible heat ratio (SHR) can be found by applying Equation 4.13 or 4.14 to find sensible cooling, Equation 4.18 or 4.19 to find the total cooling, and Equation 4.22 to find SHR.

An alternative method is to draw a line that is parallel to a line drawn from points (1) to (2) through the line-up point (80°F/67°F) on the psychrometric chart as described previously or to use the SHR protractor in the upper-left corner of a standard ASHRAE psychrometric chart.

Ventilation Air Coils and Heat Recovery Units

In some cooling applications, especially those in humid and moderate climates with high ventilation air requirements, a conventional total cooling coil cannot adequately meet the latent cooling requirement. Likewise, in cold climates, a very large component of the total heating requirement results from the need to heat the outside air that must be supplied to the building to maintain air quality. Conventional coils can be used to precool (and dehumidify) and heat ventilation air. Equations and methods of analysis for these coils are identical to those used for total cooling and heating. However, the outdoor air conditions (t_o, h_o, W_o, υ_o) would be used for the inlet air conditions (t_1, h_1, W_1, υ_1).

Heat recovery units are being applied more frequently to capture the energy in building exhaust and thereby reduce the amount of energy required to preheat and precool ventilation air. Performance is expressed in terms of effectiveness (ε), a percentage of the energy that can be recovered if the air leaving the HRU were brought to the return air temperature. Figure 4.8 shows a total heat recovery unit (wheel type) that can recover both latent and sensible heat from the exhaust airstream. In the winter, a wheel absorbs heat and moisture from the exhaust air and rotates to the ventilation airstream

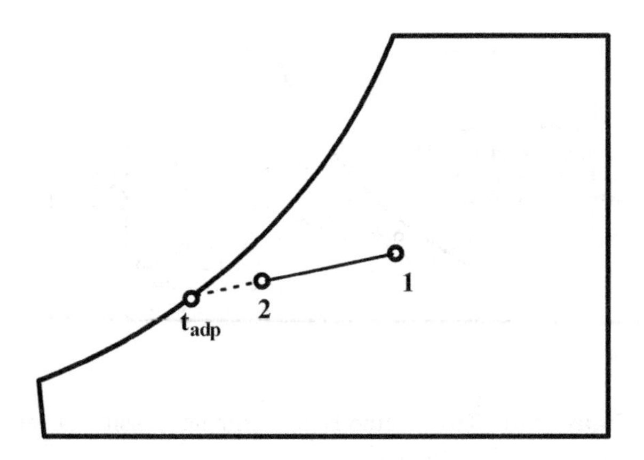

Figure 4.7 Location of apparatus dew-point temperature for bypass factor method.

where the heat and moisture are released. In the summer, the process is reversed so that the moist and hot air can release heat and moisture to the cooler and dryer exhaust air. A plate-type heat exchanger is also shown, but only sensible heat is transferred. Total HRUs are also available that substitute a membrane for the solid plate so that both sensible heat and moisture can be transferred without the moving parts of the wheel-type HRU. A variety of other types of heat recovery devices (heat pipes, run-around coils, molecular sieves) are also available.

Total, sensible, and latent effectiveness (ε_T, ε_S, ε_L) are defined by Equations 4.24a, 4.25a, and 4.26a (ASHRAE 2004). The equations are rearranged to find the conditions of the air leaving the heat recovery unit (enthalpy [h_{hru}], temperature [t_{hru}], and humidity ratio [W_{hru}]) based on the conditions of the outdoor ventilation air (t_o, h_o, W_o) and the exhausted return air (t_r, h_r, W_r) from the building, as shown in Equations 4.24b, 4.25b, and 4.26b.

Effectiveness — Heat Recovery Outlet

$$\varepsilon_T = \frac{m_s}{m_{min}} \times \frac{h_o - h_{hru}}{h_o - h_r} \rightarrow h_{hru} = h_o - \varepsilon_T \times \frac{m_{min}}{m_s} \times (h_o - h_r) \tag{4.24a,b}$$

$$\varepsilon_S = \frac{m_s}{m_{min}} \times \frac{t_o - t_{hru}}{t_o - t_r} \rightarrow t_{hru} = t_o - \varepsilon_S \times \frac{m_{min}}{m_s} \times (t_o - t_r) \tag{4.25 a,b}$$

$$\varepsilon_L = \frac{m_s}{m_{min}} \times \frac{W_o - W_{hru}}{W_o - W_r} \rightarrow W_{hru} = W_o - \varepsilon_L \times \frac{m_{min}}{m_s} \times (W_o - W_r) \tag{4.26 a,b}$$

where m_s = mass flow of outside air, m_e = mass flow of exhaust air, and m_{min} = the smaller of m_s and m_e.

Care must be exercised to minimize the added amount of fan energy for heat recovery units compared to conventional ventilation air systems. In part this is due to the friction of the HRU elements and the required air filters. In addition to the outdoor air supply fan, the exhaust fan must also overcome this added resistance, which increases the consumption compared to conventional low head, high volume axial exhaust fans. The added supply fan and motor heat is beneficial in the heating mode, but it is an additional penalty in the cooling mode. Thus, designers in warmer climates must carefully weigh the true added energy-saving benefit of HRUs. On a seasonal basis, the savings may be nonexistent compared to mechanical precooling, with lower fan power requirements since HRUs are not of significant benefit when outdoor temperatures are moderate (60°F to 85°F) and are a penalty during periods when air-side economizer operation is useful (<60°F).

To estimate the fan heat penalty (q_{OASFan}) and power consumption (kW_{HRU}) of a heat recovery unit, the following is suggested:

$$q_{OASFan}\ (\text{Btu/h}) = \frac{2545\ \text{Btu/hp} \cdot \text{h} \times w_{OASFan}\ (\text{hp}) \times \%\text{Load}/100}{\eta_{OASFanMtr}} \tag{4.27}$$

$$kW_{HRU} = \frac{0.746\ \frac{\text{kW}}{\text{hp}} \times \frac{\%\text{Load}}{100} \times w_{OASFan}\ (\text{hp})}{\eta_{OASMtr}} + \frac{0.746\ \frac{\text{kW}}{\text{hp}} \times \frac{\%\text{Load}}{100} \times w_{ExFan}\ (\text{hp})}{\eta_{ExFanMtr}} + \frac{0.746\ \frac{\text{kW}}{\text{hp}} \times \frac{\%\text{Load}}{100} \times w_{Wheel}\ (\text{hp})}{\eta_{WheelMtr}} \tag{4.28}$$

where

w_{OASFan} = rated outdoor air supply fan motor power, hp
w_{ExFan} = rated exhaust air fan motor power, hp
$w_{WheelMtr}$ = rated enthalpy wheel motor power, hp
$OASFan$ = outdoor air supply fan motor efficiency, %
$ExFan$ = exhaust air fan motor efficiency, %
$OASFan$ = enthalpy wheel motor efficiency, %
$\%Load$ = percent of actual load to motor rated load, %

Figure 4.9 is included to estimate the average efficiency of typical electric four-pole AC motors (nominal 1800 rpm). Chapter 11 contains more detailed information on motor and variable-speed drive efficiencies.

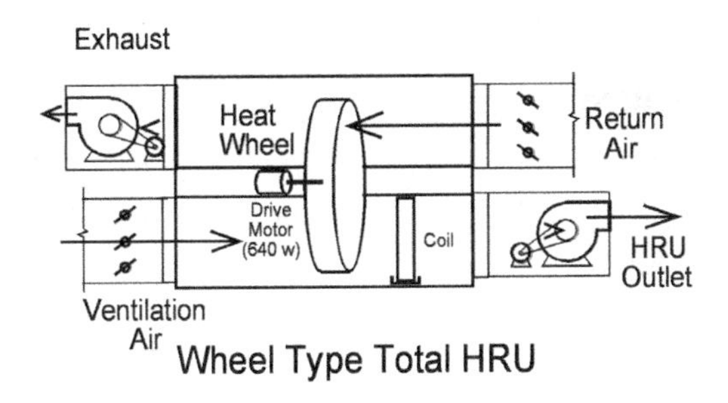

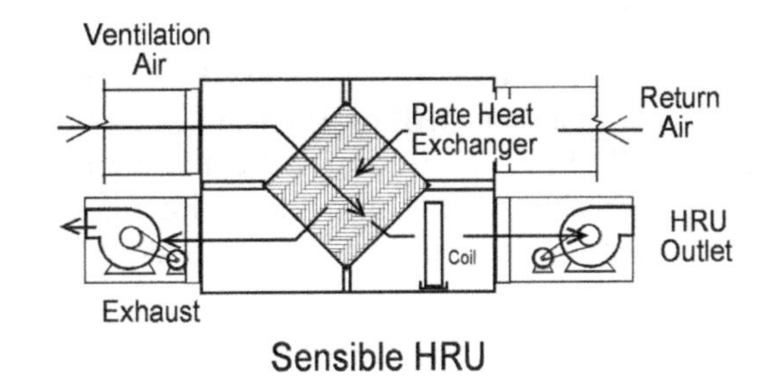

Figure 4.8 Two types of heat recovery units.

CHAPTER 4 PROBLEMS

4.1 A sling psychrometer measures the air temperatures to be 85°F dry bulb and 72°F wet bulb. Find: relative humidity, dew-point temperature, humidity ratio (in lbmv/lbma and grains), specific volume, and enthalpy. Show results on a chart and verify with the program *PsychProcess.xls* (on the accompanying CD). Assume sea level elevation.

4.2 Air flowing at 4000 cfm is heated from 70°F (RH = 40%) at rate of 95,000 Btu/h. Find the outlet air conditions (db, RH, wb, υ). Sketch the process on a psychrometric chart.

4.3 Outside air (100°F/75°F) flowing at 1000 cfm is mixed with return air (75°F/63°F) at 5000 cfm. Find the mixed air conditions (db, RH, wb, υ, and *h*). Sketch on the psychrometric chart.

4.4 A gas furnace produces 60,000 Btu/h with an airflow of 1400 cfm heated air with an inlet condition of 65°F (RH = 45%). Find the outlet air conditions (db, RH, wb, υ). Sketch the process on a psychrometric chart.

4.5 Outside air (95°F/75°F) flowing at 2500 cfm is mixed with return air (75°F/63°F) at 7500 cfm. Find the mixed air conditions (db, RH, wb, υ, and *h*). Sketch on the psychrometric chart.

4.6 A quantity of 1600 cfm of air at 80°F/67°F enters an evaporator coil with a 0.12 bypass factor and a 45°F apparatus dew point. Find the outlet air conditions (db, wb, RH, *h*), the sensible cooling capacity, the latent cooling capacity, total cooling capacity, and the SHR of the coil. Sketch on the psychrometric chart.

4.7 A 500 cfm outdoor air heat recovery unit (HRU) has a total effectiveness of 75% (both sensible and latent are equal). If the exhaust and makeup airflow rates are equal, find the conditions of the air (db, wb, *h*) leaving the HRU and entering the room when outdoor conditions are 94°F/77°F and the room air entering the HRU is 75°F/63°F. What is the capacity of this unit?

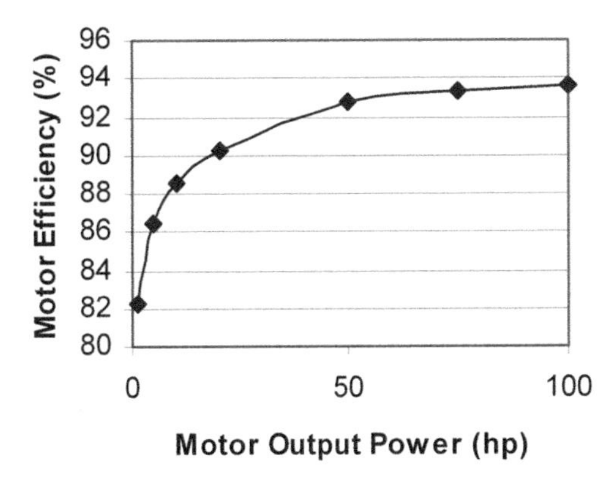

Figure 4.9 EPACT efficiency for 1 to 100 hp, four-pole motors (NEMA 1997).

4.8 A sensible heat recovery unit (HRU) with 80% efficiency draws in 1000 cfm of outside air at –10°F and exhausts an equal amount of room air at 70°F. Calculate the air temperature leaving the HRU and entering the room. What is the capacity of this unit? What is the capacity for 40°F outside air? Calculate the EER (= capacity in Btu/h ÷ power input in W) for both conditions if two fans that draw 700 W each are used.

4.9 A quantity of 2500 cfm of air at 82°F/70°F enters an evaporator coil with a 0.08 bypass factor and a 45°F apparatus dew point. Find the outlet air conditions (db, wb, RH, *h*), the sensible cooling capacity, the latent cooling capacity, total cooling capacity, and the SHR of the coil. Sketch on the psychrometric chart.

4.10 A room at 75°F/63°F has a 36,000 Btu/h total capacity with a room SHR of 0.90 and an outdoor air (95°F/75°F) requirement of 400 cfm. Find the required sensible capacity and total cooling capacity of a unit to handle the building and outdoor air loads.

4.11 Air flowing at 1500 cfm is heated from 65°F (RH = 35%) at a rate of 50,000 Btu/h. Find the outlet air conditions (db, RH, wb, υ). Sketch the process on a psychrometric chart.

4.12 Air flowing at a rate of 2000 cfm at 78°F/65°F enters a cooling unit with a total capacity (TC) of 60,000 Btu/h and a sensible heat ratio (SHR) of 0.75. Calculate the dry bulb, wet bulb, and relative humidity of the air leaving the coil. Determine the apparatus dew point and the bypass factor.

REFERENCES

ASHRAE. 2005. *2005 ASHRAE Handbook—Fundamentals*, chapter 6, Psychrometrics. Atlanta: American Society of Heating, Refrigerating and Air-Conditioning Engineers, Inc.

ASHRAE. 2004. *2004 ASHRAE Handbook—HVAC Systems and Equipment*, chapter 44, Air-to-Air Energy Recovery. Atlanta: American Society of Heating, Refrigerating and Air-Conditioning Engineers, Inc.

NEMA. 1997. *NEMA Standard MG1-12.59*, Table 12-10. Rosslyn, VA: National Electrical Manufacturers Association.

5 HVAC Equipment, Systems, and Selection

New options for HVAC equipment and systems are being introduced continuously, and the design professional must be knowledgeable about both existing and emerging technologies. This chapter will discuss a number of the most common types of equipment, beginning with the simple and progressing to more complex equipment and the systems in which it is applied. Designers are concerned with selecting equipment that will meet the performance needs of the building in the worst case and during seasonal variations. Equipment performance is often presented in manufacturers' literature at conditions that vary considerably from local design conditions. For example, unitary cooling equipment performance may be presented at near maximum recommended indoor airflow rates to achieve optimum efficiency values. In humid climates, operating air distribution systems at lower airflow rates (300–350 cfm/ton) is a viable option for improved equipment latent capacity to enhance the system's ability to maintain lower indoor relative humidity. Data from manufacturers must be available to correct for a variety of nonstandard conditions, and engineers must know how to use them effectively.

This text will provide performance data at a variety of conditions to complement the diagrams and discussions of various equipment and systems. Performance data cannot be presented for all equipment and systems in all ranges of available sizes and operating conditions. However, a spreadsheet will be discussed that is able to correct conventional equipment performance to within reasonable accuracy. This information and spreadsheet tool are provided to reduce the uncertainty in equipment selection that can lead to equipment oversizing. Equipment oversizing results in added cost not only in the equipment but throughout the system (i.e., larger ducts, larger pipes, larger electrical circuits, etc.). Furthermore, inefficiency and lack of comfort will result due to frequent on-off-on cycles (unitary) and/or operation at near-minimum efficiency (low part-load) conditions.

Multiple organizations have responsibility for developing and revising the test standards and ratings of HVAC equipment (e.g., ASHRAE 2003, chapter 56). ASHRAE develops both testing and rating standards for a broad set of equipment, and the standards are approved by ASHRAE as well as the American National Standards Institute (ANSI). The Air-Conditioning and Refrigeration Institute (ARI) has jurisdiction over equipment that involves refrigeration devices (air conditioners, refrigerators, freezers, coolers, etc.), including equipment that heats air (heat pumps) and "heat pump" water heaters. A parallel organization that represents combustion equipment is the American Gas Association (AGA), which develops standards in cooperation with ASHRAE and ANSI. These organizations cover the largest portion of the equipment discussed in this chapter. Later chapters will discuss other organizations and rating standards for other types of auxiliary HVAC equipment.

UNITARY PACKAGED EQUIPMENT—RESIDENTIAL AND SMALL COMMERCIAL

The term *unitary* suggests that all of the cooling and heating equipment is housed in a single compartment that requires no assembly at the site. Cooling-only systems have all the basic components (compressor, evaporator, condenser, expansion device, fans, motors, controls, and cabinet) assembled and charged at the factory. Heating and cooling systems may also have an integrated furnace (electric or fossil fuel) or a reversing valve so that heat pump operation is available. Many packaged systems are nonducted and discharge air directly to the conditioned space. Ducted models are also common when multiple rooms are served by a single packaged unit that is placed near the building or on the roof (rooftop unit—RTU). Two advantages of packaged units of all types are the reduced potential for refrigerant leaks, system contamination, or improper charge that may result with site assembly installations and the reduced environmental impact of smaller refrigerant charges typically required in unitary equipment.

A room air conditioner (RAC) is shown in Figure 5.1. These units are typically placed in a window and are frequently referred to as window air conditioners. Equipment is widely available in capacities ranging from 5,000 to 20,000 Btu/h. A single cabinet houses all components, with an insulated partition dividing the outdoor and indoor sections. The compressor is typically located in the outdoor section to minimize noise transmission to the room. As shown in Figure 5.1, air is drawn in by an axial condenser fan through louvers (not shown) in the exterior cabinet and is discharged away from the building through the condenser. Capillary expansion devices are common. The evaporator dominates the interior side of the unit, and the filter (not shown) covers the evaporator in a return air grille. Return air is drawn through the evaporator by a centrifugal fan and is discharged into the room at sufficiently high velocity to minimize recirculation. The condenser and evaporator fan are often driven by the same double-shaft motor as shown in the figure. The bottom of the cabinet also serves as the condensate pan, so some care is required to level the package to ensure the water exits outdoors near the condenser rather than into the room.

The primary advantages of RACs are low cost, simplicity of installation, and wide availability. Existing electrical circuits can be used for increasingly larger equipment. Because of new efficiency standards, units in the 12,000 Btu/h capacity range are able to cool 300 to 600 ft^2 of floor space while drawing less than 10 amps from a 115 VAC circuit. However, larger capacity machines are likely to require dedicated 208 or 230 VAC circuits, especially for units that also provide heating.

RAC units with heating capability are also available. Electric elements are the dominant heating source. Elements are also in units that have reversing valves to operate as heat pumps. Most manufacturers disable heat pump operation of room units when the outdoor temperature falls below 35°F and switch to electric resistance heating elements. It should be noted that electric furnaces require approximately three times the power of the cooling mode power of the heat pump.

RACs are rated according to standards developed by the Association of Home Appliance Manufacturers (AHAM) in conjunction with ANSI (AHAM 1992). Minimum efficiencies are regulated by the Department of Energy (DOE 1997) and ASHRAE Standard 90.1-2004 (ASHRAE 2004a). The method of test is described in ASHRAE Standard 16-1999 (ASHRAE 1999a) and the rating methods in ASHRAE Standard 58-1999 (ASHRAE 1999b).

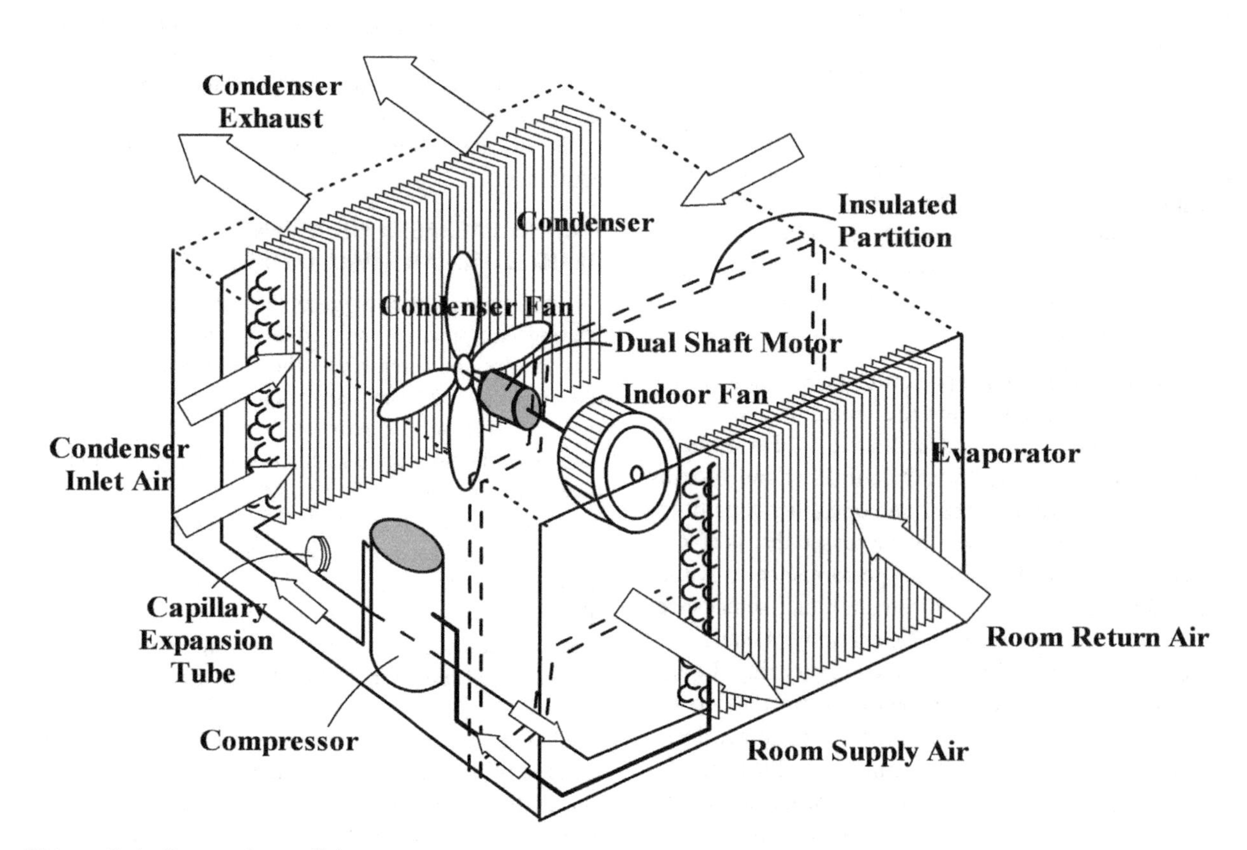

Figure 5.1 Room air conditioner.

Cooling performance, efficiency values, and rating conditions are:

- Total cooling (TC) in Btu/h (kW = Btu/h ÷ 3412)
- Energy efficiency ratio (EER) in Btu/W·h (COP_c = EER ÷ 3.412)
- Outdoor air temperature (OAT) at 95°F
- Indoor entering air temperature (EAT) at 80°F dry bulb and 67°F wet bulb

Most room units fall into classifications that require the EER to be at least 9.7 or 9.8 Btu/h (COP 2.9). Surprisingly this regulation exceeds the equivalent efficiency of much larger and more expensive HVAC systems, as will be demonstrated later.

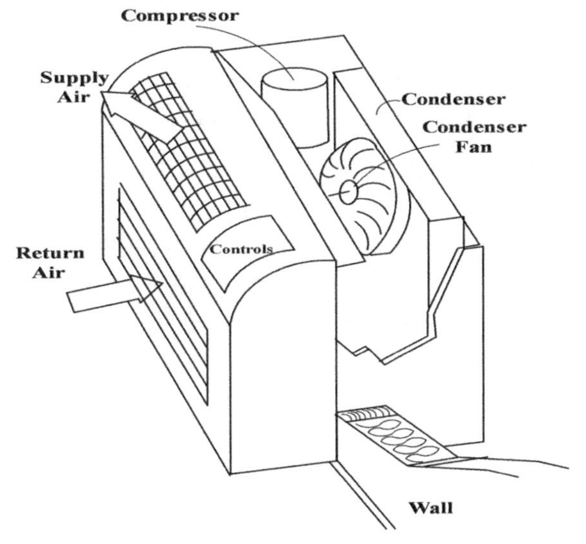

Figure 5.2 Packaged terminal air conditioner (PTAC).

Packaged terminal air conditioners (PTACs) and packaged terminal heat pumps (PTHPs) are similar to room air conditioners in configuration. They are designed to be installed inside a housing or "sleeve" that penetrates an exterior wall. A typical arrangement is shown in Figure 5.2. They are more common in commercial applications, such as motels, dormitories, and small office buildings. Design, operation, and performance are similar to the room units described above.

PTACs and PTHPs are rated according to ANSI/ARI Standard 310/380-1993 (ANSI/ARI 1993). Minimum efficiencies are regulated by ASHRAE Standard 90.1-2004 (ASHRAE 2004a). The method of test is described in ASHRAE Standard 16-1999 (ASHRAE 1999a) and the rating methods by ASHRAE Standard 58-1999 (ASHRAE 1999b).

Cooling performance, efficiency values, and rating conditions are:

- Total cooling (TC) in Btu/h (kW = Btu/h ÷ 3412)
- Energy efficiency ratio (EER) in Btu/W·h (COP_c = EER ÷ 3.412)
- Outdoor air temperature (OAT) at 95°F
- Indoor entering air temperature (EAT) at 80°F dry bulb and 67°F wet bulb

For heating:

- Total heating (TH) in Btu/h
- Coefficient of performance (W/W)
- Outdoor air temperature (OAT) at 47°F
- Indoor entering air temperature (EAT) at 70°F dry bulb

Table 5.1 presents performance data for three models of high-efficiency PTACs and PTHPs. Performance

Table 5.1 Performance of Three High-Efficiency PTACs and PTHPs

PTAC and PTHP Cooling Performance @ 230 VAC										
		Model 07			Model 09			Model 12		
OAT (°F)	EAT (db/wb)	TC (MBtu/h)	SC (MBtu/h)	Power (kW)	TC (MBtu/h)	SC (MBtu/h)	Power (kW)	TC (MBtu/h)	SC (MBtu/h)	Power (kW)
85	75/63	7.00	5.7		8.93	6.47		11.8	8.69	
	80/67	7.45	5.49		9.61	6.44		12.7	8.43	
95	75/63	6.69	5.55		8.46	6.19		11.1	8.38	
95	**80/67**	**7.10**	**5.35**	**0.62**	**9.10**	**6.02**	**0.81**	**12.0**	**8.13**	**1.12**
105	75/63	6.34	5.65		7.96	5.95		10.4	8.05	
	80/67	6.76	5.40		8.56	5.81		11.2	7.81	

PTHP Heating Performance @ 230 VAC							
OAT (°F)	EAT (°F)	TH (MBtu/h)	Power (kW)	TH (MBtu/h)	Power (kW)	TH (MBtu/h)	Power (kW)
57	70	7.10	0.60	9.20	0.79	12.3	1.09
47	**70**	**6.40**	**0.57**	**8.10**	**0.74**	**1.08**	**1.02**
37	70	5.80	0.55	7.10	0.70	9.4	0.95
~35	70	Switch Over to Electric Heat					

Bold rows indicate standard rating points.

is given for a variety of nonrated conditions. Total cooling (TC) is given for both higher (105°F) and lower (85°F) outdoor temperatures. Particular concern should be focused on the use of 80°F dry-bulb and 67°F wet-bulb indoor air. This condition is well outside the normal comfort zone for indoor air. It is a traditional rating point that assumes the return air at a comfortable condition will be mixed with warm, humid outdoor air before entering the unit. This is not the normal arrangement for PTACs and PTHPs. *Thus, 80°F/67°F should only be used as an entering air condition when the building owner specifies in writing that this should be the design point.* Table 5.1 also provides the cooling capacity (TC) at 75°F/63°F, a more acceptable indoor condition.

The sensible cooling (SC) capacities are also provided in the table. These values are important since it is critical that the sensible heat ratio of the equipment (SHR_{unit} = SC/TC) be equal to or lower than the building load sensible heat ratio ($SHR_{load} = q_s/q$), especially at off-peak conditions in humid and moderate climates.

Larger loads are often served by ducted packaged air conditioners and heat pumps (Figure 5.3). This equipment offers greater variety in heating options with fossil fuel furnaces being a popular option. Residential units are typically located near an exterior wall or on the roof and connected to the supply and return ductwork in a crawlspace, basement, or attic.

Packaged rooftop units (RTUs) are a widely used heating and cooling option. They are inexpensive, relatively easy to install, and can be placed on rooftops where they are out of view and do not take up valuable interior space. These advantages are accompanied by drawbacks since the modest price is typically accompanied by modest efficiency and service life. The rooftop location can make installation and service more difficult. Roof structures must be strengthened to support the additional weight, and precautions are required around the curb supports and service access paths to prevent roof leaks. Application of RTUs is more difficult for multi-storied buildings and sloped roofs.

The simplicity of a rooftop unit is demonstrated in Figure 5.4. The return duct can be relatively short or nonexistent. The supply duct is located in the ceiling space as shown. Since the units are packaged, refrigeration leakage and service due to improper charge are minimized.

Example Problem 5.1

a. Compute the EER for a Model 09 PTAC at ARI 310/380 conditions.
b. Select a unit to satisfy a cooling load of 8,500 Btu/h at the design conditions of 100°F outdoor air and 75°F/63°F (db/wb) indoor air conditions.
c. Determine if the unit will provide a SHR less than 0.75 when the conditions are 85°F outdoor air and 75°F/63°F (db/wb) indoor air.

Solution:

a. At the conditions of OAT = 95°F and EAT = 80°F/67°F, TC = 9.1 MBtu/h = 9100 Btu/h, and Power = 0.81 kW = 810 W. Therefore,

$$EER = \frac{9100 \text{ Btu/h}}{810 \text{ W}} = 11.2 \text{ Btu/(W·h)}$$

b. Try a Model 09:
   At the conditions of OAT = 95°F, EAT = 75°F/63°F, TC = 8.46 MBtu/h = 8,460 Btu/h
   At the conditions of OAT = 105°F, EAT = 75°F/63°F, TC = 7.96 MBtu/h = 7,960 Btu/h
   By interpolation to OAT = 100°F, TC = 8210 Btu/h (*too small*)

   Try a model 12:
   At the conditions of OAT = 95°F, EAT = 75°F/63°F, TC = 11.1 MBtu/h = 11,100 Btu/h
   At the conditions of OAT = 105°F, EAT = 75°F/63°F, TC = 10.4 MBtu/h = 10,400 Btu/h
   By interpolation to OAT = 100°F, TC = 10,750 Btu/h (*acceptable*)

c. For a Model 12 @ OAT = 85°F, EAT = 75°F/63°F, TC = 11,800 Btu/h, SC = 8,690 Btu/h

$$SHR_{unit} = SC/TC = 8.69 \text{ MBtu/h}/11.8 \text{ MBtu/h} = 0.74 \text{ (\textit{acceptable})}$$

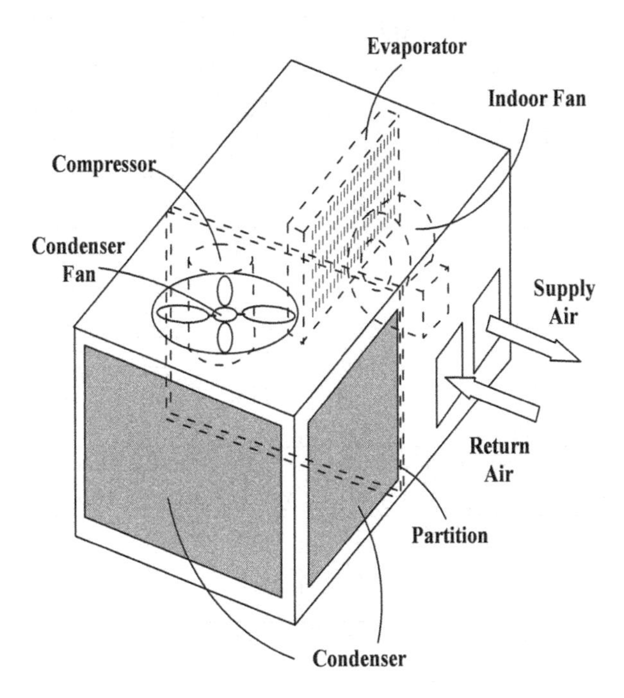

Figure 5.3 Packaged air conditioner with side discharge and return.

Table 5.2 is a sample set of performance data for a product line of rooftop units ranging in capacity from 12 to 20 tons. However, much larger units (sometimes referred to as "penthouse units") are available. Total cooling (TC), sensible cooling (SC), and unit power (kW) are presented for three flow rates and three entering air wet-bulb temperatures (wb) for each unit. The bypass factor (BF) for each flow rate is listed. All SC values are for 80°F dry-bulb EAT. Values for other dry-bulb EATs can be determined from the correction procedure given in the table for airflow (cfm) and BF. The data are in gross cooling capacity (no deduction for fan heat) rather than net capacity since fan power will vary with required external static pressure of the duct system. Therefore, the fan motor must be selected to provide required airflow and external static pressure (ESP) before TC and SC can be corrected to net values.

As stated previously, sensible cooling (SC) capacities are critical since the sensible heat ratio of the equipment (SHR_{unit} = SC/TC) must be equal to or lower than the building load's sensible heat ratio ($SHR_{load} = q_s/q$), especially at off-peak conditions in humid and moderate climates.

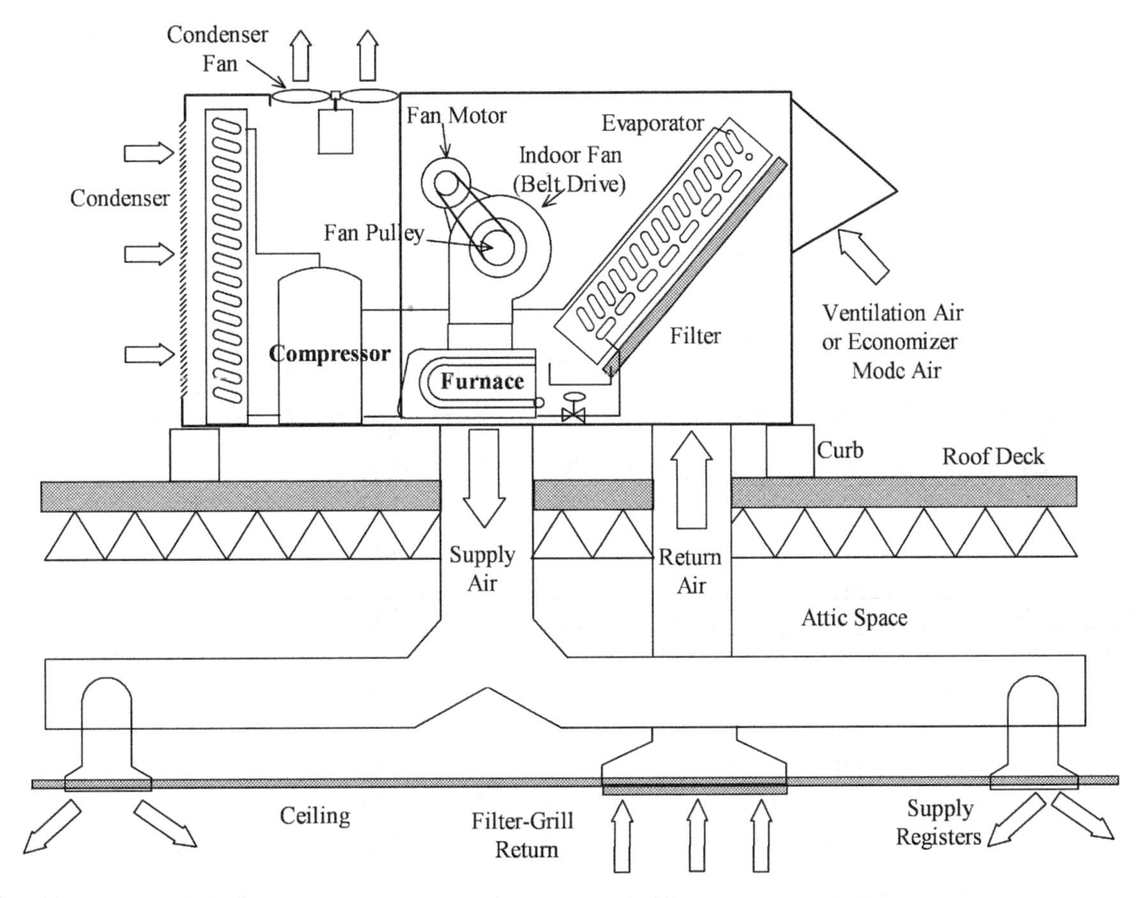

Figure 5.4 Packaged rooftop unit (RTU).

Table 5.2 Cooling Performance of Package Rooftop Units (RTUs) (C/UTC 2001a)

TC = Total Gross[1] Cooling in MBtu/h, SC = Sensible Gross[2] Cooling in MBtu/h, BF = Bypass Factor

Model 150		3600 cfm/BF = 0.03			5000 cfm/BF = 0.05			6400 cfm/BF = 0.07		
OAT		wb = 62	wb = 67	wb = 72	wb = 62	wb = 67	wb = 72	wb = 62	wb = 67	wb = 72
	TC	134	145	153	141	149	156	147	152	157
75	SC	111	93	72	130	109	77	137	119	79
	kW	7.9	8.1	8.2	8.1	8.2	8.3	8.2	8.2	8.3
	TC	125	134	141	131	138	142	136	139	142
95	SC	106	87	66	122	99	71	127	108	73
	kW	9.9	10.1	10.2	10.1	10.2	10.2	10.2	10.2	10.2
	TC	114	122	127	120	124	127	122	125	127
115	SC	100	81	60	111	93	64	114	98	66
	kW	12.3	12.5	12.6	12.4	12.5	12.6	12.5	12.5	12.6
Model 180		**4500 cfm/BF = 0.02**			**6000 cfm/BF = 0.04**			**7500 cfm/BF = 0.05**		
OAT		**wb = 62**	**wb = 67**	**wb = 72**	**wb = 62**	**wb = 67**	**wb = 72**	**wb = 62**	**wb = 67**	**wb = 72**
	TC	181	196	213	189	204	219	198	208	223
75	SC	147	124	99	173	143	113	186	163	125
	kW	11.2	11.6	12.0	11.4	11.8	12.1	11.7	11.9	12.2
	TC	168	182	197	176	190	203	185	194	206
95	SC	142	118	93	164	136	102	174	153	113
	kW	13.5	13.9	14.3	13.8	14.1	14.5	14.0	14.3	14.6
	TC	152	165	179	161	171	181	169	174	183
115	SC	134	111	87	151	128	94	158	142	101
	kW	16.1	16.5	16.9	16.4	16.7	17.0	16.7	16.9	17.1
Model 240		**6000 cfm/BF = 0.04**			**8000 cfm/BF = 0.06**			**10000 cfm/BF = 0.08**		
OAT		**wb = 62**	**wb = 67**	**wb = 72**	**wb = 62**	**wb = 67**	**wb = 72**	**wb = 62**	**wb = 67**	**wb = 72**
	TC	238	260	283	249	269	293	260	276	300
75	SC	198	166	134	229	192	151	248	216	166
	kW	15.7	16.3	16.9	16.1	16.6	17.2	16.4	16.8	17.5
	TC	222	242	264	232	252	274	243	257	281
95	SC	189	159	127	218	183	141	235	205	155
	kW	18.3	18.9	19.6	18.7	19.3	20.0	19.1	19.6	20.2
	TC	202	220	241	213	228	249	223	232	255
115	SC	180	151	119	204	173	133	216	195	146
	kW	21.2	21.8	22.5	21.6	22.2	22.8	22.1	22.4	23.0

cfm	rpm[3]	kW	bhp[3]	rpm	kW	bhp	rpm	kW	bhp	rpm	kW	bhp
M150	**ESP[3] = 0.5 in. of water**			**ESP = 1.0 in. of water**			**ESP = 1.5 in. of water**			**ESP = 2.0 in. of water**		
4000	916	1.33	1.53	1154	1.88	2.16	1361	2.47	2.85	1549	3.11	3.57
5000	1028	1.99	2.28	1234	2.57	2.95	1420	3.19	3.67	–	–	–
6000	1152	2.84	3.32	1333	3.47	3.99	–	–	–	–	–	–
M180	**ESP = 0.5 in. of water**			**ESP = 1.0 in. of water**			**ESP = 1.5 in. of water**			**ESP = 2.0 in. of water**		
4500	800	1.44	1.70	980	2.08	2.44	1134	2.76	3.24	1271	3.48	4.08
6000	947	2.59	3.03	1100	3.32	3.89	1236	4.09	4.79	1361	4.88	5.72
7500	1119	4.26	5.00	1241	5.09	5.97	–	–	–	–	–	–
M240	**ESP = 0.5 in. of water**			**ESP = 1.0 in. of water**			**ESP = 1.5 in. of water**			**ESP = 2.0 in. of water**		
6000	928	2.96	2.52	1083	3.58	4.24	1223	4.23	5.00	1351	4.86	5.75
8000	1137	4.87	5.78	1264	5.47	6.49	1382	6.10	7.23	1493	6.73	7.98
10000	1362	7.35	8.72	1468	7.95	9.43	1570	8.56	10.2	–	–	–

[1] Gross capacity does not include impact of fan heat. To compute net total and sensible capacities:
TC (MBtu/h) = TC (gross) – 3.41 × kW (fan) and SC (MBtu/h) = SC (gross) – 3.41 × kW (fan).

[2] Sensible capacities based on 80°F dry-bulb EAT; for other dry-bulb EATs: SC (MBtu/h) = SC (80) + 1.1 × (1 – BF) × (cfm/1000) × (EAT – 80).

[3] Fan performance: rpm = fan wheel revolutions per minute, ESP = external static pressure, bhp = brake horsepower.

Example Problem 5.2

For a Model 180 rooftop unit, select a fan motor size and a fan wheel pulley for a 1750 rpm motor with a 6 inch diameter drive pulley. Compute net TC, SC, kW_{Unit}, SHR, and EER for a unit at 95°F, 75°F/62°F (db/wb) EAT, 4500 cfm, and 1.0 in. ESP.

Solution:

At 4500 cfm and 1.0 in. ESP, the Model 180 fan needs a speed of 980 rpm, will draw 2.08 kW, and requires a 2.44 hp motor. The smallest standard AC motor is 3.0 hp, and the fan pulley size is

$$D_{Fan\ Pulley} = D_{Motor\ Pulley} \times (\text{rpm}_{Motor} \div \text{rpm}_{Fan}) = 6 \text{ in. } (1750 \div 980) = 10.7 \text{ inches.}$$

At 95°F OAT, 80°F/62°F, 4500 cfm: TC = 168 MBtu/h, SC = 142 MBtu/h, and kW_c=13.5.

SC is corrected from 80°F to 75°F dry-bulb EAT:

$$SC_{75} = SC_{80} + 1.1 \times (1 - BF) \times (cfm/1000) \times (EAT - 80)$$
$$= 142 + 1.1\times(1 - 0.02) \times (4500/1000) \times (75 - 80) = 118 \text{ MBtu/h}$$

Both TC and SC must be corrected for fan heat:

$$TC(net) = TC(gross) - 3.41 \times kW_{fan} = 168 - 3.41\times 2.08 \text{ kW} = 161 \text{ MBtu/h}$$

$$SC(net) = SC(gross) - 3.41 \times kW_{fan} = 118 - 3.41\times 2.08 \text{ kW} = 111 \text{ MBtu/h}$$

$$SHR = SC \div TC = 111 \div 161 = 0.69, \text{ and}$$

$$EER = TC \div (kW_{Unit} + kW_{Fan}) = 161 \div (13.5 + 2.08) = 10.3 \text{ (MBtu/kWh)} = 10.3 \text{ (Btu/Wh)}$$

UNITARY SPLIT SYSTEM EQUIPMENT—RESIDENTIAL AND SMALL COMMERCIAL

Split system air conditioners/furnaces (SSACs) and heat pumps (SSHPs) are widely used in the residential and light commercial sectors. The outdoor unit of an air-conditioner/furnace (AC/F) system typically consists of the compressor, condenser coil (condenser in cooling), and outdoor fan. The indoor unit consists of the furnace, the evaporator coil (evaporator in cooling), expansion device, and indoor fan. A variety of indoor packages are available, including upflow vertical (as shown in Figure 5.5), downflow, and horizontal flow units. In cooling, the compressor discharges refrigerant to the condenser, which is typically a one- or two-row coil. Condenser fans are typically axial designs, as shown, which are well suited to the high-volume flow rate (but low friction loss) requirement. The condensed liquid refrigerant travels to the interior coil through the smaller diameter liquid line to the expansion device (thermostatic expansion valve shown) and into the evaporator tubes. Room air is drawn through the return duct and filter and over the evaporator coil (three-row shown). The air is cooled and dehumidified, with the condensate draining from the pan and out through a trap. The cool air is discharged by the indoor fan (centrifugal design shown) into the space through the supply duct. The evaporated refrigerant returns to the outdoor unit through the larger, insulated vapor line to the compressor.

In the heating mode, the room air is returned and supplied through the same ductwork used in the cooling mode. Natural gas, propane, or oil is ignited in a set of burners, and air for combustion is ducted to the furnace from outdoors. Combustion occurs inside a heat exchanger that is sealed from the room air, which is passing on the outside of the chamber. The products of combustion are vented out of the chamber through the ceiling and roof in a double-wall metal pipe for standard efficiency furnaces (annual fuel utilization efficiency [AFUE] 80%). Higher efficiency furnaces (AFUE = 90% and above) remove sufficient heat from the combustion process to condense the water (a primary combustion product). In these units, the vent is typically a plastic pipe that can be routed through an exterior wall.

Split system heat pumps are similar to AC/Fs, but a reversing valve is added near the compressor discharge and the furnace is typically smaller (often electric) since it serves as an auxiliary system when outdoor temperatures are low. The refrigerant flow pattern is identical to the cooling mode path in the air conditioner shown in Figure 5.6 except the compressor discharge gas goes through the reversing valve (into the bottom port and out the top left port). The gas also returns to the compressor through the reversing valve (into top right port and out center top port).

In the heating mode, the reversing valve diverts the compressor discharge to the indoor coil (as noted with

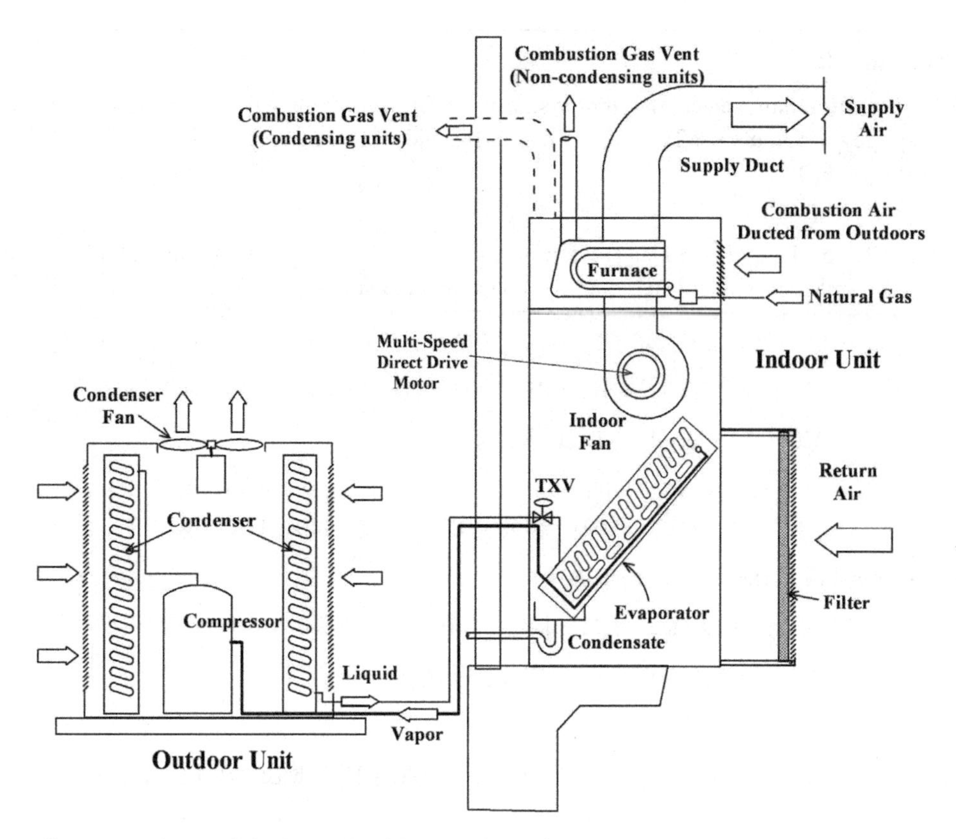

Figure 5.5 Split system air-conditioning unit with natural gas furnace.

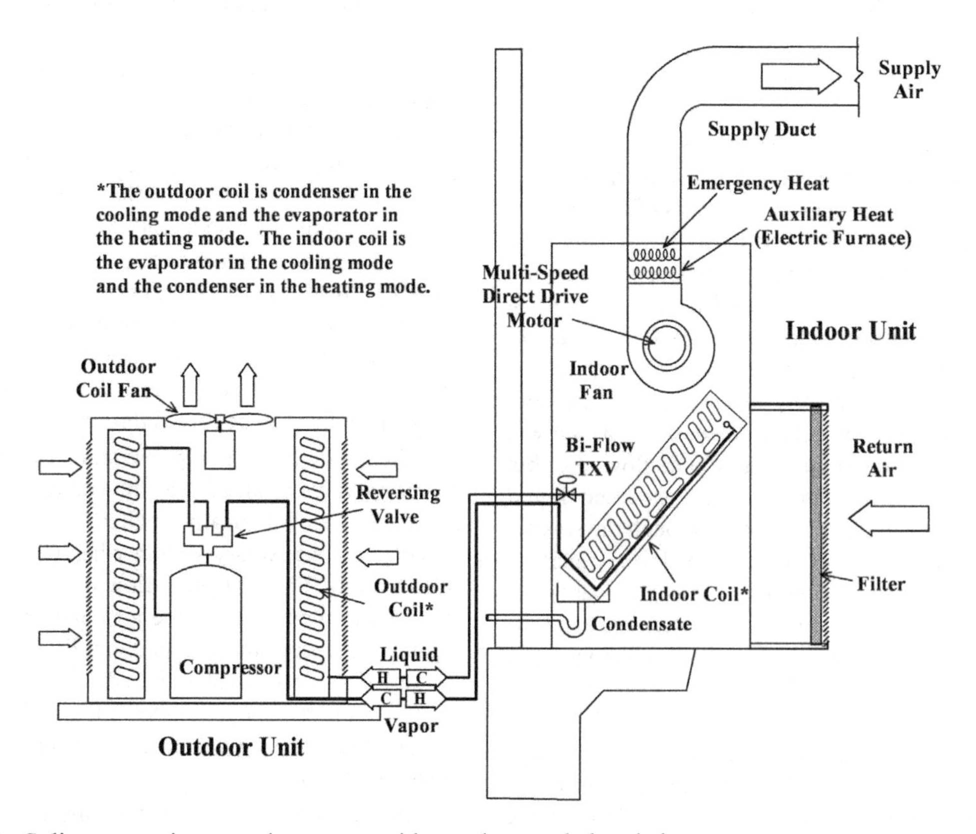

Figure 5.6 Split system air-source heat pump with supplemental electric heat.

the arrow marked "H" on the vapor line in Figure 5.6). The indoor coil is the condenser in this mode and heat is delivered to the indoor air. The condensed refrigerant flows to the expansion device (some heat pumps have one device for each mode, unlike the single bi-flow TXV shown). Refrigerant flows though the vapor line to the outdoor coil, which is now the evaporator, and then to the compressor suction through the reversing valve. Air-to-air heat pumps are equipped with an auxiliary heater to supplement the heat pump during cold periods and an emergency heater in case of unit malfunction. Controls will automatically bring on the auxiliary unit while the user must typically activate the emergency heat at the thermostat.

Packaged and split system air conditioners below 65 MBtu/h (19 kW) are rated according to ARI Standard 210/240-1994 (ARI 1994). The method of test is described in ANSI/ASHRAE Standard 37-1988 (ASHRAE 1988) and rating methods by ANSI/ ASHRAE Standard 116-1995 (ASHRAE 1995). Furnaces are rated according to ANSI/ASHRAE Standard 103-1993 (ASHRAE 1993). Minimum efficiencies for air conditioners, heat pumps, and furnaces are regulated by the Department of Energy (DOE).

Air conditioner and heat pump cooling performance, efficiency values, and rating conditions are:

- Total cooling (TC) in Btu/h (kW = Btu/h ÷ 3412)
- Seasonal energy efficiency ratio (EER) in Btu/W·h (for single-speed machines only)
- Outdoor air temperature (OAT) at 82°F
- Indoor entering air temperature (EAT) at 80°F dry bulb and 67°F wet bulb

For heat pump heating:

- Total heating (TH) in Btu/h
- Heating seasonal performance factor (HSPF) in Btu/W·h—HSPF is a corrected seasonal energy rating
- TH outdoor air temperature (OAT) at 47°F
- Indoor entering air temperature (EAT) at 65°F dry bulb

Table 5.3 is an example of the cooling and heating performance data of a high-efficiency air-source heat pump product line ranging from a unit with a nominal capacity of 24 MBtu/h (2 tons, 7 kW) to one with a 60 MBtu/h capacity (5 tons, 18 kW). Total cooling (TC), sensible cooling (SC), and system power (kW_c) are provided for a range of outdoor temperatures (OAT) and two indoor entering air dry-bulb/wet-bulb temperatures (EAT). Total heating (TH) and system power (kW_h) are provided for a lower range of temperatures. The data for each unit are provided for a single airflow rate (~400 cfm/ton) and the external static pressure (ESP in inches of water column) used in ARI Standard 210/240 (ARI 1994). The table includes correction factors for capacities and power inputs for airflow rates ±20% of the rated values. Equations are also provided to correct values to higher static pressures since the ARI Standard 210/240 values are much lower than the requirements of actual installations.

Example Problem 5.3

Compute TC, SC, kW_c, SHR, and EER for a Model 030 heat pump operating at 95°F, 75°F/63°F (db/wb) EAT, 900 cfm, and 0.4 in. ESP.

Solution:

At 95°F, 75°F/63°F, 1000 cfm: TC = 29.0 MBtu/h, SC = 23.6 MBtu/h, and kW_c = 2.58.

Applying the appropriate multipliers for 900 cfm or 90% of rated flow,

$$TC = 0.98 \times 29.0 = 28.4 \text{ MBtu/h}; \; SC = 0.95 \times 23.6 = 22.4 \text{ MBtu/h}; \; kW_c = 0.985 \times 2.58 = 2.54\,.$$

Correct for the added ESP above the rated value of 0.15 in.:

$$kW_c = 2.54 + [0.00037 \times 900 \text{ cfm} \times (0.40 - 0.15)] = 2.54 + [0.083] = 2.62 \text{ kW}$$

$$TC = 28.4 - (0.083 \times 3.41) = 28.1 \text{ MBtu/h}$$

$$SC = 22.4 - (0.083 \times 3.41) = 22.1 \text{ MBtu/h}$$

Therefore,

$$SHR = SC \div TC = 22.1 \div 28.1 = 0.79, \text{ and}$$

$$EER = TC \div kW_c = 28.1 \div 2.62 = 10.7 \text{ (Btu/W·h)}.$$

Table 5.3 Cooling and Heating Performance of Split System High-Efficiency Heat Pumps (T/ASC 2003)

		Model 024 @ 800 cfm, 0.1 in. ESP			Model 030 @ 1100 cfm, 0.15 in. ESP			Model 036 @ 1200 cfm, 0.15 in. ESP		
OAT (°F)	EAT (db/wb)	TC (MBtu/h)	SC (MBtu/h)	Power (kW)	TC (MBtu/h)	SC (MBtu/h)	Power (kW)	TC (MBtu/h)	SC (MBtu/h)	Power (kW)
85	75/63	24.1	16.0	1.71	30.3	24.1	2.31	34.6	26.3	2.66
	80/67	26.0	18.8	1.74	32.6	24.8	2.33	37.2	27.0	2.69
95	75/63	22.9	17.8	1.88	29.0	23.6	2.58	33.4	25.7	2.94
	80/67	24.7	18.3	1.90	31.2	24.2	2.60	35.8	26.4	2.97
105	75/63	21.7	17.3	2.04	27.9	23.1	2.87	32.2	25.3	3.28
	80/67	23.4	14.5	2.06	30.1	23.8	2.88	34.5	25.9	3.31
115	75/63	20.6	16.8	2.20	26.8	22.6	3.16	31.1	24.8	3.61
	80/67	22.2	17.3	2.20	29.0	23.3	3.16	33.3	24.5	3.66

Heating Performance

OAT (°F)	EAT (°F)	TH (MBtu/h)	Power (kW)	TH (MBtu/h)	Power (kW)	TH (MBtu/h)	Power (kW)
57	70	25.1	1.88	34.8	2.76	41.5	3.21
47	70	21.8	1.78	30.9	2.69	36.8	3.10
37	70	16.3	1.65	24.5	2.53	26.0	2.85
27	70	14.4	1.58	21.8	2.51	24.3	2.80
17	70	12.1	1.50	19.2	2.50	22.6	2.76

		Model 042 @ 1300 cfm, 0.15 in. ESP			Model 048 @ 1400 cfm, 0.20 in. ESP			Model 060 @ 1700 cfm, 0.20 in. ESP		
OAT (°F)	EAT (db/wb)	TC (MBtu/h)	SC (MBtu/h)	Power (kW)	TC (MBtu/h)	SC (MBtu/h)	Power (kW)	TC (MBtu/h)	SC (MBtu/h)	Power (kW)
85	75/63	42.1	31.7	3.13	48.3	35.5	3.64	55.8	41.6	4.49
	80/67	45.1	32.6	3.16	51.7	36.4	3.69	59.8	42.7	4.55
95	75/63	40.7	31.1	3.47	46.7	34.8	4.04	54.1	40.9	4.99
	80/67	43.6	32.0	3.51	50.0	35.7	4.10	53.6	42.0	5.08
105	75/63	39.1	30.4	3.88	44.8	34.0	4.52	51.8	39.9	5.57
	80/67	41.8	31.3	3.93	47.9	34.9	4.60	55.4	41.0	5.68
115	75/63	37.4	29.7	4.28	42.8	33.1	4.99	49.5	38.9	6.14
	80/67	40.1	30.6	4.34	45.8	34.0	5.10	53.0	40.0	6.29

Heating Performance

OAT (°F)	EAT (°F)	TH (MBtu/h)	Power (kW)	TH (MBtu/h)	Power (kW)	TH (MBtu/h)	Power (kW)
57	70	49.2	3.54	52.3	3.83	62.6	4.59
47	70	44.1	3.42	47.0	3.70	55.9	4.46
37	70	35.3	3.28	35.6	3.51	38.1	4.06
27	70	32.0	3.18	33.3	3.41	36.9	4.06
17	70	28.7	3.08	31.0	3.31	35.7	4.07

Multipliers for Other Airflow Rates

% of Rated Flow—Multiply Factors by Rated Values

	80%	85%	90%	95%	105%	110%	115%	120%
TC	0.97	0.975	0.98	0.99	1.005	1.01	1.015	1.02
SC	0.90	0.93	0.95	0.98	1.02	1.04	1.07	1.09
kW_c	0.975	0.98	0.985	0.99	1.01	1.02	1.03	1.04
TH	0.98	0.985	0.99	0.995	1.005	1.01	1.015	1.02
kW_h	1.05	1.04	1.02	1.01	0.99	0.98	0.97	0.96

Higher ESP Correction:

Add to Rated kW_c and kW_h CF (kW_{esp}) = 0.00037 × cfm × (ESP – ESP_{rated})

Deduct from TC and SC CF = CF (kW_{esp}) × 3.41

Add to TH CF = CF (kW_{esp}) × 3.41

It is important to note the performance trends in the data in tables and figures presented throughout this text. The most notable and significant in Table 5.3 is the increase in demand and the decline in capacity and efficiency for equipment as the outdoor air temperature rises. A 20°F rise (85°F–105°F) in OAT will result in a 10% to 12% loss in capacity and a 22% to 26% loss in efficiency. Also note the decline in capacity (6% to 8%) when the indoor air wet-bulb temperature falls from 67°F to 63°F. Also note the airflow multipliers indicate that SC capacity declines more rapidly (10%) than the total capacity (3%) when the airflow is reduced by 20%. Thus, lowering airflow may be an attractive alternative during periods of high latent loads. These same trends and others manifest themselves in all of the cooling and heating equipment options available to the designer.

Tables 5.4 and 5.5 provide the performance of a product line of standard efficiency noncondensing natural gas furnaces and a line of higher efficiency condensing furnaces. The input and output values are given in MBtu/h and efficiency in annual fuel utilization efficiency (AFUE, %). Both model lines are equipped with four-speed fan motors, which permit airflow adjustment to meet the heating load and may be switched to another speed to meet cooling coil airflow need. The color codes shown next to the fan speed indicate the color of the speed tap wire.

Table 5.4 Performance and Specifications for Natural Gas Furnaces (T/ASC 1998)

Model	Output	Input	AFUE	Fan Size	Fan Speed Taps Airflow in cfm @ 0.4/0.8 in. ESP			
	(MBtu/h)	(MBtu/h)	(%)	(hp)	Lo (Red)	M/Lo (Yel.)	M/Hi (Blu.)	Hi (Blk.)
040	32	40	80	1/3	540/360	650/450	780/590	950/760
060	47	60	80	1/3	880/700	1010/820	1160/920	1300/1020
080	63	80	80	1/3	890/770	1030/870	1180/1000	1340/1120
100	79	100	80	1/2	1190/1010	1370/1120	1550/1260	1740/1380
120	95	120	80	3/4	1390/1250	1600/1430	1820/1600	2040/1750

Table 5.5 Performance and Specifications for Condensing Natural Gas Furnaces (T/ASC 1995)

Model	Output	Input	AFUE	Fan Size	Fan Speed Taps Airflow in cfm @ 0.4/0.8 in. ESP			
	(MBtu/h)	(MBtu/h)	(%)	(hp)	Lo (Red)	M/Lo (Yel.)	M/Hi (Blu.)	Hi (Blk.)
040	38	40	92	1/5	600/220	710/310	790/420	890/520
060	56	60	92	1/3	890/690	1030/770	1160/873	1260/950
080	74	80	92	1/3	840/660	1120/920	1310/1060	1540/1230
100	93	100	92	3/4	1400/1150	1590/1280	1730/1390	1840/1480
120	112	120	92	3/4	1520/1330	1760/1500	1960/1620	2070/1790

Example Problem 5.4

Select a condensing furnace and fan speed setting in a building with a 65 MBtu/h heat loss with an indoor temperature of 70°F. Specify the outlet air temperature to be approximately 120°F if the air distribution design will be conducted so that the coil, filter, and duct losses are to be less than 0.6 in. of water.

Solution:

The Model 080 will deliver 74 MBtu/h and is the smallest unit that will provide the required 65 MBtu/h. The airflow rate needed to deliver 120°F is determined by rearranging Equation 4.14 and substituting TH for q_s:

$$\text{cfm} = \frac{\text{TH (Btu/h)}}{1.08 \times (t_2 - t_1)} = \frac{74\ (\text{MBtu/h}) \times 1000\text{Btu/MBtu}}{1.08 \times (120°\text{F} - 70°\text{F})} = 1370\ \text{cfm}$$

The airflow portion of Table 5.5 indicates the fan for the Model 080 will provide 1540 cfm at 0.4 in. ESP and 1230 cfm at 0.8 in. ESP when the motor is wired to the high speed tap. By linear interpolation, it is estimated the fan will provide 1385 cfm for an ESP of 0.6 in.

GROUND-SOURCE AND WATER-LOOP HEAT PUMPS

In moderate and cold climates, air-source heat pumps are not able to maintain comfort and efficiency throughout the heating season. Ground-source heat pumps (GSHPs) are able to provide a means of effective and efficient heating and cooling in most climates because of the relatively constant temperature of the ground and groundwater. The primary unit used in this application is a packaged water-to-air heat pump. A vertical upflow unit is shown in Figure 5.7, but other arrangements are available. The outdoor coil is eliminated and replaced with a water-to-refrigerant coil, normally a coiled version of the coaxial design shown in Figure 3.8. The compressor is also located indoors in the package. A second water coil is often added and is piped on the compressor discharge to preheat domestic hot water with the discharge gas. In the summer, the coil recovers waste heat that would normally be rejected to the environment. In winter operation, the coil will recover excess capacity of the heat pump during the many hours that it is available.

The primary water coil is connected via a water loop to the ground, groundwater, or lake water, which serves as the condenser fluid. Water is an effective cooling fluid, especially if the loop has been properly designed so the loop is able to deliver the fluid at a temperature near the earth temperature. This permits the condenser to operate at a lower pressure and temperature before the refrigerant travels to the thermostatic expansion valve (TXV) and the indoor air coil, which serves as the evaporator in the cooling mode. The system is able to operate efficiently with good dehumidification capacity.

In the winter the ground is relatively warm, and this provides an advantage when the heat pump mode is switched to heating. The water coil is now the evaporator and it operates at a high temperature and pressure because the ground/water loop is warm. The compressor receives the refrigerant from the evaporator at a favorable condition so it is able to deliver a higher temperature gas to the air coil (which is the condenser). This results in warm supply air and high efficiency.

Table 5.6 provides the performance data for a product line of water-to-air heat pumps. The format is simi-

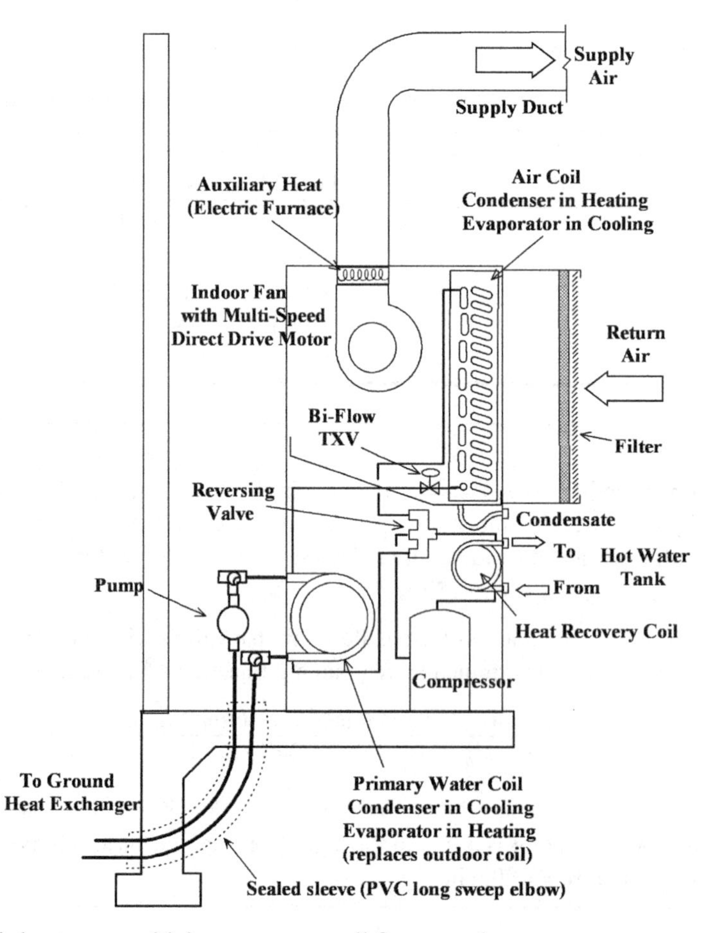

Figure 5.7 Water-to-air heat pump with heat recovery coil for ground-source systems.

Table 5.6 Cooling and Heating Performance of Water-to-Air Heat Pumps (WFI 1996)

Values for EAT = 80°F/67°F in cooling and 70°F in heating with multipliers for other conditions.

	S024 @ 800 cfm, 0.15 ESP Δh = 6.2 ft @ 4.5 gpm, 12.4 ft @ 6 gpm				S030 @ 1000 cfm, 0.2 ESP Δh = 4.6 ft @ 6 gpm, 6.9 ft @ 8 gpm				S036 @ 1200 cfm, 0.2 ESP Δh = 8.1 ft @ 7 gpm, 12.8 ft @ 9 gpm			
EWT (°F)	GPM	TC (MBtuh)	SC (MBtuh)	kW	GPM	TC (MBtuh)	SC (MBtuh)	kW	GPM	TC (MBtuh)	SC (MBtuh)	kW
60	4.5	29.5	19.8	1.41	6.0	34.0	24.1	1.76	7.0	40.2	27.4	2.16
	6.0	28.9	19.4	1.36	8.0	33.9	24.1	1.71	9.0	40.1	27.6	2.11
70	4.5	28.2	19.5	1.55	6.0	32.6	23.6	1.94	7.0	38.6	26.4	2.34
	6.0	28.0	19.3	1.50	8.0	32.6	23.7	1.89	9.0	38.6	27.0	2.29
80	4.5	27.1	19.1	1.63	6.0	31.1	23.0	2.11	7.0	37.1	25.7	2.47
	6.0	27.1	19.0	1.59	8.0	31.6	23.2	2.03	9.0	37.2	26.6	2.42
90	4.5	24.8	18.2	1.8	6.0	29.7	22.3	2.20	7.0	34.2	24.4	2.72
	6.0	25.3	18.3	1.77	8.0	29.8	22.5	2.16	9.0	34.5	25.7	2.67
EWT (°F)	GPM	TH (MBtuh)		kW	GPM	TH (MBtuh)		kW	GPM	TH (MBtuh)		kW
60	4.5	26.7		1.59	6.0	32.8		2.01	7.0	38.8		2.53
	6.0	27.4		1.6	8.0	33.2		2.02	9.0	40.0		2.56
50	4.5	23.2		1.5	6.0	28.4		1.92	7.0	33.8		2.4
	6.0	23.9		1.51	8.0	28.9		1.94	9.0	34.8		2.42
40	4.5	20.0		1.44	6.0	24.5		1.81	7.0	29.5		2.26
	6.0	20.5		1.45	8.0	24.9		1.82	9.0	30.3		2.27

	S042 @ 1400 cfm, 0.2 ESP Δh = 7.3 ft @ 8 gpm, 13.0 ft @ 11 gpm				S048 @ 1600 cfm, 0.2 ESP Δh = 14.1 ft @ 9 gpm, 22.2 ft @12 gpm				S060 @ 2000 cfm, 0.2 ESP Δh = 12.5 ft @11 gpm,18 ft @14 gpm			
EWT (°F)	GPM	TC (MBtuh)	SC (MBtuh)	kW	GPM	TC (MBtuh)	SC (MBtuh)	kW	GPM	TC (MBtuh)	SC (MBtuh)	kW
60	8	50.1	34.8	2.59	9	57.0	39.9	3.15	11	61.4	42.6	3.75
	11	49.2	33.7	2.52	12	57.4	39.7	3.05	14	61.3	43.2	3.72
70	8	45.9	33.7	2.78	9	54.5	39.2	3.4	11	59.7	42.0	3.97
	11	45.6	33.0	2.71	12	54.8	39.2	3.30	14	59.7	42.2	3.95
80	8	44.7	32.7	2.93	9	52.8	38.4	3.57	11	58.4	41.8	4.19
	11	44.7	32.3	2.86	12	53.1	38.4	3.46	14	58.3	41.8	4.16
90	8	42.2	30.8	3.22	9	49.5	36.7	3.90	11	55.7	41.5	4.63
	11	42.9	30.9	3.16	12	49.6	36.8	3.79	14	55.6	41.0	4.57
EWT (°F)	GPM	TH (MBtuh)		kW	GPM	TH (MBtuh)		kW	GPM	TH (MBtuh)		kW
60	8	45.1		2.87	9	54.0		3.5	11	62.5		4.16
	11	47.9		2.91	12	55.6		3.56	14	64.0		4.19
50	8	38.2		2.71	9	47.7		3.30	11	55.3		3.00
	11	41.3		2.76	12	48.6		3.35	14	56.7		4.02
40	8	33.1		2.56	9	41.7		3.13	11	49.3		3.87
	11	36.0		2.61	12	42.2		3.17	14	50.7		3.89

Use multipliers in lower section of Table 5.3 for other flow rates, but no ESP correction is required for this data.*

Cooling	Multipliers		SC Multipliers for DBs				Heating	Multipliers	
WB (°F)	TC	kW	70°F	75°F	80°F	85°F	DB (°F)	TC	kW
61	0.91	0.96	0.84	1.02	1.10	–	60	1.02	0.92
64	0.95	0.98	0.73	0.93	1.10	–	65	1.01	0.96
67	1.0	1.0	0.59	0.81	1.0	1.17	70	1.0	1.0
71	1.10	1.04	–	–	0.76	0.96	75	0.99	1.04

* Data for this product line includes power of the fan motor. Other products may need fan and pump power correction according to ISO 13256-1 (ISO 2000).

lar to that for air heat pumps, but water loop temperatures replace OAT. The performance is provided for two liquid flow rates. All data in the table are for entering air temperatures (EATs) of 80°F/67°F in cooling and 70°F in heating. Multipliers for TC, SC, kWc, TH, and kWh are provided for other EATs. Airflow rate multipliers from Table 5.3 can be applied for other airflow rates. No correction factor is needed for external static pressure (ESP) in this data set. No correction for antifreeze solutions of 30% or less are needed in the range of temperatures given in the table. However, manufacturers may de-rate performance for temperatures below 40°F. One of the supplemental spreadsheet programs on the CD that accompanies this text is a performance correction of the water-to-air heat pump whose performance is given in Table 5.6. Design of the ground loops, groundwater systems, and lake water systems are discussed in ASHRAE (2003, chapter 32) and Kavanaugh and Rafferty (1997). Figure 5.8 shows several heat pump types.

Figure 5.9 depicts a water-loop heat pump (WLHP) system, which is a variation of the ground-source heat pump that is often used in commercial buildings. A common loop is tied to water-to-air and water-to-water heat pumps located throughout the building. A fluid cooler or cooling tower is used to dissipate the heat from the units in cooling, which is the dominant mode in commercial buildings in the US. In heating, the units must be supplemented with a boiler. This can be a negative factor in cold climates in poorly insulated buildings that have high heating requirements since energy must be supplied to the units and the boiler. However, the requirement in many modern buildings is small compared to the cooling load. The WLHP can also take advantage of low-temperature waste heat. Another option is to replace the boiler with a ground loop or groundwater loop that is sized to meet only the heating

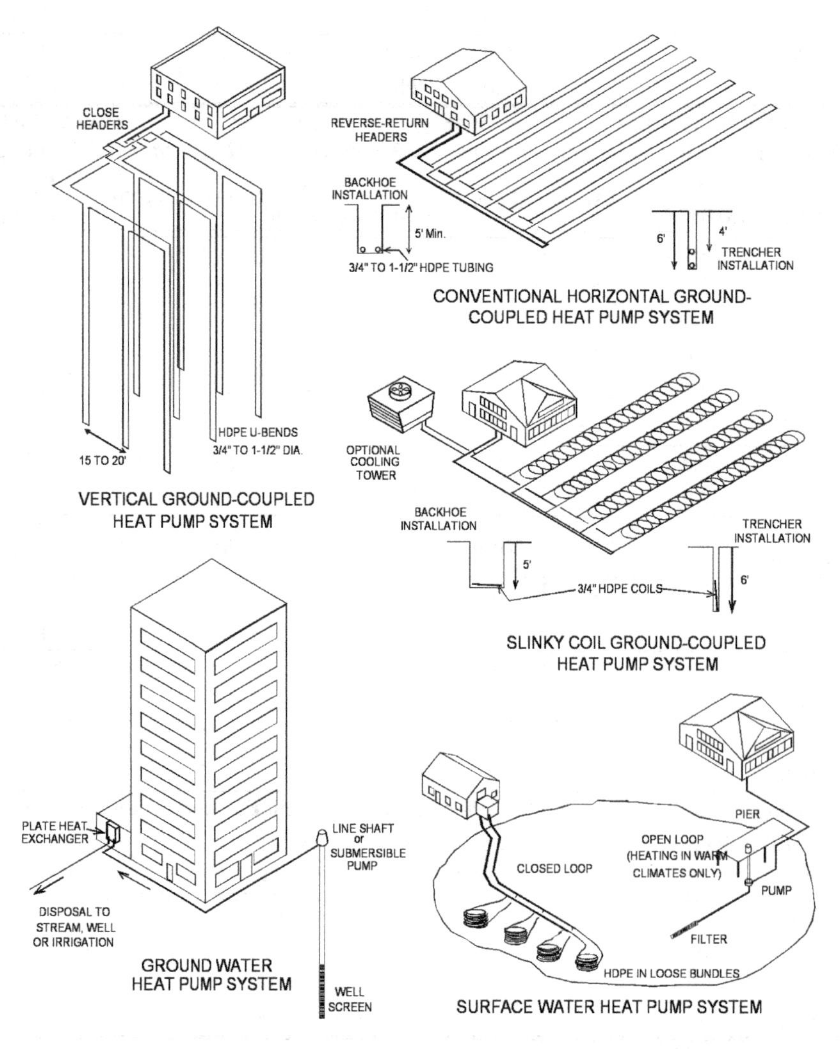

Figure 5.8 Ground-source heat pump (GSHP) types (Kavanaugh and Rafferty 1997).

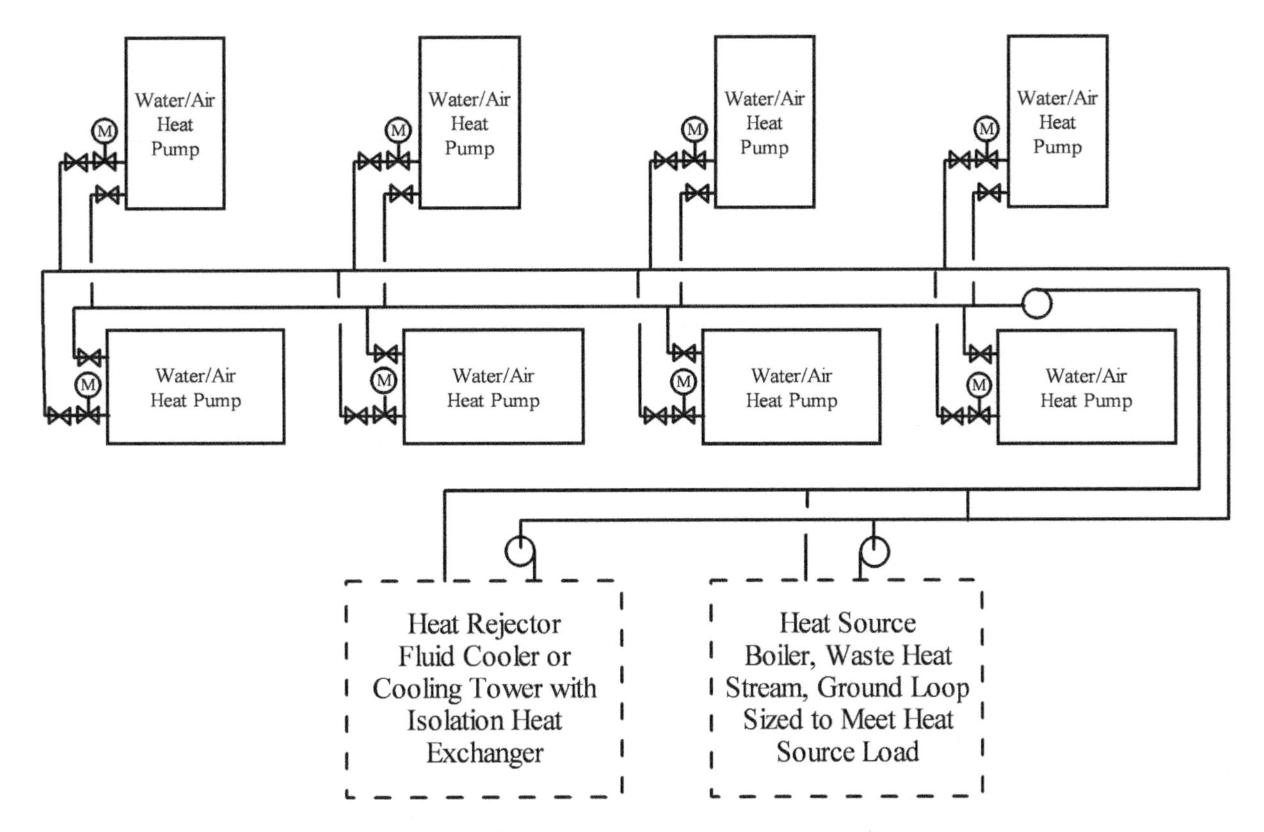

Figure 5.9 Water-loop heat pump (WLHP) system.

requirement of the building. This is typically much smaller than the size required to satisfy the cooling load. In cooling, the ground/water loop is used in parallel with the fluid cooler or cooling tower. These systems can also take advantage of the many hours when outdoor conditions are moderately cold and some zones require cooling (interior and/or high internal loads) and other zones require heating (perimeter and/or high outdoor air loads). The heat added to the water loop by the units in cooling tends to be balanced by heat removed by the units in heating. This concept is considered a form of heat recovery. The potential also exists for free cooling interior zones (without heat pump cooling mode operation) if the water-loop temperature can be controlled in the mid-50°F or lower range.

Water-to-air heat pumps are the first unitary product line to be converted from ARI standards to International Organization for Standardization (ISO) ratings. The new standard for water-to-air heat pumps is ISO 13256-1 (ISO 2000), which makes accommodations for three common applications that were formerly covered by three individual ARI standards (ARI 320, 325, and 330). The three applications are water-loop heat pump (WLHP), groundwater heat pump (GWHP), and ground-loop heat pump (GLHP). The ISO standard resolved the debate over the most appropriate value for ESP for the rating by setting the required ESP to 0. Thus, the user must add the value for fan power input. This fan heat is converted from watts to Btu/h and deducted from the total capacity in cooling or added to the total heating capacity. A similar procedure was followed with the pump wattage. Equations to perform this task are presented at the bottom of Table 5.7.

CHILLED WATER SYSTEMS

Chilled water systems (CWSs) are a widely used HVAC option in medium and large-sized commercial and industrial applications. The chiller is the central component of the system. It is a compressor (typically scroll, reciprocating, screw, or centrifugal) that has a water-to-refrigerant evaporator. Chillers with water-to-refrigerant condensers are classified as water-cooled. Most often the condenser water is cooled with a cooling tower as shown in Figure 5.10 (condenser water is sprayed over fill [baffles] and exposed directly to the outdoor air) or a closed-circuit fluid cooler (water circulated through the inside of tubes with the air and a water spray on the outside of the tubes). Cooling towers are able to reduce the condenser water to within about 5°F to 10°F of the outdoor air wet-bulb temperature because the high-velocity air induces high evaporation rates from the large surface area created by the many tiny droplets. In typical climates, this corresponds to a temperature leaving the tower and entering the condenser around

Table 5.7 ISO Standard 13256-1 Rating Temperatures (ISO 2000)

	WLHP (°F)	GWHP (°F)	GLHP (°F)
EWT_c	86	59	77
EWT_h	68	50	32
EAT_c (db)	80.6	80.6	80.6
EAT_c (wb)	66.2	66.2	66.2
EAT_h	68	68	68

ISO 13256-1 total cooling (TC) capacity, total heating (TH) capacity, and power input ratings are based on an air-side ESP of 0.0 in. of water and a water-side external head of 0.0 ft of water. Correction factors (CFs) should be applied to the values of EER and COP that appear in the table above.

To determine the correction factor for fan power (CF_{fan}):

$$CF_{fan}\text{ (W)} = 0.118 \times \text{cfm} \times \text{ESP (in. of water)} \div [\text{fan-motor eff. (\%)}] \text{ (30\% is the default)}$$

To determine the correction factor for pump power (CF_{pump}):

$$CF_{pump}\text{ (W)} = 0.189 \times \text{gpm} \times \text{h (ft of water)} \div [\text{pump-motor eff. (\%)}] \text{ (30\% is the default)}$$

To determine actual power for either cooling or heating:

$$Power_c\text{ (W)} = Power_c\text{ (ISO)} + CF_{fan} + CF_{pump};\ Power_h\text{ (W)} = Power_h\text{ (ISO)} + CF_{fan} + CF_{pump}$$

To determine total (or sensible) cooling capacity and efficiency:

$$\text{TC (Btu/h)} = \text{TC (ISO)} - 3.41 \times CF_{fan} \rightarrow \text{EER} = \text{TC} \div Power_c \text{ Btu/W·h}$$

To determine total heating capacity and coefficient of performance:

$$\text{TH (Btu/h)} = \text{TH(ISO)} + 3.41 \times CF_{fan} \rightarrow \text{COP} = (\text{TC} \div 3.41) \div Power_h \text{ W/W}$$

Example Problem 5.5

A water-to-air heat pump has ISO 13256-1 ratings @ 1150 cfm and 9 gpm with h = 7.9 ft water.

(WLHP) TC = 33.8 MBtu/h, EER = 14.9, TH = 40.4 MBtu/h, COP = 4.6

(GLHP) TC = 35.8 MBtu/h, EER = 16.4, TH = 16.7 MBtu/h, COP = 3.4

Determine the performance of the unit (TC, kW_c, and EER) for the conditions of EWT = 80°F and EAT = 80°F/67°F in cooling. The required ESP is 0.5 in. of water, and a 200 W pump is required to circulate water.

Solution:

Correct for EWT = 80°F
@ WLHP conditions (EWT = 86°F), TC = 33.8 MBtu/h, EER = 14.9 MBtu/Wh
@ GLHP conditions (EWT = 77°F), TC = 35.8 MBtu/h, EER = 16.4 MBtu/Wh
via interpolation to EWT = 80°F, TC_{80} = 35.1 MBtu/h, EER_{80} = 15.9 MBtu/Wh

Thus, $kW_{c,80} = TC_{80} \div EER_{80} = 35.1 \div 15.9 = 2.21$ kW.
Correct for EAT (wb) = 67°F @ ISO 13256-1 condition of EAT (wb) = 66.2°F.
CF for TC = 0.99, and CF for kW_c = 0.995 (by interpolation from Table 5.6 between EAT [wb] = 64°F and 67°F).

Thus, $TC_{80,67} = 35.1 \div 0.99 = 35.5$ MBtu/h, and $kW_{c80,67} = 2.21 \div 0.995 = 2.22$ MBtu/h.

Correct for fan and pump power.
$CF_{fan} = 0.118 \times 1150 \text{ cfm} \times 0.5 \text{ in.} \div 0.30 = 226$ W.
Pump is not corrected since value is given in problem statement (200 W).
$Power_c = 2.22 \text{ kW} \times 1000 \text{ W/kW} + 226 + 200 = 2646$ W.

$$TC_{80,67,net} = TC_{80,67}\text{ (Btu/h)} - 3.41 \times CF_{fan} = 35.5 \text{ MBtu/h} \times 1000 \text{ Btu/MBtu} - 3.41 \times 226 \text{ W}$$

$$TC_{80,67,net} = 34{,}700 \text{ Btu/h}$$

$$EER_{80,67,net} = TC_{80,67,net} \div Power_c = 34{,}700 \div 2646 = 13.1 \text{ Btu/Wh}$$

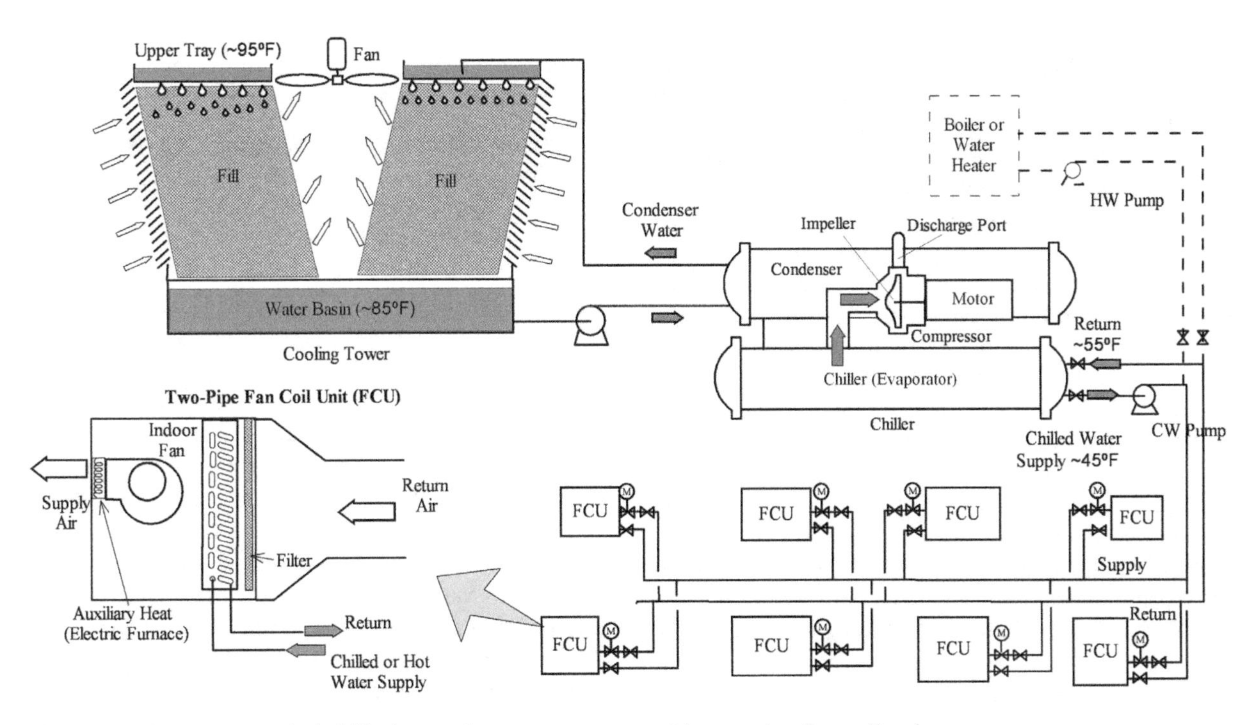

Figure 5.10 Water-cooled chilled water/hot water system with two-pipe fan-coil units.

85°F at design conditions. Chillers are able to operate at higher efficiency because of the lower temperatures.

Air-cooled chillers with air-to-refrigerant condensers are also available, but they do operate at much lower efficiency than water-cooled chillers during warm weather. They require less maintenance and are more common in medium-sized commercial applications in spite of poor efficiencies.

Chilled water from the evaporator is circulated throughout the building. Supply water temperatures can be as high as 48°F when the dehumidification load is light or as low as 32°F when thermal storage is incorporated. The chiller can operate at higher efficiency with 48°F supply water, but greater water flow, pump energy, and pipe sizes are required to meet a given load. The chilled water (CW) pump size, energy demand, and piping sizes will be much smaller with the colder supply water, but the chiller will operate at much lower efficiency and capacity. This presents a classic design optimization problem of selecting the optimum flow rate that maximizes efficiency when both chiller power and pump power are considered. Water distribution systems will be discussed in more detail in chapter 10.

Fin tube water-to-air coils are used in the space for cooling and in some cases heating. Small unitary cabinets that serve a single zone are classified as fan-coil units (FCUs). The FCUs shown in Figure 5.10 are two-pipe units (chilled or water supply pipe and return pipes). In applications with minimal heating load, electric coils located in the cabinet may be used. This high-demand heating option may be an acceptable heating method in warm climates or in applications that have low requirements. However, a more common two-pipe application is a system that provides hot water from a boiler or fossil fuel water heater. The system is switched over from the chiller to the heater to provide hot water in the piping loop during the winter months. While this does generate hot water from a non-electric resistance source, it can result in limits to comfort conditions during the fall and spring. Heating may be needed in some zones while cooling is simultaneously required in others. In some zones, heating is required in the morning while cooling is needed in the afternoon. A two-pipe changeover system cannot satisfy these types of loads without some occupant discomfort. The piping network shown in Figure 5.10 is a direct return design. Note that the FCUs at the greatest distance from the chiller will tend to get less water than those near the chiller. Testing and balancing (TAB) is required to throttle balancing or the use of automatic flow control valves on the units near the chiller to ensure the more remote FCUs have adequate flow. Flow can be balanced more simply by installing a "reverse-return" header, which is a third main line that picks up the return from the most distant coils first. This option does result in additional piping cost.

Figure 5.11 is a diagram of a four-pipe FCU that has both a cooling coil and a heating coil. This equipment provides greater levels of comfort since change-over from heating to cooling is almost instantaneous. Simultaneous heating and cooling in different zones can also be accomplished with a four-pipe system. In some cases, it may be necessary to "reheat" the chilled air when dehumidification requirements are abnormally high while sensible cooling loads are low. The heating coil of four-pipe FCUs can supply this sensible load to maintain adequate relative humidity levels in the zone. The primary disadvantage of a four-pipe system, in addition to the added coil in each unit and the central boiler or water heater, is the second piping loop.

Fan-coil units are typically equipped with fans that supply a constant air volume (CAV) flow. They typically remain on continuously during occupancy in order to induce ventilation air. In some cases, they will cycle off with load if they are not required to provide ventilation.

Variable air volume (VAV) systems are incorporated to take advantage of a substantial amount of energy savings made possible by reducing the airflow rate in proportion to load. In theory, fan power can be reduced by the cube of the required airflow since friction loss declines as a function of volume flow (or velocity) squared and fan power is the product of volume flow rate and friction head loss. However, this theory will overpredict savings since its application assumes constant efficiency in all components and no added restrictions in the air distribution system.

An additional factor in considering the size and type of cooling and heating coils is the reluctance of building owners to invest in adequate maintenance required for a large number of small fan-coil units. This, coupled with the desire to capture energy savings with VAVs, often sways the decision toward large coils located in air-handling units (AHUs). AHUs serve several zones, which are connected via a large duct system that is supplied with chilled primary air as shown in Figure 5.12. The flow rate in the main air handler is varied by changing the speed of the supply fan motor. Fan speed control is accomplished with an inverter drive that converts the 60 Hz AC signal to a modified sine wave with variable frequency. Air temperature control in the zones is achieved by varying the volume flow of primary air with a modulating air damper or valve in a VAV terminal box. As the primary flow is modulated during mild loads, it may become necessary to maintain air distribution through supply registers with small fans located in the VAV terminal that draw in an equivalent amount of room air. There are two types of fan-powered variable air volume (FPVAV) terminals: parallel and series. When the primary air volume is reduced by the thermostat, the

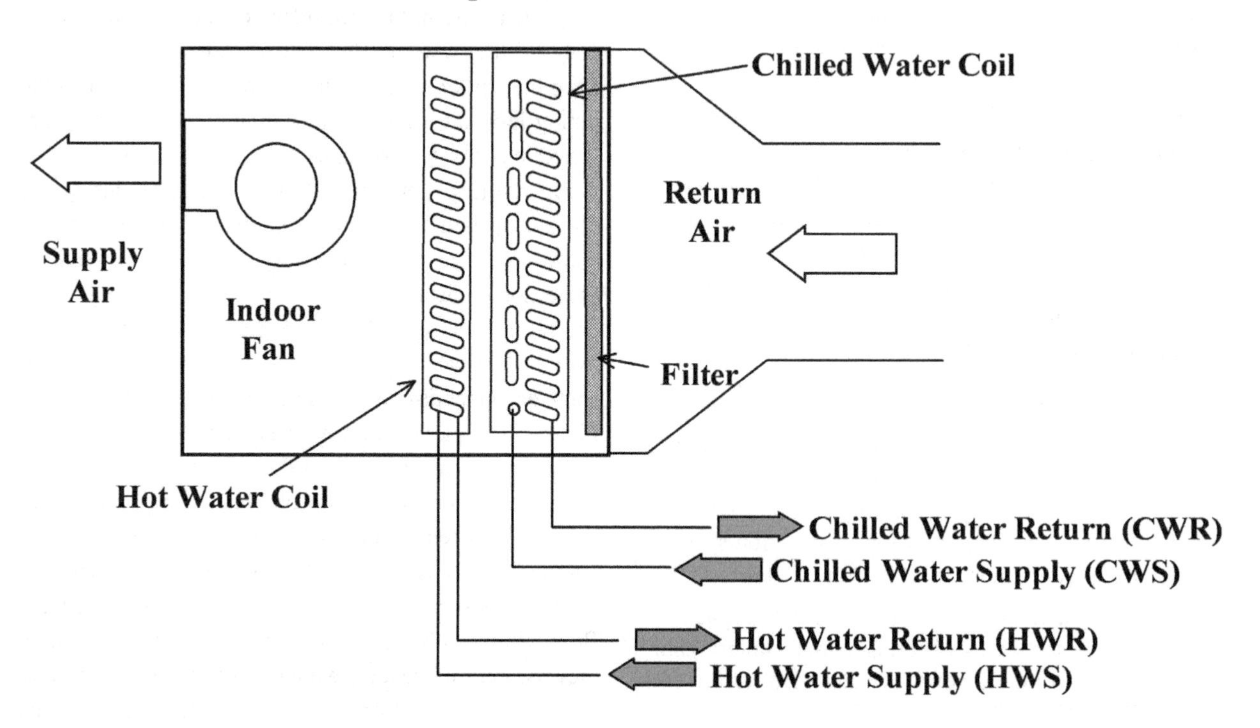

Figure 5.11 Four-pipe fan-coil unit.

fans in parallel FPVAV terminals provide the balance of the air required for adequate distribution. Series FPVAV terminal fans operate continuously with the full volume of zone airflow; when the primary air is reduced, a greater portion of the air is recirculated by the fan.

A volume of air approximately equivalent to the supply air volume must be returned to the central air-handling unit. This often requires a return air fan that may also be variable speed in some cases. Air can also be exhausted from the main return (as shown in Figure 5.12) or it can be expelled locally (i.e., restroom exhaust fans, etc.) before the return air duct. Traditionally, ventilation air has been supplied in the return of the central AHU. However, increasing attention is being devoted to the need to supply sufficient ventilation air to each and every zone even at part-load. This is problematic when the ventilation air is provided with the primary air that is being throttled with load.

There is a variety of solutions to deal with the challenge of high ventilation air loads. In some cases this involves the use of a dedicated outdoor air system (DOAS) that is decoupled from the primary cooling and heating system (Coad 1996). The ventilation air is precooled and preheated before being distributed with heat recovery units (see chapter 4) or conventional coils. In some cases, precooled or preheated air can be injected into the primary system as shown in Figure 5.12. However, rates must be adjusted for ventilation air system effectiveness according to ASHRAE Standard 62.1-2004 (ASHRAE 2004b) as discussed in chapter 6.

The engineer should be concerned about the increased complexity and energy consumption and demand of the added components. It is possible to defeat the primary purpose of VAV (energy savings) by the addition of fans, especially those with low-efficiency types in FPVAV terminals and those that provide high external static pressures (>2.0 inches of water). The total demand of the entire VAV system should be compared to the demand of a simple system with low ESP fan-coil units. High ESP FPVAV systems should be avoided with low-efficiency chillers (COP < 4.0 [kW/ton > 0.88]), such as air-cooled designs. Table 5.8 summarizes the requirements for minimum COP for chillers from ASHRAE Standard 90.1-2004. These values are con-

Table 5.8 ASHRAE Standard 90.1 Minimum Chiller COP, EER, and Maximum kW/ton (ASHRAE 2004a)

Conditions according to ARI Standard 550 (ARI 1998d): 44°F LWT, 85°F/95°F Condenser Water or 95°F OAT									
Chiller Type	< 150 tons (530 kW)			≥ 150, < 300 tons			≥ 300 tons (1060 kW)		
	COP	kW/ton	EER	COP	kW/ton	EER	COP	kW/ton	EER
Air-Cooled w/Condenser	2.8	1.26	9.6	2.8	1.26	9.6	2.8	1.26	9.6
Air-Cooled w/o Condenser	3.1	1.14	10.6	3.1	1.14	10.6	3.1	1.14	10.6
Water Cooled Reciprocating	4.2	0.84	14.3	4.2	0.84	14.3	4.2	0.84	14.3
Water Cooled Scroll and Screw	4.45	0.79	15.2	4.9	0.72	16.7	5.5	0.64	18.8
Water Cooled Centrifugal	5.0	0.70	17.1	5.55	0.63	18.9	6.1	0.58	20.8

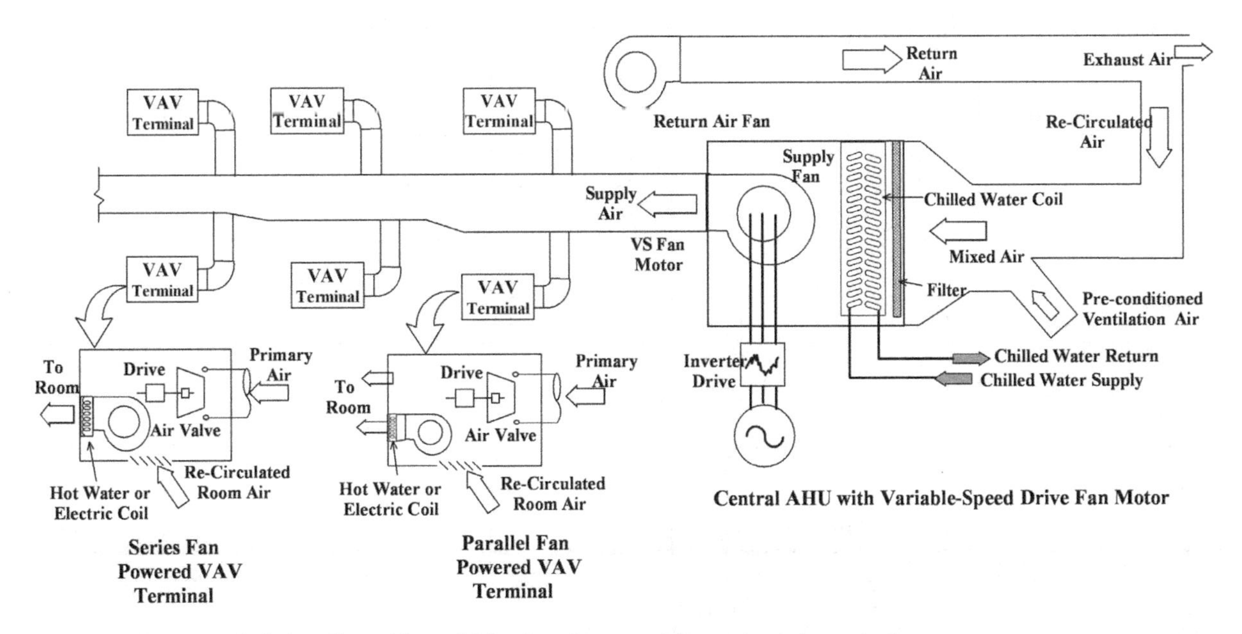

Figure 5.12 Central air handler with variable air volume and fan-powered terminals.

verted to EER and maximum allowable kW/ton, an inverse efficiency commonly used by manufacturers in the US. The spreadsheet program *HVACSysEff.xls* is available on the CD accompanying this book to compare the system COP, kW/ton, and EER of all HVAC types when all components are included.

Table 5.9 provides the capacity (ton), demand (kW), and efficiency (EERBtu/W·h) of a product line of small air-cooled chillers with scroll compressors. Data are given for three outdoor air temperatures (OATs) and three chiller leaving water temperatures (LWTs). Figure 5.13 provides information regarding the evaporator head loss as a function of water flow rate for the five chiller models shown in Table 5.9. As noted, the values in Table 5.9 correspond to a drop in chilled water temperature of 10°F, which results from a chilled flow rate of 2.4 gpm/ton. Therefore, nominal evaporator flow rate can be approximated by multiplying 2.4 by the chiller capacity in tons. A typical arrangement for an air-cooled chiller is shown is Figure 5.14.

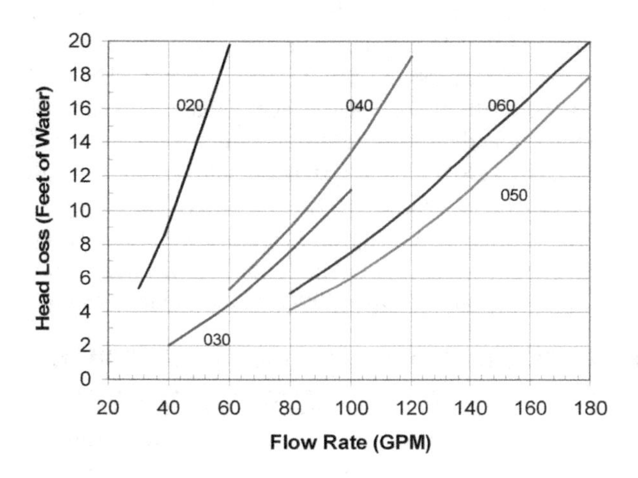

Figure 5.13 Evaporator head loss for Table 5.9 (AC) and Table 5.10 (WC) chillers.

Table 5.10 provides the capacity (tons), demand (kW), and efficiency (EER ≡ Btu/W·h) of a product line of small water-cooled chillers with scroll compressors. Data are given for three condenser entering water temperatures (EWTs) and three chiller leaving water temperatures (LWTs). Figure 5.13 can be used to determine the evaporator head loss for the air-cooled chiller product line. Figure 5.15 provides information regarding the condenser head loss as a function of water flow rate for the water-cooled chillers. As noted, the values in Table 5.10 correspond to a drop in chilled water temperature of 10°F and a rise in condenser water of 10°F. This results from a chilled flow rate of 2.4 gpm/ton and a condenser water flow rate of 3.0 gpm/ton. Therefore, nominal evaporator flow rate can be approximated by multiplying 2.4 by the chiller capacity in tons, and condenser flow rate can be approximated by multiplying 3.0 by the chiller capacity in tons.

Table 5.9 Air-Cooled Chiller Capacity,* Demand,* and Efficiency (T/ASC 2002)

	LWT	OAT = 75°F			OAT = 95°F			OAT = 115°F		
	°F	tons	kW	EER†	tons	kW	EER	tons	kW	EER
Model 020	40	18.0	17.5	12.4	16.2	21.3	9.2	14.2	26.1	6.5
	45	19.7	18.0	13.2	17.8	21.8	9.8	15.6	26.7	7.0
	50	21.5	18.4	14.0	19.4	22.3	10.4	17.1	27.3	7.5
Model 030	40	28.5	28.1	12.2	25.7	33.9	9.2	22.7	40.8	6.7
	45	31.2	28.6	13.1	28.2	34.3	9.8	24.9	41.6	7.2
	50	34.0	29.2	14.0	30.8	35.0	10.5	26.3	42.1	7.5
Model 040	40	35.3	34.1	12.4	31.7	41.5	9.2	27.8	51.0	6.5
	45	38.5	35.0	13.2	34.7	42.5	9.8	30.5	52.1	7.0
	50	42.0	35.9	14.0	37.8	43.5	10.4	33.3	53.2	7.5
Model 050	40	44.1	43.3	12.2	39.7	51.9	9.2	35.0	63.0	6.7
	45	48.3	44.3	13.1	43.5	53.1	9.8	38.5	64.3	7.2
	50	52.7	45.3	14.0	47.5	54.3	10.5	42.1	65.7	7.7
Model 060	40	58.5	56.6	12.4	52.6	68.5	9.2	46.2	83.4	6.7
	45	63.7	58.1	13.2	57.3	70.2	9.8	50.6	85.4	7.1
	50	69.1	59.6	13.9	62.3	72.0	10.4	55.0	87.5	7.5

† EER (Btu/W·h) = 12 × TC (tons) ÷ W (kW) = 12,000 × TC (tons) ÷ W (watts)
gpm (Evap.) ≈ TC (Btu/h) ÷ [500 × Δt_{evap} (°F)]
* Performance based on 100% water and 10°F chilled water temperature drop.
For 30% ethylene glycol: 70% water, multiply capacity (tons) by 0.97 and kW by 0.985.
For 30% propylene glycol: 70% water, multiply capacity (tons) by 0.955 and kW by 0.98.

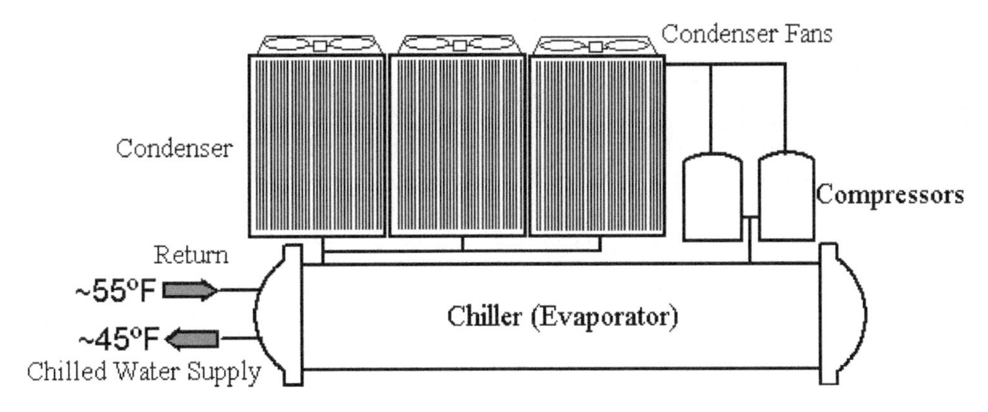

Figure 5.14 Air-cooled chiller.

Table 5.10 Water-Cooled Scroll Compressor Chiller Capacity,* Demand,* and Efficiency (T/ASC 1999)

	LWT °F	Condenser EWT = 75°F			Condenser EWT = 85°F			Condenser EWT = 95°F		
		tons	kW	EER†	tons	kW	EER	tons	kW	EER
Model 020	40	19.2	13.8	16.7	18.6	15.4	14.6	17.7	17.1	12.4
	45	21.1	13.9	17.9	20.4	15.4	14.9	19.5	17.2	13.4
	50	23.2	14.1	19.7	22.4	15.6	17.2	21.4	17.4	14.8
Model 030	40	28.3	20.7	16.4	27.0	22.9	14.1	25.6	25.4	12.1
	45	30.9	20.8	17.9	29.7	23.0	15.5	28.2	25.6	13.2
	50	34.1	21.1	19.5	32.6	23.3	16.8	31.0	25.0	14.4
Model 040	40	38.0	27.3	16.8	36.5	30.2	14.5	34.6	33.6	12.4
	45	41.8	27.5	18.2	40.1	30.4	15.8	38.1	34.0	13.5
	50	45.9	27.9	19.8	43.9	30.8	17.1	41.9	34.3	14.6
Model 050	40	47.0	34.1	16.6	44.9	37.7	14.3	42.7	41.0	12.2
	45	51.8	34.3	18.1	50.0	38.0	15.7	46.6	42.3	13.4
	50	56.8	34.8	19.6	54.3	38.5	16.9	51.7	42.8	14.5
Model 060	40	57.7	42.1	16.4	55.0	46.5	14.2	52.3	51.7	12.1
	45	63.3	42.5	17.9	60.5	47.0	15.5	57.5	52.2	13.2
	50	69.1	43.0	19.3	66.0	47.5	16.7	62.9	52.8	14.3

† EER (Btu/W·h) = 12 × TC (tons) ÷ W (kW) = 12,000 × TC (tons) ÷ W (watts)
gpm (Evap.) ≈ TC (Btu/h) ÷ (500 × Δt_{evap}) (10°F)
gpm (Cond.) ≈ [TC (Btu/h) + 3.41 × w (W)] ÷ [500 × Δt_{cond} (°F)]
* Performance based on 100% water and 10°F chilled water temperature drop and 10°F condenser water temperature rise.
For 30% ethylene glycol: 70% water, multiply capacity (tons) by 0.97 and kW by 0.985.
For 30% propylene glycol: 70% water, multiply capacity (tons) by 0.96 and kW by 0.98.

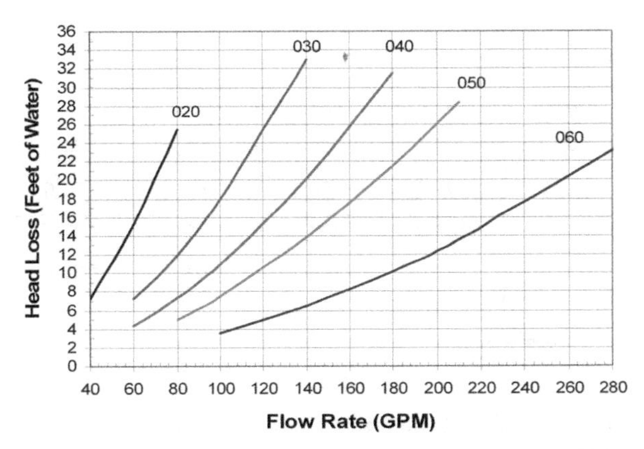

Figure 5.15 Water-cooled scroll compressor chiller condenser head loss.

The performance of a larger chiller product line with screw compressors is shown in Table 5.11. Head loss characteristics of the evaporator and condenser are shown in Figure 5.16. The efficiency and required water flow rates are not shown but can be calculated by the equations in the table. Data are based on a 10°F temperature difference across the evaporator and condenser.

Chilled water systems offer a good deal of flexibility in that building zones with vastly different load characteristics can be accommodated by the choice of fan-coil or air-handler specifications. For example, a computer room with a constant high sensible load would be a good choice for a high flow rate, constant volume air-handling unit. The high airflow rate would ensure high capacity but low dehumidification capability, which

Example Problem 5.6

a. Compute the capacity, demand, EER, kW/ton, COP, required condenser flow rate and head loss, and required chilled water flow rate and head loss for a Model 050 water-cooled scroll compressor chiller with 85°F condenser EWT and 45°F chilled LWT.
b. Compare the capacity, demand, EER, kW/ton, and COP values of a Model 050 air-cooled chiller for an outdoor air temperature (OAT) of 95°F.

Solution:

a. From Table 5.10 for a Model 050 @ 85°F condenser EWT and 45°F chilled LWT,

$$TC = 50.0 \text{ tons}, kW_c = 38.0, EER = 15.7 \text{ Btu/W·h}$$

$$kW/ton = 38.0 \div 50.0 = 0.76, COP_c = EER \div 3.412 = 15.7 \div 3.412 = 4.6$$

From Table 5.10 footnote equation and Figure 5.15 Model 050 curve @ 146 gpm, h = 15 ft:

$$gpm_{Cond} = [50 \text{ tons} \times 12{,}000 \text{ Btu/ton·h} + 3.412 \times 38 \text{ kW} \times 1000\text{W/kW}] \div [500 \times (95 - 85°F)] = 146 \text{ gpm}$$

From Table 5.10 footnote equation and Figure 5.13 Model 050 curve @ 120 gpm, h = 8.5 ft:

$$gpm_{Evap} = [50 \text{ tons} \times 12{,}000 \text{ Btu/ton·h}] \div [500 \times (55 - 45°F)] = 120 \text{ gpm}$$

b. From Table 5.9 for a Model 050 @ 95°F OAT and 45°F chilled LWT,

$$TC = 43.5 \text{ tons}, kW_c = 53.1, EER = 9.8 \text{ Btu/W·h}$$

$$COP_c = EER \div 3.412 = 9.8 \div 3.412 = 3.1$$

Table 5.11 Water-Cooled Screw Compressor Chiller Capacity* and Demand*

	LWT	Condenser EWT = 80°F		Condenser EWT = 85°F		Condenser EWT = 90°F		Condenser EWT = 95°F	
		tons	kW	tons	kW	tons	kW	tons	kW
Model 096	40	89.1	62.5	85.4	66.0	82.0	69.7	79.3	74.1
	44	97.6	63.7	93.6	67.3	89.6	71.1	85.9	75.0
	48	106.4	65.6	102.5	68.5	98.1	72.5	93.9	76.51
Model 120	40	116.6	81.0	111.8	86.0	107.3	91.2	103.7	97.0
	44	127.6	82.3	122.5	87.3	117.4	92.6	112.5	98.1
	48	139.1	84.2	134.0	88.7	128.5	93.9	123.0	99.5
Model 160	40	147.6	104.3	144.9	113.5	143.1	123.7	142.4	135.5
	44	159.0	103.6	153.6	113.0	153.4	123.0	151.4	136.0
	48	170.4	104.0	168.1	112.1	165.2	122.4	162.0	133.3
Model 210	40	199.4	136.1	195.8	148.0	193.0	161.0	191.1	175.4
	44	214.7	136.9	211.2	147.6	207.3	160.5	204.2	174.6
	48	229.3	137.1	227.0	148.2	223.1	159.9	218.9	174.0
Model 270	40	252.3	179.6	248.3	191.9	245.2	208.3	244.2	227.5
	44	271.2	175.8	267.0	190.8	262.7	207.0	259.4	224.8
	48	289.9	176.7	286.5	189.4	282.0	205.8	277.5	223.6

EER (Btu/W·h) = 12 × TC (tons) ÷ W (kW) = 12,000 × TC (tons) ÷ W (watts)
gpm (Evap.) ≈ TC (Btu/h) ÷ [500 × Δt_{evap} (°F)]
gpm (Cond.) ≈ [TC (Btu/h) + 3.41 × w (W)] ÷ [500 × Δt_{cond} (°F)]
*Performance based on 100% water and 10°F chilled water temperature drop (Δt_{evap}) and 10°F condenser water temperature rise (Δt_{cond}).
For 30% ethylene glycol: 70% water, multiply capacity (tons) by 0.97 and kW by 0.985.
For 30% propylene glycol: 70% water, multiply capacity (tons) by 0.96 and kW by 0.98.

Example Problem 5.7

A 200-ton chiller uses forty 2000 cfm series FPVAV terminals. Determine the fan heat penalty in MBtu/h and tons for these terminals.

Solution:

Figure 5.17 indicates a 2000 cfm series FPVAV terminal will draw 1000 W (or 1 kW). Forty terminals will require 40 kW. This is input power to the fan motors, and it will be converted into heat in three stages (heat due to motor inefficiency, heat due to fan inefficiency, and conversion of kinetic energy of air into heat via air distribution system friction). Thus, the entire input power to the motor will be converted to heat. The results are expressed in the desired units with the use of conversion factors.

$$q_{Fan} = 40 \text{ kW} \times 3.412 \text{ MBtu/kWh} = 136.5 \text{ MBtu/h} \div 12 \text{ MBtu/ton·h} = 11.4 \text{ tons}$$

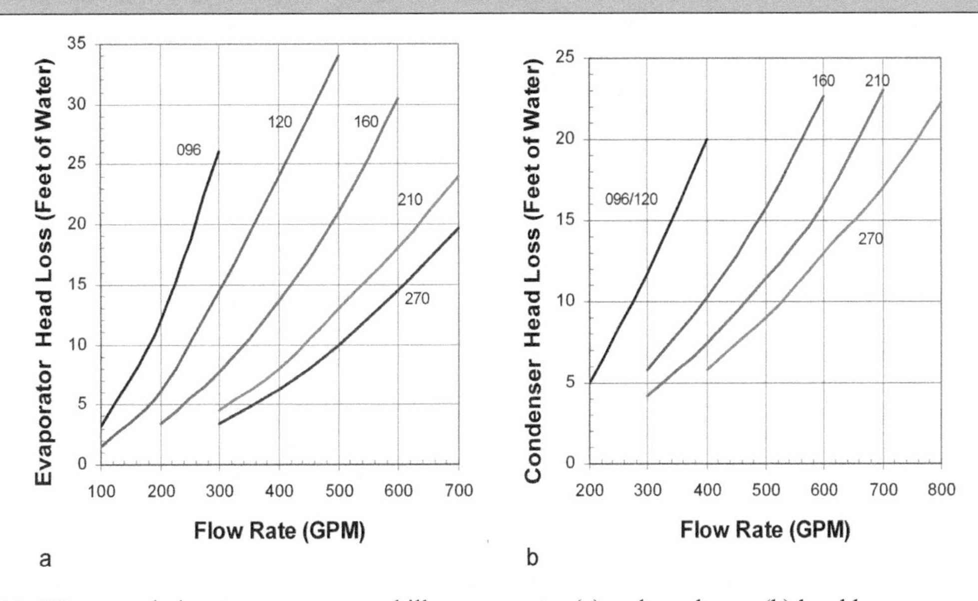

Figure 5.16 Water-cooled screw compressor chiller evaporator (a) and condenser (b) head losses.

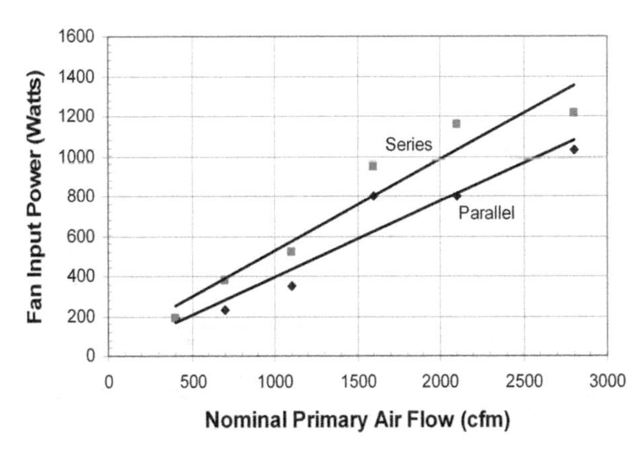

Figure 5.17 VAV fan power for parallel and series FPVAV terminals (C/UTC 2001b).

is a good match to the load. The added cost and complexity of a VAV system would be unnecessary because of the constant nature of the load.

This zone might be located in the same building as a classroom or training room that has a very high occupant density. Code compliance will dictate a high required ventilation air rate, which, coupled with moisture generation from the occupants, will result in a high latent load (low SHR) in many climates. Chilled water could be delivered to a fan-coil unit with a low airflow rate to match the internal latent load and to an outdoor air preconditioning coil to remove humidity from the outdoor air.

Figure 5.17 is manufacturer's fan power data for a product line of series and parallel FPVAV terminals. These data are necessary in order to deduct the summarized fan heat for all the terminals from the cooling capacity of the coil in the main air-handling unit.

Table 5.12 provides the total capacity (TC), sensible capacity (SC), and water-side head loss (h) for a product line of chilled water coils. The coils have four rows and 12 fins per inch (FPI). Airflow rate (cfm), water flow rate (gpm), entering air temperature (EAT), and entering (chilled) water temperature (EWT) are varied. Recall that the EWT for the coil corresponds to the leaving water temperature (LWT) of the chiller.

Table 5.12 Total and Sensible Capacities (MBtu/h)—4-Row, 12-FPI Water Coils (MA 1998)

	24-HW-4		EAT = 85°F/71°F		EAT = 80°F/67°F		EAT = 75°F/63°F	
EWT	gpm/h	cfm	TC	SC	TC	SC	TC	SC
42°F	6 gpm	600	30.4	17.5	25.7	16.0	20.4	14.2
	4.1 ft	800	35.2	20.9	29.0	18.9	23.8	17.0
	9.5 gpm	600	33.9	18.9	27.7	17.0	22.5	15.0
	10.5 ft	800	39.0	22.3	32.8	20.5	26.0	18.0
45°F	6 gpm	600	27.9	16.6	22.7	14.8	17.5	13.0
	4.1 ft	800	31.5	19.6	26.4	17.9	20.5	15.7
	9.5 gpm	600	30.7	17.7	25.0	15.8	19.5	13.8
	10.5 ft	800	35.9	21.2	29.0	18.9	23.1	16.7
48°F	6 gpm	600	24.9	15.4	19.8	13.7	15.3	12.1
	4.1 ft	800	28.1	18.4	22.0	16.3	17.4	14.5
	9.5 gpm	600	27.9	16.5	21.8	14.5	16.4	12.5
	10.5 ft	800	32.1	19.8	25.2	17.4	19.7	15.3
	Model 36-HW-4		**EAT = 85°F/71°F**		**EAT = 80°F/67°F**		**EAT = 75°F/63°F**	
EWT	**gpm/h**	**cfm**	**TC**	**SC**	**TC**	**SC**	**TC**	**SC**
42°F	8 gpm	900	47.0	27.0	39.3	24.5	31.5	21.8
	4.3 ft	1200	54.5	32.2	44.8	29.0	36.8	28.2
	13 gpm	900	52.2	29.1	42.6	25.9	34.6	23.1
	11.1 ft	1200	62.3	35.1	50.8	31.4	41.6	28.1
45°F	8 gpm	900	43.2	25.5	34.6	22.7	28.0	20.3
	4.3 ft	1200	49.1	30.1	41.0	27.5	31.8	24.1
	13 gpm	900	47.0	27.1	38.9	24.3	30.4	21.3
	11.1 ft	1200	56.2	32.7	45.1	29.2	36.2	25.9
48°F	8 gpm	900	38.2	23.7	30.3	21.0	23.8	18.5
	4.3 ft	1200	43.7	28.3	35.4	25.4	27.0	22.2
	13 gpm	900	43.2	25.5	33.7	22.3	25.6	19.3
	11.1 ft	1200	49.5	30.2	40.8	27.5	30.5	23.5
	Model 48-HW-4		**EAT = 85°F/71°F**		**EAT = 80°F/67°F**		**EAT = 75°F/63°F**	
EWT	**gpm/h**	**cfm**	**TC**	**SC**	**TC**	**SC**	**TC**	**SC**
42°F	10 gpm	1200	61.6	35.6	51.5	32.3	41.6	28.8
	4.5 ft	1600	71.2	42.4	58.8	38.4	48.2	34.5
	16 gpm	1200	68.5	38.3	56.5	34.4	46.0	30.7
	11.5 ft	1600	81.6	46.2	57.2	41.7	55.0	37.3
45°F	10 gpm	1200	56.5	33.6	45.7	30.0	36.7	26.8
	4.5 ft	1600	64.2	39.8	51.6	35.6	41.5	31.7
	16 gpm	1200	62.3	35.8	51.2	32.2	39.9	28.1
	11.5 ft	1600	73.6	43.3	59.2	38.5	47.4	34.2
48°F	10 gpm	1200	50.2	31.4	40.1	27.9	30.2	24.1
	4.5 ft	1600	57.1	37.1	46.3	33.7	35.4	29.3
	16 gpm	1200	56.6	33.7	44.3	29.4	33.7	25.5
	11.5 ft	1600	65.5	40.1	51.3	35.4	40.3	31.4
	Model 60-HW-4		**EAT = 85°F/71°F**		**EAT = 80°F/67°F**		**EAT = 75°F/63°F**	
EWT	**gpm/h**	**cfm**	**TC**	**SC**	**TC**	**SC**	**TC**	**SC**
42°F	13 gpm	1500	76.8	44.1	64.2	40.2	51.6	35.6
	3.9 ft	2000	88.4	42.3	72.8	47.3	59.6	42.7
	21 gpm	1500	85.6	47.6	69.9	42.5	56.8	37.9
	10 ft	2000	100.8	57.1	82.9	51.4	67.8	46.0
45°F	13 gpm	1500	70.6	41.7	57.1	37.3	45.8	33.4
	3.9 ft	2000	79.4	49.1	65.0	44.9	51.3	39.2
	21 gpm	1500	77.1	44.1	63.5	39.9	49.6	34.8
	10 ft	2000	91.2	53.4	75.2	48.4	59.0	42.5
48°F	13 gpm	1500	63.0	38.9	49.5	34.5	38.8	30.4
	3.9 ft	2000	75.5	46.0	57.3	41.6	44.1	36.4
	21 gpm	1500	70.6	41.7	55.4	36.6	41.8	31.7
	10 ft	2000	81.2	49.7	65.8	44.7	50.2	38.8

SELECTION PROCEDURE FOR CHILLED WATER COILS, AIR CONDITIONERS, AND HEAT PUMPS

The following is a set of recommended steps to select equipment that meets the total cooling requirement of the building and to specify operating conditions to ensure that proper dehumidification simultaneously occurs.

1. *Sum the total cooling requirements*—Building envelope, internal loads, ventilation air, etc. Express answers in Btu/h, MBtu/h (Btu/h ÷ 1000), and tons of cooling (tons = MBtu/h ÷ 12). Details are presented in chapter 8.
2. *Find sensible and latent components of loads.* Sum sensible cooling requirements and calculate the sensible heat ratio of the load (SHR_{load} = sensible load ÷ total load). Latent load is the difference between the total and sensible loads. Buildings with high latent loads (25% or more) present moisture-removal challenges.

3a. *Find a unit to meet the required net total capacity (TC) of the load.* Enter the equipment tables using the design outdoor (dry-bulb) temperature (or the entering water temperature [EWT for water cooled/source equipment]), the indoor air wet-bulb temperature, and the nominal rated airflow. Find a unit that has a capacity equal to or slightly greater than the TC of the load. It is also wise to note the power input of the unit at these conditions (kW_{wb} or $kW_{odb,wb}$ [W_{wb} or $W_{odb,wb}$]). This power should include the power for the outdoor fan.

b. *For units that provide capacity in gross total and sensible cooling, estimate fan heat and deduct from the gross total and sensible values to find net total and sensible capacities.* Manufacturers will have a set of fan curves or tables (see Table 5.2) to find the indoor fan power required to deliver the estimated airflow at different external static pressure (ESP) levels. This is the pressure required to overcome friction in the ductwork, dampers, registers, grilles, and external filters. Residential systems may require as little as 0.5 in. of water, small commercial systems (10 to 40 tons) usually require 1 to 2 in., and larger systems may require 2 to 6 in. of water. Fan power is converted to heat by:

$$Watts_{fan} = Fan\ hp_{req'd} \times 746 \div \eta_{Motor} \quad (5.1a)$$

$$kW_{fan} = Fan\ hp_{req'd} \times 0.746 \div \eta_{Motor} \quad (5.1b)$$

$$q_{fan}\ (Btu/h) = 3.412 \times W_{fan} \quad (5.2a)$$

$$q_{fan}\ (MBtu/h) = 3.412 \times kW_{fan} \quad (5.2b)$$

Fan input is converted to heat in three stages (heat due to motor inefficiency, heat due to fan inefficiency, conversion of kinetic energy of air into heat via air distribution system friction). If the fan motor is located outside the conditioned space, the motor heat contribution can be ignored. This can be accomplished by multiplying the fan input power (W_{fan}, kW_{fan}) in Equations 5.2a and 5.2b by the fan motor efficiency (η_{Motor}).

After TC and SC are corrected, step 2 should be repeated to recalculate SHR_{unit} and compared with the value of SHR_{load}.

Note: Units with total capacities of 65,000 Btu/h or less may be rated in net total capacity, which means the fan heat has already been deducted. So in these instances step 3b is unnecessary.

4. *Specify the required airflow rate and make sure SHR_{unit} is slightly less than or equal to SHR_{load}.* If $SHR_{unit} > SHR_{load}$, the dehumidification (latent) capacity of the unit will be inadequate. So the airflow should be lowered and step 3a (and 3b) is repeated. If $SHR_{unit} \ll SHR_{load}$, the airflow should be raised and step 3a (and 3b) is repeated. A slightly lower SHR_{unit} is often acceptable since the unit is being sized at design conditions, which typically do not correspond with maximum latent load. Therefore, a slightly higher SHR_{unit} indicates an improved capability to handle higher latent loads during off-peak conditions. A general guide to reduce the number of iterations is presented here when return air is 80°F/67°F. SHRs will be slightly higher when return air is 75°F/63°F.

Estimated airflow (Q_m) ≈ 300 to 375 cfm/ton × tons ($SHR_{load} < 0.72$)

400 cfm/ton × tons ($0.72 < SHR_{load} < 0.80$)

425 to 475 cfm/ton × tons ($SHR_{load} > 0.80$)

5. *Find demand in kW and EER*:

$$EER\ (Btu/W{\cdot}h) = TC\ (Btu/h) \div (W_{Total}) = TC\ (Btu/h) \div (W_{Unit} + W_{InFan} + W_{OutFan\ or\ Pump}) \quad (5.3a)$$

$$EER\ (MBtu/kWh) = TC\ (MBtu/h) \div (kW_{Unit} + kW_{InFan} + kW_{OutFan\ or\ Pump}) \quad (5.3b)$$

HEATING MODE SELECTION PROCEDURE FOR UNITARY HEAT PUMPS AND FURNACES

The following is a set of recommended steps to select equipment that meets the total heating requirement of the building.

1. *Sum the total heating requirements* (q_{Total})—Building envelope and ventilation air. Express answers in Btu/h and/or MBtu/h (Btu/h ÷ 1000).

2. *Estimate the required airflow rate.* Air-source heat pumps are normally sized in the cooling mode and the airflow is determined by that procedure. However, lower flows can be used to minimize "cold blow" discomfort if controls have this option.

   Water-to-air heat pumps are also frequently sized in the cooling mode and the airflow is determined by that procedure. In cold climates, flows may be determined to meet the heating requirement and should be held at or below 400 cfm/heating equivalent ton to minimize "cold blow" discomfort. However, this may lead to oversizing in the cooling cycle in colder climates, so designers should consider cooling mode SHRs when determining a fixed flow.

   Furnace standard controls permit the use of a different speed tap on the indoor blower motor so that airflow can be different in heating than in cooling to permit comfortable air discharge temperatures (110°F to 125°F) from furnaces. Recall that $t_{supply} \approx t_{return} + q \div (1.08 \times \text{cfm})$.
3. *Determine heating capacity and size auxiliary heat if necessary.* Air-source heat pump performance data are used to determine the heating capacity power requirement for compressor and fans for the outdoor design condition and indoor air temperature. Air heat pumps must be stopped, reversed, and defrosted periodically. If the manufacturer does not note that the performance is based on this consideration, a deduction (usually 10%) should be given to the unit. Some manufacturers lump total input power into a single value; others break it down for the compressor and fans. If the corrected total heating capacity (TH) is less than the load (q_{Total}), auxiliary heat (usually electric) must be added.

$$q_{Aux} = q_{Total} - \text{TH (less deduct for defrost penalty)} \quad (5.4)$$

The required size of the electric auxiliary furnace is:

$$\text{kW}_{Aux} = q_{Aux}\ (\text{MBtu/h}) \div 3.412\ (\text{MBtu/kWh}) \quad (5.5)$$

System COP is:

$$\text{COP} = [\text{TH} + q_{Aux}]\ (\text{MBtu/h}) \div [3.412\ (\text{kW}_{comp} + \text{kW}_{IDFan} + \text{kW}_{ODFan} + \text{kW}_{Aux})] \quad (5.6a)$$

$$= [\text{TH} + q_{Aux}]\ (\text{Btu/h}) \div [3.412\ (\text{W}_{comp} + \text{W}_{IDFan} + \text{W}_{ODFan} + \text{W}_{Aux})\ (\text{W})] \quad (5.6b)$$

Water-to-air heat pump performance data are used to determine the heating capacity and power requirements based on the water-loop temperature and indoor air temperature. A water flow rate of 3 gpm/ton is typical, but 2 gpm/ton can be used in some cases. Water-to-air heat pumps do not have a defrost deduction. Some manufacturers lump total input power into a single value; others break it down for the compressor and fans. If the corrected heating capacity (TH) is less than the load (q_{Total}), auxiliary heat (usually electric) must be added as with air heat pumps, although this is only necessary in colder climates.

System COP is:

$$\text{COP} = \text{TH} + q_{Aux}\ (\text{MBtu/h}) \div [3.412\ (\text{kW}_{comp} + \text{kW}_{IDFan} + \text{kW}_{Pump} + \text{kW}_{Aux})] \quad (5.7a)$$

$$= \text{TH} + q_{Aux}\ (\text{Btu/h}) \div [3.412\ (\text{W}_{comp} + \text{W}_{IDFan} + \text{W}_{Pump} + \text{W}_{Aux})] \quad (5.7b)$$

Furnaces can be selected to meet the entire heating load and be able to handle a cfm near or less than the cooling mode cfm. See Example Problem 5.4.

COOLING TOWERS AND FLUID COOLERS

Cooling towers and fluid coolers are used in many HVAC and industrial processes. Cooling towers have direct exchange between the outdoor air and cooling water, while fluid coolers have enclosed tubing with a water spray and air circulation on the outside surfaces. Water has superior heat transfer characteristics compared to air, and the evaporative cooling phenomenon permits the leaving water temperature to approach the outside air wet-bulb temperature. The result is that the condensing temperature of water-cooled equipment can be significantly lower than that of air-cooled equipment. The performance will be enhanced dramatically, as indicated by the COP values in ASHRAE Standard 90.1-2004 that only require a COP of 2.8 for air-cooled equipment, but water-cooled chillers can have COPs above 6.0 (ASHRAE 2004a). However, cooling towers require significant maintenance procedures and precautions to prevent *Legionella* and other water-induced air quality problems (ASHRAE 2004c, chapter 36; ASHRAE 2000).

Figure 5.18 provides a description of several types of cooling towers and fluid coolers. Figure 5.19 is a selection chart from a small cooling tower manufacturer. The user inputs the desired temperature range (Range = $t_{wi} - t_{wo}$) on the chart and follows a vertical line to the approach temperature (Approach = $t_{wo} - t_{awb}$). From this intersection, a horizontal line is drawn to the entering wet-bulb line (t_{awb}). At this point, a vertical line is drawn downward to intersect a horizontal line for water flow rate (gpm). The intersection point is then used to determine the cooling tower model number, as shown in Figure 5.19.

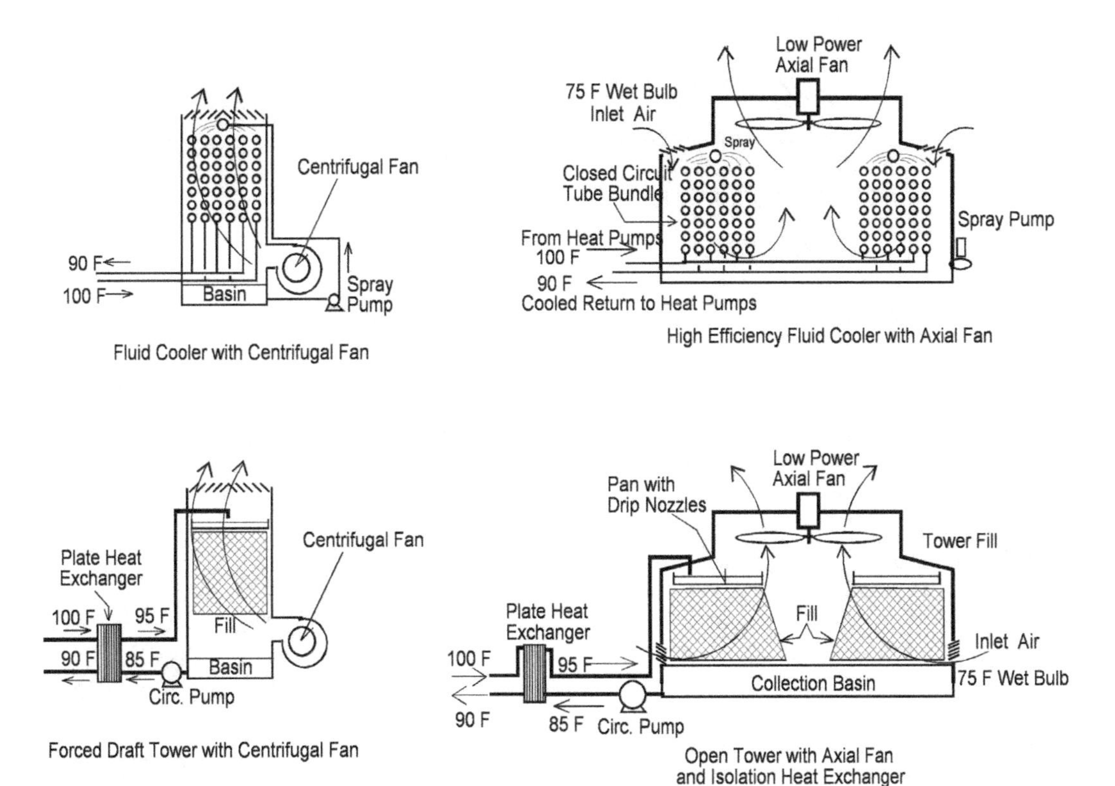

Figure 5.18 Cooling tower and fluid cooler types.

ECONOMIZERS

In many applications, there are periods when the outdoor air temperature is low and the building may require cooling. Water-side or air-side economizers are options to assist the mechanical cooling system or to provide total cooling (ASHRAE 2004c, chapter 5). The water-side economizer consists of a water coil located upstream of the primary cooling coil. Low-temperature water can be provided from a cooling tower, a fluid cooler, a surface water heat pump coil, or a groundwater loop to precool or totally cool the supply air. The air-side economizer uses cool outside air to either assist mechanical cooling or provide total cooling if the outside air is sufficiently cool and dry. A mixing box is typically required to allow 100% of the supply air to be drawn from the outside. Figure 5.4 provides an example of an air-side economizer on a rooftop unit (RTU).

Economizers can substantially reduce compressor energy consumption. Air-side economizers require precautions to ensure the added amount of outside air does not create additional building pressurization and humidity control problems. Water-side economizers have minimal impact in these areas, but they create added loads on the cooling tower and water treatment system, which must be designed for winter operation.

CHAPTER 5 PROBLEMS

5.1 Find the total cooling capacity (gross), sensible cooling capacity (gross), and input kW for a Model 150 rooftop unit when the outdoor temperature is 95°F, indoor temperature is 80°F/67°F, and airflow is 5000 cfm.

5.2 Find the required fan power to deliver 5000 cfm at an ESP of 1.2 in. w.g. for the unit selected in Problem 5.1.

5.3 Repeat Problem 5.1 for an indoor temperature of 74°F/62°F.

5.4 Correct the results of Problem 5.3 for fan heat to obtain total capacity (net), sensible cooling capacity (net), and resulting sensible heat ratio (SHR).

5.5 A building in St. Louis, Missouri, has a sensible heat gain of 23,000 Btu/h and a total load of 33,000 when outdoor conditions are 97°F/76°F and mixed indoor air conditions entering the cooling coil are 80°F/67°F. Select a cooling unit from Table 5.3 to meet the load and SHR requirement. Specify the required cfm and resulting EER.

5.6 Repeat Problem 5.5 for an indoor condition of 75°F/63°F.

5.7 Determine if the unit selected in Problem 5.6 can meet an SHR of 0.68 for an outdoor condition of

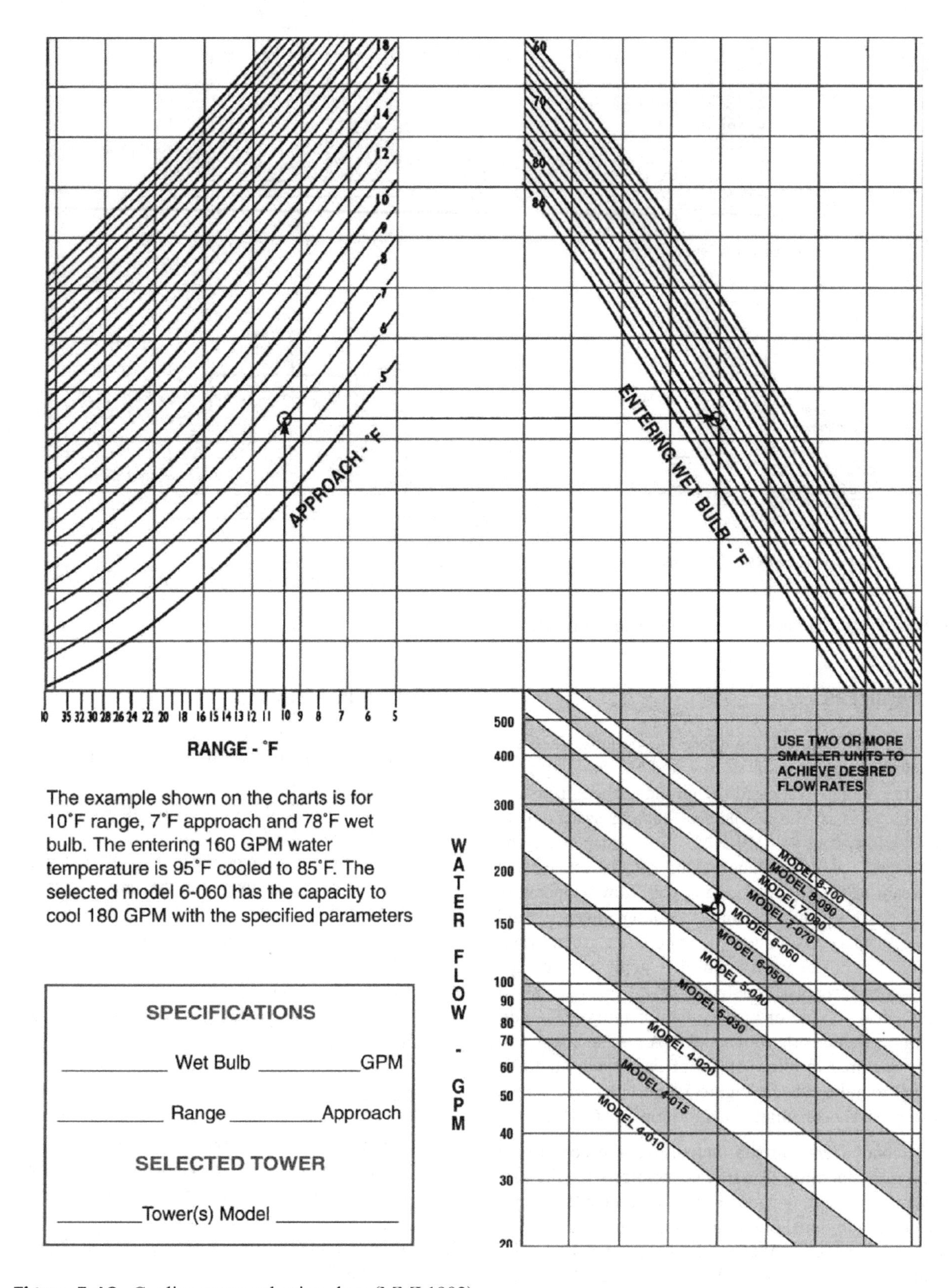

Figure 5.19 Cooling tower selection chart (MMI 1992).

85°F and indoor condition of 75°F/63°F at the design cfm. Can it meet the SHR at a lower cfm?

5.8 The building heat loss is 37,000 Btu/h when the indoor temperature is 70°F and the outdoor temperature is 20°F. Use the heating data of the Problem 5.6 heat pump to determine the unit's capacity (with a 10% defrost cycle deduct) and size the electric resistance supplementary backup if necessary. Find the system COP.

5.9 Meet the requirements of Problem 5.8 by selecting a natural gas furnace for an indoor temperature of 70°F.

5.10 Repeat Problems 5.6 and 5.8 using a water-to-air heat pump with a 90°F entering water temperature in cooling and a 45°F entering water temperature in heating. Assume a pump power requirement of 160 W (this replaces the outdoor fan of an air unit) and indoor fan is included in the total kW.

5.11 A building has a sensible heat gain of 140 MBtu/h and a total load of 190 MBtu/h when outdoor conditions are 95°F/75°F and mixed indoor air conditions entering the cooling coil are 78°F/64.5°F. Select a rooftop cooling unit from Table 5.2 to meet the load. Specify the required cfm, SHR_{unit}, and fan motor size to deliver 1.2 in. of water external static pressure (ESP) and the resulting EER.

5.12 A building zone has a total sensible heat gain of 105,000 Btu/h (walls, roof, windows, internal, people) and a latent gain of 20,000 Btu/h. The required outdoor air ventilation rate is 800 cfm. Indoor conditions are 75°F/63°F and outdoor conditions are 95°F/75°F, and outside air is mixed with the return air before entering the unit. Select a rooftop unit to cool this zone. The fan must deliver 1.0 in. water of external static pressure (ESP). Recall the capacities given are gross. You must convert them to total net capacities by deducting the fan heat.

5.13 A water-cooled chiller must provide water at 45°F to ten fan coil units that require 45 MBtu/h (net) each with fans that draw 600 W each. The condenser water is cooled with a cooling tower that can provide 85°F with an entering water temperature (EWT) of 95°F.

a. Select a chiller to meet this load.

b. Calculate the required chilled water flow in gpm for a 55°F chiller entering temperature (base answer on chiller capacity).

c. Calculate the required condenser water flow based on 3.0 gpm per ton of chiller capacity.

d. Determine the head loss in feet of water across the evaporator and condenser.

e. Determine the chiller gross kW/ton (gross) and EER (Btu/W·h).

f. Determine system net kW/ton and EER if two pumps (chilled water and condenser water) draw 2.0 kW and 2.25 kW, respectively.

5.14 A four-zone building has the loads shown below. The room air entering the coils is 80°F/67°F and chilled water at 45°F is supplied. Select fan coil units (assuming a 10% deduction for fan heat) and specify airflow and water flow while attempting to maintain a coil outlet temperature of 55°F ±2.0°F.

	10 a.m. Cooling Loads (MBtu/h)		3 p.m. Cooling Loads (MBtu/h)	
	Sensible	Total	Sensible	Total
Zone 1	30	40	42	60
Zone 2	45	60	35	45
Zone 3	25	35	38	54
Zone 4	30	38	40	55
Total				

5.15 Select a chiller (or chillers) to meet the combined loads of the coils in problem 5.14. Specify unit model number, required water flow, and gross kW/ton and EER.

REFERENCES

AHAM. 1992. *ANSI/AHAM Standard RA C-1-1992, Room Air Conditioners.* Washington, DC: Association of Home Appliance Manufacturers.

ARI. 1993. *ANSI/ARI Standard 310/380, Packaged Terminal Air Conditioners and Heat Pumps.* Arlington, VA: Air Conditioning and Refrigeration Institute.

ARI. 1994. *ANSI/ARI Standard 210/240, Unitary Air-Conditioning and Air-Source Heat Pump Equipment.* Arlington, VA: Air Conditioning and Refrigeration Institute.

ARI. 1998a. *ARI Standard 320, Water-Source Heat Pumps.* Arlington, VA: Air-Conditioning and Refrigeration Institute.

ARI. 1998b. *ARI Standard 325, Ground Water-Source Heat Pumps.* Arlington, VA: Air-Conditioning and Refrigeration Institute.

ARI. 1998c. *ARI Standard 330, Ground Source Closed-Loop Heat Pumps.* Arlington, VA: Air-Conditioning and Refrigeration Institute.

ARI. 1998d. *ARI Standard 550, Centrifugal and Rotary Screw Water-Chilling Packages.* Arlington, VA: Air-Conditioning and Refrigeration Institute.

ASHRAE. 1993. *ANSI/ASHRAE Standard 103-1993, Method of Testing for Annual Fuel Utilization Efficiency of Residential Central Furnaces and Boilers.* Atlanta: American Society of Heating, Refrigerating and Air-Conditioning Engineers, Inc.

ASHRAE. 1995. *ANSI/ASHRAE Standard 116-1995, Methods of Testing for Rating Seasonal Efficiency of Unitary Air Conditioners and Heat Pumps.* Atlanta: American Society of Heating, Refrigerating and Air-Conditioning Engineers, Inc.

ASHRAE. 1998. *ANSI/ASHRAE Standard 37-1988, Methods of Testing for Rating Unitary Air-Conditioning and Heat Pump Equipment* (superseded). Atlanta: American Society of Heating, Refrigerating and Air-Conditioning Engineers, Inc.

ASHRAE. 1999a. *ANSI/ASHRAE Standard 16-1999, Method of Testing for Rating Room Air Conditioners and Packaged Terminal Air Conditioners.* Atlanta: American Society of Heating, Refrigerating and Air-Conditioning Engineers, Inc.

ASHRAE. 1999b. *ANSI/ASHRAE Standard 58-1999, Method of Testing for Rating Room Air Conditioners and Packaged Terminal Air Conditioner Heating Capacity.* Atlanta: American Society of Heating, Refrigerating and Air-Conditioning Engineers, Inc.

ASHRAE. 2000. *ASHRAE Guideline 12, Minimizing the Risks of Legionellosis Associated with Building Water Systems.* Atlanta: American Society of Heating, Refrigerating and Air-Conditioning Engineers, Inc.

ASHRAE. 2001. *ANSI/ASHRAE/IESNA Standard 90.1-2001, Energy Standard for Buildings Except Low-Rise Residential Buildings.* Atlanta: American Society of Heating, Refrigerating and Air-Conditioning Engineers, Inc. Table 6.2.1C.

ASHRAE. 2003. *2003 ASHRAE Handbook—HVAC Applications*, chapter 32, Geothermal energy; chapter 56, Codes and standards. Atlanta: American Society of Heating, Refrigerating and Air-Conditioning Engineers, Inc.

ASHRAE. 2004a. *ANSI/ASHRAE/IESNA Standard 90.1-2004, Energy Standard for Buildings Except Low-Rise Residential Buildings.* Atlanta: American Society of Heating, Refrigerating and Air-Conditioning Engineers, Inc.

ASHRAE. 2004b. *ANSI/ASHRAE Standard 62.1-2004, Ventilation for Acceptable Indoor Air Quality.* Atlanta: American Society of Heating, Refrigerating and Air-Conditioning Engineers, Inc.

ASHRAE. 2004c. *2004 ASHRAE Handbook—HVAC Systems and Equipment*, chapter 5, Decentralized cooling and heating; chapter 36, Cooling towers. Atlanta: American Society of Heating, Refrigerating and Air-Conditioning Engineers, Inc.

Coad, W.J. 1996. Indoor air quality: A design parameter. *ASHRAE Journal* 38(6):39–47.

C/UTC. 2001a. *Light Commercial Products and Systems*, Single packaged rooftop units. Carrier/United Technologies Corporation, Syracuse, NY.

C/UTC. 2001b. *Light and Heavy Commercial Products and Systems*, Fan powered variable air volume terminals. Carrier/United Technologies Corporation, Syracuse, NY.

DOE. 1997. *Federal Register*, September 24. US Department of Energy, Washington, DC.

ISO. 2000. *ISO Standard 13256-1, Water-to-Air and Water-to-Brine Heat Pumps.* Geneva: International Organization for Standardization.

Kavanaugh, S.P., and K.D. Rafferty. 1997. *Ground-Source Heat Pumps: Design of Geothermal Heating and Cooling System for Commercial and Industrial Buildings.* Atlanta: American Society of Heating, Refrigerating and Air-Conditioning Engineers, Inc.

MA. 1998. Horizontal Chilled Water Blower Coil Units-BHW Series. Magic Aire Division, United Electric Company. Wichita Falls, TX.

MMI. 1992. Innovative cooling tower. MMI Cooling Tower Division, Fort Wayne, IN.

T/ASC. 1995. Upflow/Horizontal Gas Fired Furnace. Publication No. 22-1640-05-398 (EN), Trane-American Standard Co., LaCrosse, WI.

T/ASC. 1998. Upflow/Horizontal Condensing Gas Fired Furnace. Publication No, 22-1674-02, Trane-American Standard Co., LaCrosse, WI.

T/ASC. 1999. Cold Generator® Scroll Liquid Chillers. No. SLC-DS-1, Trane-American Standard Co., LaCrosse, WI.

T/ASC. 2001. Packaged Terminal Air Conditioners and Heat Pumps. Publication No. PTAC-PRC001-EN, Trane-American Standard Co., LaCrosse, WI.

T/ASC. 2002. Air-Cooled Liquid Chillers. No. CG-PRC007-EN, Trane-American Standard Co., LaCrosse, WI.

T/ASC. 2003. Split System Heat Pump Product and Performance Data: XL14i. Publication No. 22-1750-01-0603(EN), Trane-American Standard Co., Tyler, TX.

WFI. 1996. Spectra™ Horizontal and Vertical Water-Source Heat Pumps. WF700 Water Furnace International, Fort Wayne, IN.

6 Comfort, Air Quality, and Climatic Data

THERMAL COMFORT AND AIR CONDITIONS

Many factors contribute to the perception of satisfaction with the thermal environment. Temperature and moisture are primary conditions that engineers attempt to control in order to create thermal comfort. However, air speed (V), skin wettedness (w), non-uniform conditions, radiation effects, clothing (clo), and activity level (met) are additional conditions that affect thermal comfort. This brief discussion is restricted primarily to the impact of temperature and humidity. More detail can be found in the *2005 ASHRAE Handbook—Fundamentals*, "Thermal Comfort" (ASHRAE 2005a, chapter 8) and *ANSI/ASHRAE Standard 55, Thermal Environmental Conditions for Human Occupancy* (ASHRAE 2004a). Unfortunately, other conditions (odors, acoustics, etc.) may impact overall occupant perception of comfort.

Thermal comfort is achieved by maintaining body temperature (typically between 98°F and 99°F) by dissipation of heat. An average adult male (~154 pounds and 68 inches tall) at rest produces heat at a rate of 350 Btu/h (sensible + latent). The surface area of this person is 19.4 ft^2. The heat transfer rate per unit area required is 18.4 Btu/h·ft^2, which is the base metabolic rate of 1 met. The average female has a surface area of 17.2 ft^2. The surface area (A_D) of other humans can be approximated by (DuBois and Dubois 1916)

$$A_D = 0.108 m^{0.425} l^{0.725}, \tag{6.1}$$

where

A_D = surface area, ft^2;
m = body mass, lb; and
l = height, in.

The heat rate of other non-average humans at rest can be approximated by multiplying 18.4 Btu/h·ft^2 (1 met) by the surface area (A_D). Table 6.1 lists the metabolic heat rates for other activities.

Body temperature is regulated by control of the heat rate at the skin via heat convection and radiation ($C + R$) and evaporation (E_{sk}). When the body temperature rises, dilation of the skin blood vessels allows increased flow to carry internal heat to the skin for transfer to the environment. At higher body temperatures, sweating also occurs to enhance heat dissipation. When air conditions are good for evaporation (lower relative humidity), skin can remain relatively dry. When the surrounding air is humid or very warm, the moisture must spread out on the skin surface and perspiration is evident. The fraction of skin covered by moisture is referred to as *skin wettedness* (w). As sweating continues, a residue of salt will accumulate. This salt lowers the vapor pressure of water, which impedes evaporation. Washing this residue with water will provide some relief though reduction of this residue and wettedness.

When the body temperature falls below the desired setpoint, blood vessels are constricted to conserve body heat. With further decline in body temperature, muscle tension is used to generate additional heat, which may result in shivering when muscle groups are opposed.

Table 6.1 Typical Human Metabolic Heat Generation Rates (ASHRAE 2005a, chapter 8)

Activity	Btu/h·ft^2	met
Sleeping	13	0.7
Seated, quiet	18	1.0
Reading or writing	18	1.0
Typing	20	1.1
Standing	22	1.2
Filing (standing)	26	1.4
Lifting or packing	39	2.1
Walking at 3 mph	48	2.6
Dancing	44–81	2.4–4.4
Calisthenics	55–74	3.0–4.0
Basketball	90–140	5.0–7.6
Housecleaning	37–63	2.0–3.4
Light machine work	37–44	2.0–2.4
Pick and shovel work	74–88	4.0–4.8

However, the insulating effects of clothing will eventually be necessary to prevent further lowering of body temperature as air temperature declines (ASHRAE 2005a, chapter 8).

Engineers must consider these physiological effects when selecting indoor air design conditions. For most humans at typical metabolic rates, air temperatures in the 68°F to 80°F range are best suited for maintaining body temperatures without conscious adaptive physiological activities. Figure 6.1 is the center section of the psychrometric chart, which provides two comfort zones that include the effects of temperature, humidity, and clothing. In the summer, lighter clothing is worn so occupants are satisfied with temperatures in the 74°F to 80°F range at 50% relative humidity. Lower temperature is needed when the relative humidity is higher, and slightly higher temperature is acceptable if the humidity is near 30%. Air temperatures in the 68°F to 75°F range are acceptable in the winter since occupant clothing will be heavier. It should be noted that winter indoor air humidity is more likely to be near the 30% relative humidity level since dry outdoor ventilation and infiltration air will be displacing moist room air. Below this humidity, occupants will experience discomfort from dryness and conditions are better suited to annoying static electricity discharges. Thus, humidification is often necessary in colder climates.

Figure 6.1 shows a comfort chart from the *2001 ASHRAE Handbook—Fundamentals* (ASHRAE 2001a) and an updated figure for the acceptable range of operative temperatures and humidity (ASHRAE 2004a). In the comfort chart the shaded areas represent the ASHRAE "comfort envelopes." The upper limit of the summer comfort zone is the 68°F wet-bulb line, and the 64°F wet-bulb line is for the winter comfort zone. An additional line is provided to indicate the upper limit of 65% relative humidity to comply with *ANSI/ASHRAE Standard 62.1-2004, Ventilation for Acceptable Indoor Air Quality* (ASHRAE 2004b). This limit addresses air quality issues resulting from microbial growth that are more likely to occur at elevated relative humidity values.

The revised version of ASHRAE Standard 55 expands the range of acceptable temperature and humidity as shown in Figure 6.1 (ASHRAE 2004a). However, these limits do not consider the possibility of condensation on building surfaces that can result with the higher humidity ratio (0.012 lb/lb) and the adverse effects of low humidity (skin drying, irritation of mucus membranes, dryness of the eyes, and static electricity generation). It is suggested that designers use the "comfort envelopes" appearing in Figure 6.1 modified to comply with the 65% relative humidity limit.

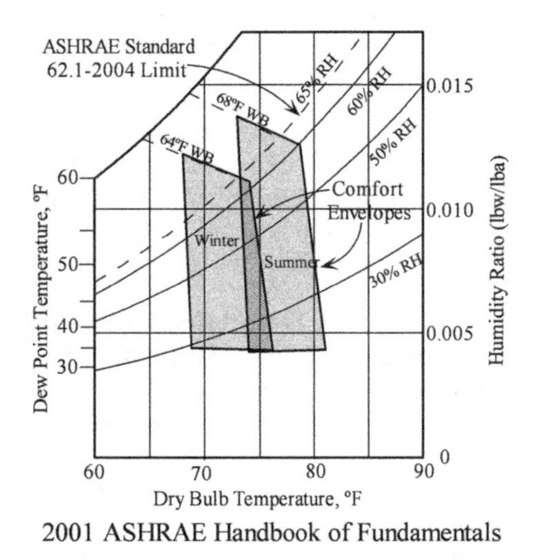

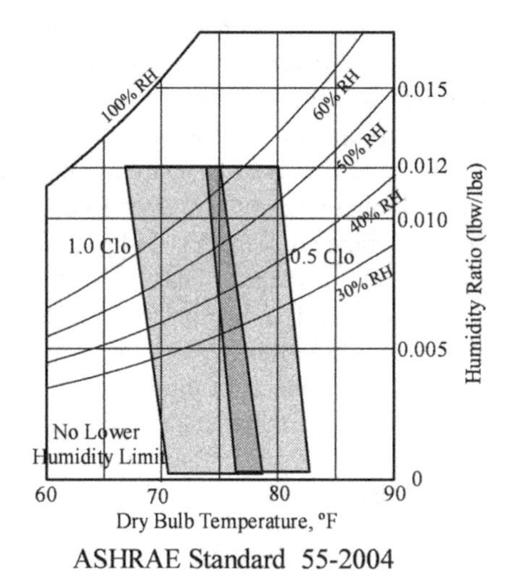

Figure 6.1 ASHRAE comfort chart and Standard 55-2004 (ASHRAE 2004a) temperature and humidity ranges.

Example Problem 6.1

Compute the heat rate of a 6 ft, 10 in., 260 lb basketball player during peak competition.

Solution:

Using Equation 6.1 and $q = (q/A_D) \times A_D$,

$$A_D = 0.108(260\ \text{lb})^{0.425}[(6\ \text{ft} \times 12\ \text{in./ft}) + 10\ \text{in.}]^{0.725} = 28.01\ \text{ft}^2$$

$$q\ (\text{Btu/h}) = 140\ (\text{Btu/h/ft}^2) \times 28.01\ \text{ft}^2 = 3920\ \text{Btu/h}$$

These comfort zones will not ensure complete satisfaction of all occupants and activity levels, so limits must be placed to recognize other factors that affect the perception of thermal comfort. For the zones shown in Figure 6.1, limitations include (ASHRAE 2005a, chapter 8):

- Satisfaction of 90% of the occupants if the environment is thermally uniform
- Satisfaction of 80% of the occupants if variations in the local thermal environment cause some body extremities or sides to be too hot or too cold
- Mean air velocities < 0.3 fps for an indoor space temperature (t_a) = 68°F, < 0.5 fps for t_a = 74°F, or < 0.6 fps for t_a = 79°F.

VENTILATION FOR ACCEPTABLE INDOOR AIR QUALITY

ASHRAE Standard 62 has received a great deal of attention in recent years. Significant revisions to the 1989 version were warranted since more than 40 pages of interpretations accompanied the document. The extensive nature of the proposed standard (62-1989R) made complete revision unwieldy. The standard was placed on "continuous maintenance" to permit revisions to be developed and approved on a piecemeal basis in the form of addenda. An interim version was published in 1999 with several small changes. The most significant modification in the 1999 version was the removal of the portion of the footnote in the ventilation airflow table (Table 2, "Outdoor Air Requirements for Ventilation") that permitted "a moderate amount of smoking." The current version is Standard 62.1-2004 (ASHRAE 2004b). Several significant addenda to the 2001 version were approved after undergoing the rigorous procedure of public review (three times in some cases), approval by the Standard 62 Project Committee and the ASHRAE Standards Committee, approval by the ASHRAE Board of Directors, consideration of appeals, and, finally, approval by the American National Standards Institute (ANSI). These approved addenda are embedded in the 2004 version, but the following discussion addresses the document that was Addendum n to ASHRAE Standard 62-2001 (ASHRAE 2001b). Part of Addendum n is now Section 6.2 of the 2004 standard and is a revised procedure for calculating the outdoor ventilation airflow rate.

The importance of Standard 62.1-2004 is due to the fact that the amount of ventilation air is a significant factor in terms of air quality, comfort, and energy consumption.

- High rates are needed to remove indoor contaminants and secondarily to limit indoor air CO_2 concentrations that are elevated by human respiration.
- In cold climates, poorly controlled injection of high percentages of cold air may cause discomfort and problems with equipment performance.
- In warm climates with high and moderate outdoor humidity levels, the injection of high percentages of outdoor air will result in latent loads that cannot be handled by simple conventional cooling equipment.
- Higher percentages of ventilation air often result in much higher energy consumption, even when heat recovery units are used due to elevated fan energy requirements of this equipment, poor efficiency during moderate conditions, or potentially counterproductive performance when economizer mode cooling is possible.

The 2004 version of the standard was written in a format that could be easily adapted to building codes. However, pre-existing codes, alternative codes, and other standards are used by code authorities. These prescriptions will supersede the use of Standard 62.1-2004 to determine ventilation system design. Many details, alternative procedures, and other aspects of the complete standard must also be consulted in order to provide "ventilation for acceptable indoor air quality." The issues that are presented in more detail in the standard include separation of air intakes and exhaust outlets, demand-controlled ventilation, recommended maintenance activity and frequency, drain pan design, outdoor air quality, and building envelope quality.

Addendum n (Section 6.2)

Many comments were received during the Standard 62.1 revision process concerning the amount of ventilation air required. Addendum n is the result of many hours of debate, background research, and negotiation. This 62.1 revision replaces the cfm per person values that appeared in Table 2 of previous versions of the standard with a procedure to calculate the required minimum airflow delivered to the breathing zone (V_{bz}). This permits consideration of the number and type of occupants and the size and type of space.

$$V_{bz} = R_P P_z + R_a A_z \tag{6.2}$$

where

A_z = zone floor area, ft^2

P_z = zone population (maximum, calculated average, or default if unknown)

R_P = outdoor airflow rate per person from Table 6.2 in this book

R_a = outdoor airflow rate per unit area from Table 6.2 in this book, cfm/ft^2

Note: To avoid confusion with ASHRAE Standard 62.1, this section uses the terminology *airflow* and symbol V rather than the normal ASHRAE practice of *volumetric flow rate* and symbol Q (ASHRAE 2005a, chapter 37).

Table 6.2 includes default occupancy values per unit area for each type of space listed. These default values of (P_z/A_z) can be applied to Equation 6.2 to arrive at default values of airflow rate per person. In most cases the resulting value of minimum required ventilation airflow *to the breathing zone* (V_{bz}) is less than the values for minimum *outdoor air requirements* prescribed in Table 2 of previous versions of the standard.

$$V_{bz}\text{ (cfm)} = \left(R_P + \frac{R_a}{(P_z/A_z)_{Default}}\right) \times P_z \tag{6.3}$$

Standard 62.1-2004 recognizes the fact that outdoor air intake airflow (V_{ot}) may not be delivered to the breathing zone if the ventilation system is ineffective. Therefore, procedures are incorporated to correct for the effectiveness of the air distribution. The zone air distribution effectiveness (E_z) accounts for how well the ventilation supply air is delivered to the breathing zone (4.5 ft above the floor). The outdoor airflow (V_{oz}) that must be supplied to the zone is

$$V_{oz} = V_{bz}/E_z, \tag{6.4}$$

Table 6.2 Minimum Ventilation Rates in Breathing Zone—Abbreviated Version

Complete listing found in Table 6-1 of ASHRAE Standard 62.1-2004 (ASHRAE 2004b).

Occupancy Category	People Outdoor Air Rate R_P cfm/person	Area Outdoor Air Rate R_a cfm/ft^2	Default Values Peo./1000 ft^2	cfm/person
Education				
Day care	10	0.18	25	17
Classroom (ages 5–8)	10	0.12	25	15
Classroom (ages 9+)	10	0.12	35	13
Lecture classroom	7.5	0.06	65	8
Food & Beverage				
Restaurant dining	7.5	0.18	70	10
Cafeteria/fast food	7.5	0.18	100	9
Hotels, Dorms				
Bed/Living rooms	5	0.06	10	11
Lobbies/pre-function	5	0.06	20	8
Assembly	5	0.06	120	6
Office Buildings				
Office space	5	0.06	5	17
Reception area	5	0.06	30	7
Telephone/data entry	5	0.06	60	6
Public Assembly				
Conference	5	0.06	50	6
Auditorium	5	0.06	150	5
Library	5	0.12	10	17
Museum	7.5	0.06	40	9
Retail				
Sales	7.5	0.12	15	16
Mall—common area	7.5	0.06	40	9
Supermarket	7.5	0.06	8	15
Sports				
Gym, playing areas	–	0.30	30	–
Spectator areas	7.5	0.06	150	8
Aerobics room	20	0.06	40	22
Dance floor	20	0.06	100	21
Game arcade	7.5	0.18	20	17

where

E_z = 1.2 floor supply of cool air and ceiling return, provided low-velocity displacement ventilation achieves unidirectional flow and thermal stratification;

E_z = 1.0 ceiling supply of cool air; ceiling supply of warm air with floor return; floor supply of warm air with floor return; ceiling supply of warm air less than 15°F above room air temperature with ceiling return, provided the diffuser jet velocity of 150 fpm reaches the breathing zone; or floor supply of cool air with ceiling return, provided a jet velocity of 150 fpm reaches the breathing zone;

E_z = 0.8 ceiling supply of warm air 15°F greater than room air temperature and ceiling return or makeup supply air drawn in on the opposite side of the room from the exhaust and/or return;

E_z = 0.7 floor supply of warm air and ceiling return; and

E_z = 0.5 makeup supply air drawn in near to the exhaust and/or return location.

For single-zone systems where one air handler supplies outdoor and recirculated air to only one zone (Figure 6.2), the outdoor air intake flow (V_{ot}) is equal to the zone outdoor airflow (V_{oz}).

$$V_{ot} = V_{oz} \tag{6.5}$$

When one air handler supplies only outdoor air to one or more zones, the outdoor air intake flow (V_{ot}) is equal to the sum of the zone outdoor airflows. This type of system is referred to as a *100% outdoor air system* or, in the case where the ventilation air system is decoupled from the primary air system, a *dedicated outdoor air system* (DOAS), as shown in Figure 6.2.

$$V_{ot} = \Sigma V_{oz} \tag{6.6}$$

A DOAS separates the ventilation air system from the primary HVAC system. Typically, the ventilation air is conditioned (cooled, dehumidified, heated, humidified) to near indoor conditions and is delivered to the space in a separate distribution system or partially integrated into the primary system. This permits simple control (Coad 1996).

In many cases the occupant densities in zones fluctuate. In these cases the breathing zone outdoor airflow may be determined from the average occupancy rather than the peak. However, the time period (T) during the average occupancy can be determined by Equation 6.7, which considers the volume of the zone (Vol_z).

$$T\text{ (min)} = \frac{3 \times Vol_z\text{ (ft}^3\text{)}}{V_{bz}\text{ (cfm)}} \tag{6.7}$$

Also, occupants may move from normally occupied zones to normally unoccupied zones (i.e., meeting rooms). The outdoor air intake flow may be reset in cases where the occupancy is scheduled by time of day, a direct count of occupants, or estimates of occupancy or ventilation rate per person (i.e., CO_2 sensors).

Multiple zone systems that deliver a mixture of outdoor air and recirculated air to several zones must be corrected for the occupant diversity (D) in the zones and ventilation system efficiency (E_V). The occupant diversity is used to compute the uncorrected outdoor intake (V_{ou}).

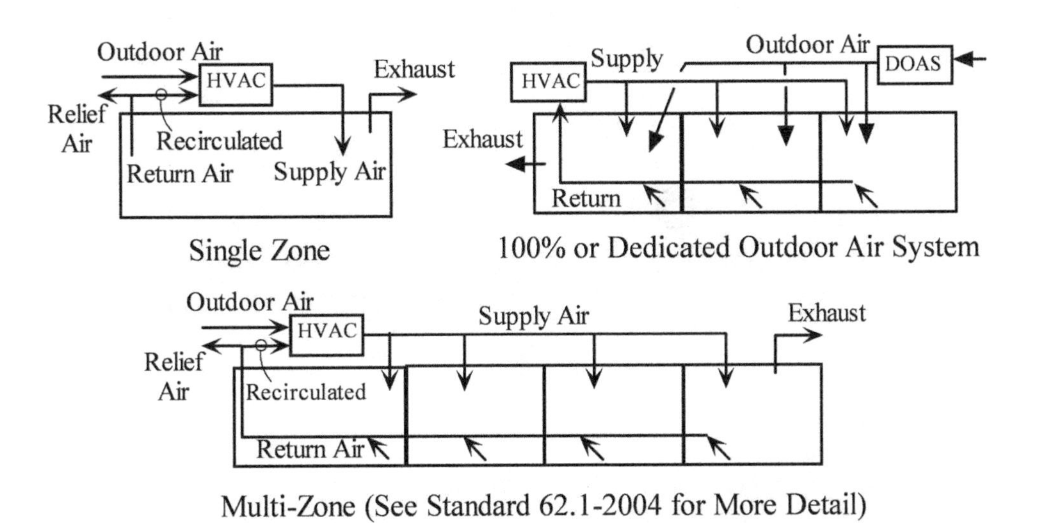

Figure 6.2 Single-zone, 100% outdoor air, and multi-zone ventilation air systems.

$$V_{ou} = D\Sigma_{allzones}R_pP_Z + \Sigma_{allzones}R_aA_z \qquad (6.8)$$

where

D = $P_s/\Sigma_{allzones}P_Z$, and

P_s = the total population in the area served by the multi-zone system.

Multiple-zone systems can provide only a single outdoor air-to-supply air fraction. However, each zone has an individual requirement for this fraction. Thus, zones with higher outdoor air requirements may not receive adequate ventilation air since they are receiving the average fraction. Previous versions of ASHRAE Standard 62 have employed the multiple-space equation to increase the overall fraction based on the critical or highest zone outdoor air fraction. Addendum n provides a similar procedure. The zone primary outdoor air fraction (Z_p) is computed for every zone from the ratio of the zone ventilation airflow rate (V_{oz}) to the primary airflow rate (V_{pz}).

$$Z_p = V_{oz}/V_{pz} \qquad (6.9)$$

The maximum value of Z_p is found and used to determine the system ventilation efficiency (E_v).

Example Problem 6.2

Verify the default cfm/person value for a hotel bedroom.

Solution:

Using Equation 6.3,

$$V_{bz}\text{ (cfm)} = \left(5\text{ cfm/person} + \frac{(0.06\text{ cfm})/\text{ft}^2}{10\text{ people}/1000\text{ ft}^2}\right) \times P_z = 11\text{ cfm/person} \times P_z\,.$$

Example Problem 6.3

Compute the required ventilation air intake flow for the seven-zone office complex shown in Figure 6.3 using a dedicated outdoor air system. Repeat the calculation for a system in which the ventilation air is provided with the primary air through a central VAV air handler at 100% design airflow and at 30% part-load airflow. The primary air is supplied at the flow rates shown in the figure through ceiling diffusers with return near the floor.

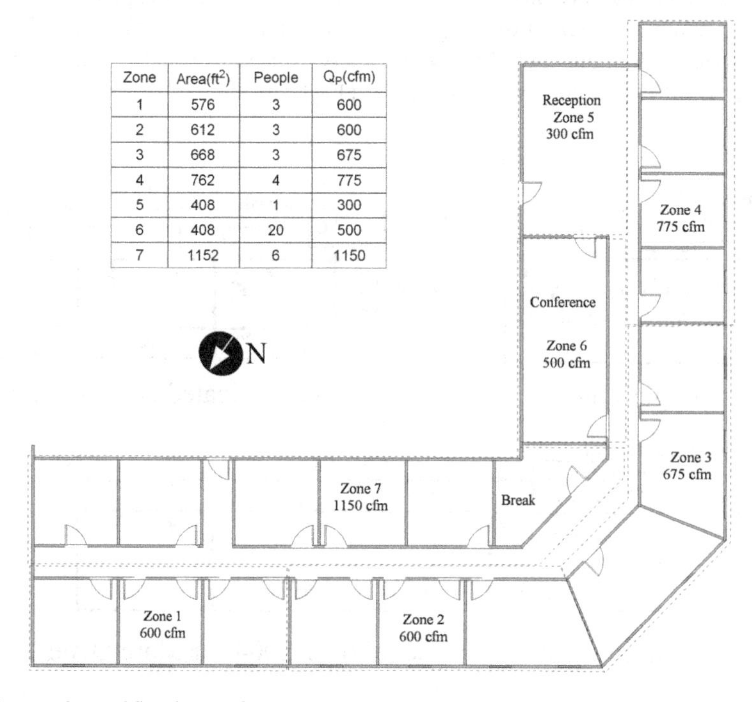

Zone	Area(ft²)	People	Qp(cfm)
1	576	3	600
2	612	3	600
3	668	3	675
4	762	4	775
5	408	1	300
6	408	20	500
7	1152	6	1150

Figure 6.3 Floor plan and specifications of a seven-zone office complex with conference room.

$$E_v = 0.9 \text{ if } Z_p(\max) \leq 0.25$$

$$E_v = 0.8 \text{ if } Z_p(\max) \leq 0.35$$

$$E_v = 0.7 \text{ if } Z_p(\max) \leq 0.45$$

$$E_v = 0.6 \text{ if } Z_p(\max) \leq 0.55$$

Use Appendix G of ASHRAE Standard 62.1-2004 if $Z_p(\max) > 0.55$

The corrected value of outdoor air intake flow (V_{ot}) for a multiple-zone system is

$$V_{ot} = V_{ou}/E_V . \quad (6.10)$$

The *outdoor air intake flow* computed with Equation 6.10 uses the same nomenclature (V_{ot}) as the *corrected total outdoor airflow rate* computed using the multiple space equation in previous versions of the standard (Equation 6-1, ASHRAE Standard 62-1999).

Dedicated Ventilation Air System (DOAS) Solution

Equation 6.2 is applied to zones 1 through 7 using the values in the figure. The value for R_P is 5 cfm/person and R_a is 0.06 cfm/ft^2 for all categories (office, reception, conference room). For zone 1,

$$V_{bz1} = (5 \text{ cfm/person} \times 3 \text{ people} + 0.06 \text{ cfm/ft}^2 \times 576 \text{ ft}^2) = 50 \text{ cfm} .$$

The zone air distribution effectiveness (E_z) for both the ceiling supply of cool air and ceiling supply of warm air with floor return is 1.0. From Equation 6.4,

$$V_{oz1} = V_{bz1}/E_Z = 50 \text{ cfm}/1.0 = 50 \text{ cfm} .$$

When Equation 6.6 is applied to all seven zones, the outdoor air intake flow is:

$$V_{ot} = \Sigma V_{oz} = (5 \times 3 + 0.06 \times 576) \div 1.0 + (5 \times 3 + 0.06 \times 612) \div 1.0 + (5 \times 3 + 0.06 \times 668) \div 1.0 + (5 \times 4 + 0.06 \times 768) \div 1.0 + (5 \times 1 + 0.06 \times 408) \div 1.0 + (5 \times 20 + 0.06 \times 408) \div 1.0 + (5 \times 6 + 0.06 \times 1152) \div 1.0 = 475 \text{ cfm}$$

This arrangement will result in some overventilation, especially when the conference room is unoccupied. A recommended alternative would be to install a stand-alone ventilation air system for the conference room that is controlled via occupancy or CO_2 concentration. The remaining six zones have similar loads, occupancy patterns, and outdoor air fractions and can be effectively handled by a common ventilation air system with or without occupancy or CO_2 concentration control. Results of this alternative are

$$V_{ot}(\text{Zones 1-5, 7}) = 351 \text{ cfm}$$

and $$V_{ot}(\text{Zone 6}) = 124 \text{ cfm} .$$

A second option specifically permitted by Standard 62.1-2004 is dynamic reset of the airflows "as operating conditions change." In this example, occupants will move from their offices to the conference room and reset could be scheduled for regular daily meetings or with CO_2 sensors. When the conference room is unoccupied the outdoor air intake flow is

$$V_{ot} = \Sigma V_{oz} = (5 \times 3 + 0.06 \times 576) \div 1.0 + (5 \times 3 + 0.06 \times 612) \div 1.0 + (5 \times 3 + 0.06 \times 668) \div 1.0 + (5 \times 4 + 0.06 \times 768) \div 1.0 + (5 \times 1 + 0.06 \times 408) \div 1.0 + (0 \times 20 + 0.06 \times 408) \div 1.0 + (5 \times 6 + 0.06 \times 1152) \div 1.0 = 375 \text{ cfm}$$

However, resetting flow to the unoccupied office areas may be somewhat more problematic when occupancy scheduling is not consistent or universal. Therefore, the option of two ventilation systems is likely to be a simpler and more effective alternative.

Multiple-Zone Recirculating Ventilation Air System Solution

Equation 6.2 is used to determine V_{bz} and Equation 6.4 is used to determine V_{oz} for each of the seven zones. For a multiple-zone system, the zone primary outdoor air fraction (Z_p) must be computed from Equation 6.9 for each zone. Results of these equations and the intermediate computations are shown in Table 6.3 for both the conditions of 100% primary airflow and 30% primary airflow.

The diversity (D) is used to compute the uncorrected outdoor airflow intake (V_{ou}) using Equation 6.8. In this case, the diversity of 0.50 is multiplied by the summation of column 2 in Table 6.3 and added to the summation of column 3.

$$V_{ou} = 0.50 \times 200 \text{ cfm} + 276 \text{ cfm} = 376 \text{ cfm}$$

At 100% primary airflow, the maximum value of Z_p is less than 0.25, so the system ventilation efficiency (E_v) is 0.9 and the outdoor air intake flow is

$$V_{ot} = V_{ou} \div E_v = 376 \div 0.9 = 469 \text{ cfm} .$$

At 30% of primary airflow, the maximum value of Z_p is 0.83, which requires a computation using Appendix G of Standard 62.1-2004 and additional information. The system ventilation efficiency (E_v) will be 0.60 or

Table 6.3 Multiple-Space Breathing Zone Airflow (V_{bz}), Zone Outdoor Airflow (V_{oz}), and Outdoor Air Fractions (Z_p) at 100% and 30% of Primary Airflow (V_p)

1	2	3	4	5	6	7	8	9	10	11
Zone	R_p × Peo	R_a × Area	V_p@100%	V_{bz}	V_{oz}	Z_p	V_p@30%	V_{bz}	V_{oz}	Z_p
1	15	35	600	50	50	0.083	180	50	50	0.275
2	15	37	600	52	52	0.086	180	52	52	0.287
3	15	40	675	55	55	0.082	203	55	55	0.272
4	20	46	775	66	66	0.085	233	66	66	0.284
5	5	24	300	29	29	0.098	90	29	29	0.328
6	100	24	500	124	124	**0.249**	150	124	124	**0.830**
7	30	69	1150	99	99	0.086	345	99	99	0.287
Σ	200	276	4600				1380			

Table 6.4 Multiple Space Breathing Zone Airflow (V_{bz}), Zone Outdoor Airflow (V_z), and Outdoor Air Fractions (Z_p) at 100% and 30% of Primary Airflow (V_p) for Six Office Zones

Zone	R_p × Peo	R_a × Area	V_p @ FL	V_{bz}	V_{oz}	Z_p	V_p at PL	V_{bz}	V_{oz}	Z_p
1	15	35	600	50	50	0.083	180	50	50	0.275
2	15	37	600	52	52	0.086	180	52	52	0.287
3	15	40	675	55	55	0.082	203	55	55	0.272
4	20	46	775	66	66	0.085	233	66	66	0.284
5	5	24	300	29	29	**0.098**	90	29	29	**0.328**
7	30	69	1150	99	99	0.086	345	99	99	0.287
Σ	100	251	4100				1380			

less. Assuming an efficiency of 0.50, the outdoor air intake flow is

$$V_{ot} = V_{ou} \div E_v = 376 \div 0.5 = 752 \text{ cfm} .$$

The increase in the required ventilation rate for 30% maximum primary airflow further enhances the attractiveness of an individual ventilation air system for the high-occupancy conference room. Table 6.4 lists the values for the intermediate calculation for zone airflows and outdoor air fractions for a ventilation system that serves the six office zones.

In this case, the diversity of 1.0 is multiplied by the summation of column 2 in Table 6.4 and added to the summation of column 3.

$$V_{ou} = 1.0 \times 100 \text{ cfm} + 251 \text{ cfm} = 351 \text{ cfm}$$

At 100% primary airflow, the maximum value of Z_p is less than 0.25, so the system ventilation efficiency (E_v) is 0.9 and the outdoor air intake flow is

$$V_{ot} = V_{ou} \div E_v = 251 \div 0.9 = 390 \text{ cfm} .$$

The airflow to zone 6 would be the same 124 cfm as in the dedicated ventilation option. However, this flow would not occur simultaneously with the airflow to the office zones when properly controlled.

At 30% of primary airflow, the maximum value of Z_p is 0.328, which yields a system ventilation efficiency (E_v) of 0.80; the outdoor air intake flow is

$$V_{ot} = V_{ou} \div E_v = 251 \div 0.8 = 439 \text{ cfm} .$$

Note: The ASHRAE *62.1 User's Manual* was recently published (ASHRAE 2005b). It provides more detailed recommendations and explanations of the standard and includes a spreadsheet for performing ventilation computations for multi-zone systems.

Additional Standard 62.1-2004 Topics

The reader is encouraged to consult the complete standard for additional prescriptions and design procedures. The following summarizes the standard's benefits.

- A listing of definitions is provided (Section 3).
- As outdoor air quality must also be investigated, sources for this information are provided in Section 4 with a listing of the required documentation.
- Section 5 describes ventilation systems and equipment. The section provides guidance and references for component specifications and installation practices. The minimum separation distance for air intakes is specified in Table 5-1. Separation of exhaust outlets and outdoor air intakes is discussed in Appendix F.

- In addition to the ventilation rate procedure described in Section 6.2, the indoor air quality (IAQ) procedure is described in Section 6.3. This alternate procedure allows the ventilation system to be "designed to maintain specific contaminants at or below certain limits... to achieve the design target level of perceived indoor air quality acceptability by building occupants and/or visitors" (p. 15).
- As mentioned previously, Section 6.2 also permits the outdoor air intake flow to be reset in cases where the occupancy is scheduled by time of day, a direct count of occupants, or estimates of occupancy or ventilation rate per person (i.e., with CO_2 sensors). Appendix C provides the rationale for basing rates on CO_2 concentration. An example CO_2 concentration computation is given that indicates acceptable dilution is achieved for sedentary occupants if the indoor concentration is 700 ppm (parts per million) above the outdoor CO_2 concentration, which is typically in the 300–500 ppm range.
- Section 6.4 prescribes minimum exhaust air rates in Table 6-4. Exhaust air can be any combination of outdoor air, recirculated air, or transfer air (air being moved from one indoor space to another).
- Section 7 describes the protection required during construction and start-up prescriptions.
- Section 8 lists the minimum maintenance activities and frequencies.

Residential Ventilation Rates

The recent publication of *ASHRAE Standard 62.2-2004, Ventilation and Acceptable Indoor Air Quality for Low-Rise Residential Buildings* (ASHRAE 2004c), has mandated mechanical ventilation systems for residential occupancy. Prior to the publication of Standard 62.2, Standard 62.1 covered residential applications with the requirement of 0.35 air changes per hour (ACH) but not less than 15 cfm total, a 100 cfm intermittent or 25 cfm continuous exhaust fan (or operable windows) for kitchens, and a 50 cfm intermittent or 20 cfm continuous exhaust fan (or operable windows) for bathrooms. The new standard requires (ASHRAE 2004c)

$$Q_{fan}\,(\text{cfm}) = 0.01A_{Floor}\,(\text{ft}^2) + 7.5(N_{Bedrooms} + 1)\,.$$

The assumption is that the occupancy is two people in one bedroom and one person in the rest. If the occupancy is greater, the rate must be increased by 7.5 cfm per person above ($N_{Bedrooms}$ + 1).

Currently, residential structures in humid climates (> 4500°F-days infiltration degree-days, as defined by ANSI/ASHRAE Standard 119-1988 [RA 2004] [ASHRAE 2004d]) are exempt from the mechanical ventilation requirement.

CLIMATIC DATA

Design Conditions

Information regarding the outdoor design conditions and desired indoor conditions are the starting point for heating and cooling load calculation programs. Tables 1A and 1B of the chapter "Climatic Design Information" from the *2001 ASHRAE Handbook—Fundamentals* (ASHRAE 2001a, chapter 27) provide the necessary reference. Columns 2a and 2b in Table 1A are the outdoor design temperatures commonly used for the heating mode. The subheading "99.6%" represents the temperatures whose frequency of occurrence is exceeded 99.6% of the time (or the value to which the outdoor temperature is equal or lower than 0.4% of the time). The next column presents this same information for the 99% value, which will be a slightly higher temperature than the value appearing in the 99.6% column. The most recent version of the ASHRAE Handbook presents a significantly expanded version of this information, which is available in an electronic format (ASHRAE 2005a, chapter 28).

The cooling design conditions are listed on the facing page in Table 1B (ASHRAE 2001a) in columns 2a through 2f for the maximum outdoor air dry-bulb (DB) temperature. These conditions appear in sets of dry-bulb temperature and the mean coincident wet-bulb (MWB) temperature (under the heading "Cooling DB/MWB") since both values are needed to determine the sensible and latent (dehumidification) loads in the cooling mode. The 0.4%, 1%, and 2% represent the temperatures that are exceeded 0.4%, 1%, and 2% of the time on an annual basis. The 0.4% values are 95°F (DB) and 77°F (MWB) for Tuscaloosa, Alabama.

In moderate and humid climates in buildings that have significant ventilation air requirements (> 10% of supply), the maximum load often occurs during periods of maximum wet-bulb (WB) conditions due to the higher enthalpy of the outdoor air (which corresponds closely to high wet-bulb temperatures). Columns 3a through 3f of Table 1B (under the heading "Evaporation WB/MDB") present these values and the corresponding coincident dry-bulb (MDB) temperature. The 0.4% values are 80°F (WB) and 90°F (MDB) for Tuscaloosa.

Often the maximum dehumidification load will occur during the periods when the outdoor air dew-point temperature (DP) is greatest. Three sets of DP values (0.4%, 1%, and 2%), with the corresponding humidity ratio (HR) and coincident dry-bulb temperature, are presented in Table 1B (under the heading "Dehumidification DP/MDB and HR"). These conditions will typically result in elevated indoor relative humidity if the HVAC system does not have adequate latent capacity. Thus, sensible and latent loads at these conditions should be analyzed. The final required input from Table 1B is the design-day daily range (DR) of the dry-bulb temperature. This value is used to correct the cooling load temperature difference (CLTD) values as discussed in chapter 8 of this text. The value is 19.6°F for Tuscaloosa.

Complete analysis dictates that the designer use both the DB/MWB and WB/MDB conditions to check for maximum total cooling load and repeat the computation for the DP/MDB and HR conditions to determine total load, sensible load, and sensible heat ratio (SHR_{Load}). The HVAC equipment should meet the total load at both DB/MWB and WB/MDB conditions and have a SHR_{Unit} less than or equal to SHR_{Load}.

Table 6.5 is a summary of the climatic data appearing in the ASHRAE Handbook for several selected sites in the continental US. The latitude, elevation, daily range, and the 99.6% (heating) and 0.4% (cooling) values are given. However, an additional column for wet bulb is added for the dehumidification conditions since cooling load calculations typically require wet bulb as an input rather than dew-point temperature or humidity ratio in grains.

Table 6.5 provides only a few of the sites listed in the Handbook chapter, and several columns of data are omitted for all sites. The Handbook chapter contains climatic design information for over 1000 locations in the US, 250 sites in Canada, and 1700 international locations.

Table 6.5 Climatic Design Data for Selected Locations

See Tables 1A, 1B, 2A, 2B, 3A, and 3B, *2001 ASHRAE Handbook—Fundamentals*, chapter 27, for 3000+ additional locations.

A	B	C	D	E		F		G				H
Location (City, State)	Lat. (°N)	Elev. (ft)	Heating (99.6%) (°F)	Cooling [1] (0.4%)		Evaporation[2] (0.4%)		Dehumidification[3] (0.4%)				DB Range (°F)
				DB (°F)	MWB (°F)	WB (°F)	MDB (°F)	DP (°F)	HR (Gr[4])	DB (°F)	WB (°F)	
Albuquerque, NM	35	5320	13	96	60	65	83	61	98	68	63	25
Atlanta, GA	34	1033	18	93	75	77	88	74	133	82	76	17
Baltimore, MD	39	154	11	93	75	78	88	75	132	83	77	19
Birmingham, AL	34	630	18	94	75	78	89	75	135	83	77	19
Boise, ID	44	2870	2	96	63	66	90	58	79	72	63	30
Boston, MA	42	30	7	91	73	75	87	72	119	80	74	15
Brownsville, TX	26	20	36	95	78	80	89	78	146	83	79	17
Chicago, IL	42	673	–6	91	74	77	88	74	130	84	77	20
Cleveland, OH	41	804	1	89	73	76	85	73	125	82	75	19
Dallas, TX	33	597	17	100	74	78	92	75	132	82	77	20
Denver, CO	40	5330	–3	93	60	65	81	60	96	69	63	27
Lake Charles, LA	30	33	29	93	78	80	88	78	148	84	80	16
Los Angeles, CA	34	105	43	85	64	70	78	67	99	75	70	11
Miami, FL	26	13	46	91	77	80	87	78	144	83	79	11
Minneapolis, MN	45	837	–16	91	73	76	88	73	124	83	76	19
Nashville, TN	36	590	10	94	76	78	89	75	134	83	77	19
Omaha, NE	41	1332	-8	95	75	78	90	75	136	85	77	20
Oklahoma City	35	1300	9	99	74	77	91	73	129	83	76	21
Phoenix, AZ	34	1089	34	110	70	76	97	71	118	82	74	23
Raleigh, NC	36	440	16	93	76	78	88	75	134	82	77	19
Sacramento, CA	39	23	31	100	69	72	96	62	84	82	69	33
St. Louis, MO	39	564	2	95	76	79	90	76	138	85	78	18
Salt Lake, UT	41	4230	6	96	62	66	85	60	92	73	64	28
Seattle, WA	47	450	23	85	65	66	83	60	78	71	64	18
Tallahassee, FL	30	69	25	95	77	80	89	77	142	83	79	19
Tuscaloosa, AL	33	171	20	95	77	80	90	77	142	84	79	20

[1] These design conditions typically result in the highest sensible cooling load and highest total load when ventilation requirements and infiltration are low.
[2] These design conditions will result in higher total cooling loads when the ventilation air requirements and infiltration are high in humid and moderate climates.
[3] These design conditions typically result in cooling loads with the lowest sensible heat ratio (SHR_{Load}).
[4] Gr = grains of moisture (7000 grains = 1 lb).

Seasonal Climatic Data

In order to estimate annual energy consumption (or savings), seasonal climatic data are required. This information is available in a variety of formats and levels of detail. The simplest presentation is "degree-day" or "degree-hour" data (NCDC 1992a). This information is a single value that represents the summation of the average daily (or hourly) indoor-to-outdoor temperature difference. The indoor temperature is typically offset a few degrees to recognize the impact of internal heat depressing the temperature at which heating is required. For example, a base temperature of 65°F is often used for an indoor temperature of 70°F. Heat will be lost from the room when the outdoor temperature is 65°F, but heating will not be required since the loss is offset by the interior heat generated by occupants, equipment, and lighting. A day with an average temperature of 35°F will contribute 30°F days to the heating degree-day total. The annual heating degree-day total is the summation of the daily degree-days for all days with an average temperature below 65°F (AFM 1988). A building with higher levels of insulation and/or large internal heat sources will have a lower base temperature. Therefore, heating degree-days are available for base temperatures much lower than 65°F (NCDC 1992b). Cooling degree-day and degree-hour data are also available (AFM 1988) for a variety of base temperatures (NCDC 1992b). Much more detailed hourly weather data are available for detailed energy analysis simulations. Data include dry-bulb and wet-bulb temperatures, wind speed and direction, and solar radiation (ASHRAE 1997; NCDC 1999). Weather data with a level of detail between the degree-day data and the comprehensive hourly data are the "Bin Data." Outdoor dry-bulb temperature data are arranged in temperature bins of typically 5°F increments, as shown in Table 6.6. The number of hours per year (or month) during which the outdoor temperature falls into a 5°F increment is referred to as the number of *bin hours*. For example, there are typically seven hours per year in Birmingham, Alabama, where the temperature is between the 5°F increment of 95°F to 99°F (97°F is the average temperature of the bin). Thus, there are seven bin hours in the 97°F temperature bin.

The bin data can be reported in terms of the totals for all periods of the day, or they can be subdivided into four- or eight-hour increments. Table 6.6 uses four-hour increments. Of the seven bin hours in the 97°F bin, six occur between noon and 4 p.m. and one occurs between 4 p.m. and 8 p.m. This allows a more detailed approach to bin calculations since loads change with time of day and occupancy.

Table 6.6 data also include mean coincident wet-bulb temperatures (WB), which are indicators of the humidity level at the location. Note the data are in terms of wet-bulb temperatures rather than hours. These data are necessary for more detailed energy calculations that consider latent loads.

Table 6.6 Birmingham, Alabama, Bin Dry-Bulb Temperatures, Four–Hour Increments with Wet-Bulb Temperatures

Bin DB	Annual Hours of Occurrence											
Temp.	Midnt–4 a.m.		4–8 a.m.		8 a.m.–Noon		Noon–4 p.m.		4–8 p.m.		8 p.m.–Midnt	
(°F)	Hours	WB (°F)	Hours	WB (°F)	Hours	WB (°F)	Hours	WB (°F)	Hours	WB (°F)	Hours	WB (°F)
97.0	0	0.0	0	0.0	0	0.0	6	75.0	1	77.0	0	0.0
92.0	0	0.0	0	0.0	41	74.0	139	74.0	30	77.0	0	0.0
87.0	0	0.0	1	75.0	158	74.0	180	72.0	109	73.0	1	76.0
82.0	2	76.0	40	73.0	178	71.0	202	69.0	186	72.0	33	74.0
77.0	99	72.0	143	72.0	176	68.0	165	65.0	165	69.0	207	72.0
72.0	227	70.0	211	69.0	138	64.0	119	62.0	171	66.0	208	68.0
67.0	206	64.0	158	64.0	120	60.0	150	56.0	140	61.0	173	63.0
62.0	144	60.0	153	60.0	123	53.0	135	50.0	125	55.0	132	59.0
57.0	144	55.0	123	54.0	132	49.0	118	48.0	128	51.0	151	53.0
52.0	162	48.0	139	49.0	159	46.0	113	46.0	133	46.0	160	48.0
47.0	117	44.0	114	44.0	102	41.0	69	40.0	108	42.0	100	44.0
42.0	106	40.0	133	39.0	65	37.0	41	37.0	84	37.0	108	39.0
37.0	84	35.0	72	35.0	32	32.0	11	31.0	43	33.0	80	34.0
32.0	88	30.0	80	30.0	18	29.0	8	28.0	24	28.0	65	30.0
27.0	52	26.0	58	25.0	10	25.0	0	0.0	7	23.0	28	26.0
22.0	22	22.0	26	21.0	8	20.0	4	22.0	6	20.0	12	20.0
17.0	7	16.0	5	17.0	0	0.0	0	0.0	0	0.0	2	17.0
12.0	0	0.0	4	12.0	0	0.0	0	0.0	0	0.0	0	0.0

Table 6.7 Boston, Massachusetts, Bin Dry-Bulb Temperatures, Four-Hour Increments

Bin Temp. (°F)	Annual Hours of Occurrence					
	Midnt–4 a.m.	4–8 a.m.	8 a.m.–Noon	Noon–4 p.m.	4–8 p.m.	8 p.m.–Midnt
97.0	0	0	2	5	0	0
92.0	0	0	10	23	1	0
87.0	0	1	20	67	27	0
82.0	0	7	78	94	67	9
77.0	25	27	118	85	94	49
72.0	55	80	114	134	124	116
67.0	160	155	134	120	120	151
62.0	156	150	128	119	115	156
57.0	199	174	124	137	152	154
52.0	90	116	90	96	107	90
47.0	113	83	92	95	102	113
42.0	146	143	155	188	168	163
37.0	157	142	135	105	168	164
32.0	176	179	110	94	80	129
27.0	62	76	65	52	81	81
22.0	71	50	40	32	29	48
17.0	14	39	24	3	15	18
12.0	18	18	12	9	2	7
7.0	12	10	4	2	7	8
2.0	6	10	5	0	1	4

Table 6.8 Chicago, Illinois, Bin Dry-Bulb Temperatures, Four-Hour Increments

Bin Temp. (°F)	Annual Hours of Occurrence					
	Midnt–4 a.m.	4–8 a.m.	8 a.m.–Noon	Noon–4 p.m.	4–8 p.m.	8 p.m.–Midnt
92.0	0	0	15	64	18	0
87.0	0	0	64	113	45	0
82.0	0	7	131	113	90	21
77.0	25	51	114	116	125	81
72.0	120	125	131	111	151	167
67.0	156	155	88	65	80	143
62.0	159	130	78	73	72	103
57.0	111	112	93	109	96	101
52.0	112	92	88	89	100	104
47.0	92	102	97	96	83	107
42.0	120	104	72	94	131	115
37.0	106	105	133	134	131	111
32.0	182	182	156	129	135	173
27.0	93	98	66	68	94	92
22.0	75	75	60	40	44	60
17.0	63	56	30	21	33	40
12.0	25	32	25	13	11	19
7.0	5	14	11	8	16	12
2.0	15	16	7	4	5	11
–3.0	1	4	1	0	0	0

Table 6.9 Dallas, Texas, Bin Dry-Bulb Temperatures, Four-Hour Increments

Bin Temp. (°F)	Annual Hours of Occurrence					
	Midnt–4 a.m.	4–8 a.m.	8 a.m.–Noon	Noon–4 p.m.	4–8 p.m.	8 p.m.–Midnt
102.0	0	0	0	20	7	0
97.0	0	0	17	129	64	0
92.0	0	0	81	146	116	8
87.0	6	2	156	132	145	96
82.0	111	74	147	152	152	168
77.0	231	220	158	156	146	189
72.0	170	200	135	158	144	140
67.0	99	105	129	120	136	116
62.0	159	132	129	122	128	156
57.0	137	140	131	93	122	138
52.0	122	112	99	78	100	104
47.0	122	117	77	76	94	129
42.0	106	125	93	35	53	111
37.0	109	92	56	26	24	57
32.0	68	107	43	13	22	36
27.0	11	19	7	4	7	9
22.0	9	15	2	0	0	3

Table 6.10 Los Angeles, California, Bin Dry-Bulb Temperatures, Four-Hour Increments

Bin Temp. (°F)	Annual Hours of Occurrence					
	Midnt–4 a.m.	4–8 a.m.	8 a.m.–Noon	Noon–4 p.m.	4–8 p.m.	8 p.m.–Midnt
102.0	0	0	4	4	0	0
97.0	0	0	4	4	0	0
92.0	0	1	4	4	0	0
87.0	0	0	11	1	5	0
82.0	4	5	15	14	8	7
77.0	3	8	91	81	7	4
72.0	8	35	300	372	91	26
67.0	178	224	334	348	312	187
62.0	359	328	359	434	432	436
57.0	374	280	256	185	470	490
52.0	355	347	74	13	130	262
47.0	152	183	8	0	5	46
42.0	27	45	0	0	0	2
37.0	0	4	0	0	0	0

Tables 6.7 through 6.10 are bin dry-bulb temperature data for Boston, Chicago, Dallas, and Los Angeles arranged in four-hour increments without the coincident wet-bulb temperatures. Data in Tables 6.7 through 6.10 were developed from ASHRAE RP-385 (ASHRAE 1994). This source has an exhaustive set of data that includes monthly totals of dry-bulb temperature bin data, coincident wet-bulb temperatures, wind speed and direction, and solar radiation. A discussion of the energy calculation procedure using bin data is presented in chapter 9.

CHAPTER 6 PROBLEMS

6.1 Compute the heat rate of a 5 ft, 10 in., 160 lb male machinist at work.

6.2 Repeat Problem 6.1 for a 5 ft, 6 in., 120 lb performing ballerina.

6.3 What range of indoor temperature and humidity is best to satisfy occupants in the summer? In the winter? Why is there a difference?

6.4 Why are people more comfortable in the winter with a lower thermostat setting?

6.5 Find the required ventilation air for a 1500 ft^2 college classroom with 40 students. The ventilation air is delivered through ceiling vents and returned though a grille near the floor.

6.6 Repeat Problem 6.5 if the return is in the ceiling and the HVAC unit is in cooling.

6.7 Repeat Problem 6.6 if the unit is in heating and the delivery temperature is 100°F.

6.8 You are required to design a ventilation air system for a 3000 ft^2 library with supply and return in the ceiling, but no occupancy is provided. Specify the required ventilation airflow rate.

6.9 Determine the required ventilation air rate for a 3000 ft^2, five-bedroom, three-bathroom home.

6.10 A building with four zones has the airflow requirements below. Determine the required ventilation air rate for a multi-zone ventilation air system.

Supply air:

Zone 1 = 800 cfm, Zone 2 = 1200 cfm,
Zone 3 = 700 cfm, Zone 4 = 1500 cfm

Ventilation air:

Zone 1 = 200 cfm, Zone 2 = 275 cfm,
Zone 3 = 150 cfm, Zone 4 = 500 cfm

6.11 An office with six zones is served with a single rooftop unit that provides 1.0 cfm/ft^2 of supply air through the ceiling. The return is also in the ceiling. Ventilation air is supplied at the rooftop unit return. Compute the required ventilation air rate in the summer and winter given the following table.

Zone	Use	People	Area (ft^2)
1	Reception	5	700
2	Office	2	400
3	Office	8	800
4	Office	4	700
5	Conference	10	500
6	Office	1	400

6.12 Find the following for Chicago:

Elevation
Dry-bulb temperature at 0.4% cooling design
Mean wet-bulb temperature at 0.4% cooling design
Temperature at 99.6% heating design condition
0.4% design wet-bulb
Dry-bulb temperature at 0.4% WB condition
Daily range on cooling design day

6.13 Repeat Problem 6.12 for Tuscaloosa, Alabama.

6.14 Explain the meaning of 99.6% and 0.4% design conditions in Table 6.3.

REFERENCES

AFM. 1988. *Engineering Weather Data*. AFM-88-29, TM 5-785, NAVFAC P-89. Washington, DC: US Departments of the Air Force, the Army, and the Navy.

ASHRAE. 1994. Bin and Degree Hour Weather Data for Simplified Energy Calculations. Atlanta: American Society of Heating, Refrigerating and Air-Conditioning Engineers, Inc.

ASHRAE. 1997. *Weather Year for Energy Calculations*, Version 2. Atlanta: American Society of Heating, Refrigerating and Air-Conditioning Engineers, Inc.

ASHRAE. 2001a. *2001 ASHRAE Handbook—Fundamentals*. Atlanta: American Society of Heating, Refrigerating and Air-Conditioning Engineers, Inc.

ASHRAE. 2001b. *ANSI/ASHRAE Standard 62-2001, Ventilation for Acceptable Indoor Air Quality*. Atlanta: American Society of Heating, Refrigerating and Air-Conditioning Engineers, Inc.

ASHRAE. 2004a. *ANSI/ASHRAE Standard 55-2004, Thermal Environmental Condition for Human Occupancy*. Atlanta: American Society of Heating, Refrigerating and Air-Conditioning Engineers, Inc.

ASHRAE. 2004b. *ANSI/ASHRAE Standard 62.1-2004, Ventilation for Acceptable Indoor Air Quality*. Atlanta: American Society of Heating, Refrigerating and Air-Conditioning Engineers, Inc.

ASHRAE. 2004c. *ASHRAE Standard 62.2-2004, Ventilation and Acceptable Indoor Air Quality for Low-Rise Residential Buildings*. Atlanta: American Society of Heating, Refrigerating and Air-Conditioning Engineers, Inc.

ASHRAE. 2004d. *ANSI/ASHRAE Standard 119-1988 (RA 2004), Air Leakage Performance for Detached Single-Family Residential Buildings*. Atlanta: American Society of Heating, Refrigerating and Air-Conditioning Engineers, Inc.

ASHRAE. 2005a. *2005 ASHRAE Handbook—Fundamentals*, chapter 8, Thermal comfort; chapter 27, Climatic design information; chapter 28, Climatic design information; chapter 37, Abbreviations and symbols. Atlanta: American Society of Heating, Refrigerating and Air-Conditioning Engineers, Inc.

ASHRAE. 2005b. *62.1 User's Manual.* Atlanta: American Society of Heating, Refrigerating and Air-Conditioning Engineers, Inc.

Coad, W.J. 1996. Indoor air quality: A design parameter. *ASHRAE Journal* 38(6):39–47.

DuBois, D., and E.F. Dubois. 1916. A formula to estimate the approximate surface area if height and weight are known. *Archives of Internal Medicine* 17:863–71.

NCDC. 1992a. Monthly normals of temperature, precipitation, heating and cooling degree-day, Climatography of the US #81. National Climatic Data Center, Asheville, NC.

NCDC. 1992b. Annual degree-days to selected bases, Climatography of the US #81. National Climatic Data Center, Asheville, NC.

NCDC. 1999. Engineering Weather Data. National Climatic Data Center, Asheville, NC.

7 Heat and Moisture Flow in Buildings

This chapter discusses basic residential and nonresidential construction practices, components, heat transfer, airflow, and moisture movement in the building envelope. The term *building envelope* refers to the outer elements of the structure, "including foundations, walls, roof, windows, doors, and floors" (ASHRAE 1991). There are a variety of building codes and standards, such as those from ICC (2003) and ASHRAE (2004), that provide guidance and prescription for items summarized in this section. The practice discussed here reflects practices in the continental US, which vary from practices in other locations. This chapter summarizes information available in greater detail in the references and also contains thermal property data on basic building materials and assemblies and building material moisture migration properties. The information in this chapter is closely related to chapter 8, "Cooling and Heating Load Calculations and Analysis."

FOUNDATIONS, WALLS, WINDOWS, AND ROOFS

Foundations

Figure 7.1 is a diagram of basic residential (and light commercial) construction framework, which is often referred to as "stick-built" construction. As shown in the figure, foundation options include poured-in-place concrete slab-on-grade or concrete block perimeters with wood floor joists or trusses (see Figure 7.2

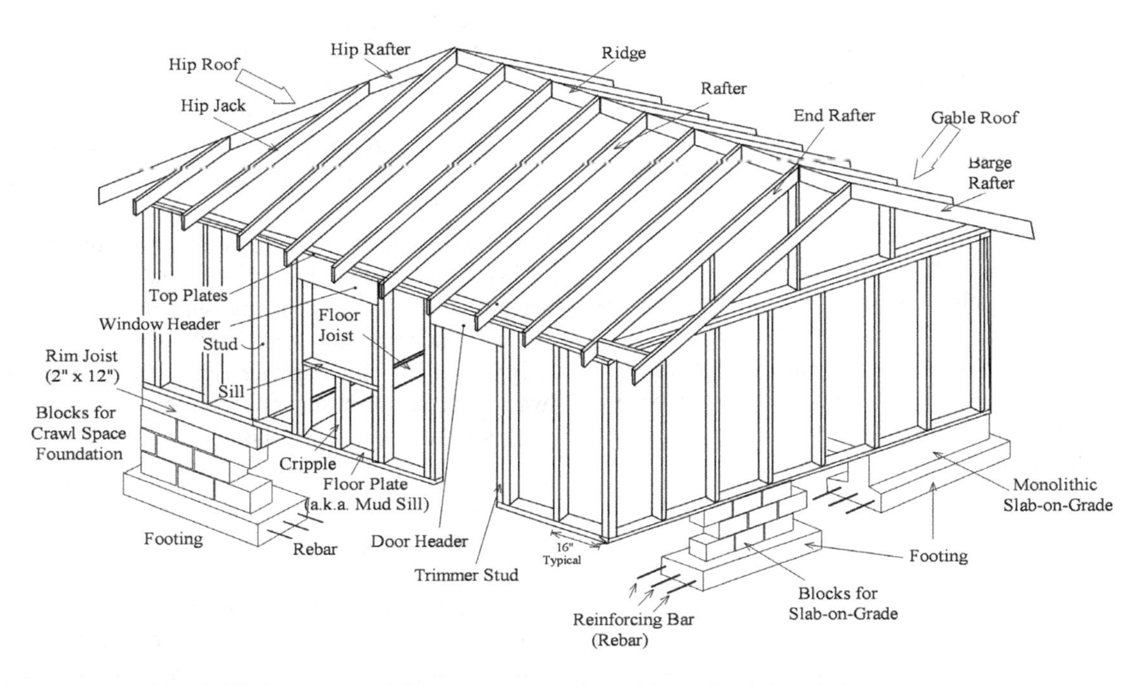

Figure 7.1 Residential/light commercial frame construction with roof and foundation types.

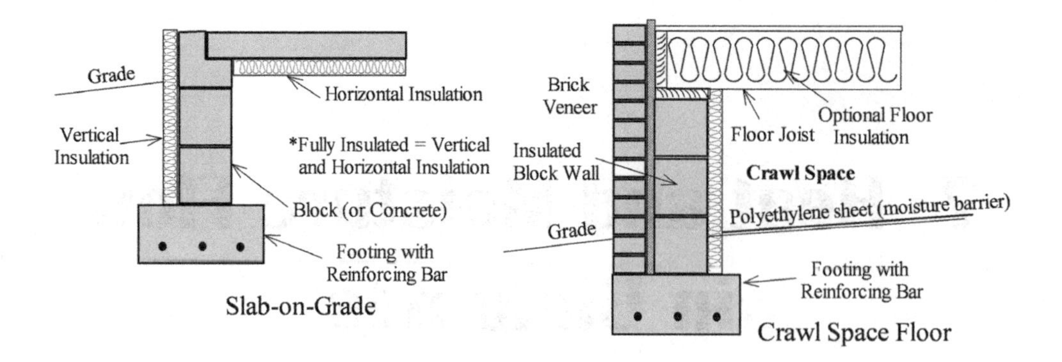

Figure 7.2 Slab-on-grade and crawlspace foundations with insulation locations.

also). In both practices a concrete footing is used with reinforcing bars ("rebar") located near the base of the footing to provide tensile strength to supplement the compressive strength of concrete. The concrete slab-on-grade foundations can be a continuous "monolithic" slab in which the footing, vertical walls, and horizontal slab are poured at the same time. Another practice is to pour the concrete footing, lay vertical cinder blocks, fill the base of the foundation with soil that can be compacted, and then pour the slab.

Crawlspace and basement foundations also require a footing. Blocks are laid (or concrete walls are poured) to elevate the floor off the existing grade around the perimeter of the building. Interior walls or columns are also used since the cost of trusses or floor joists that can span the entire building width are very high. Traditional practice in the US is to use nominal 2 × 10 in. or 2 × 12 in. dimension lumber (typically pine) for floor and rim joists. (Note: 2 × 10s are actually 1½ × 9¼ in. and 2 × 12s are 1½ × 11¼ in.). The construction industry in many applications is converting to "engineered lumber" products such as wood floor trusses, I-beams with oriented strand board (OSB) webs, or laminated beams. Plywood or OSB subfloors are fastened to the floor joist or beams as a final step before wall construction.

In all foundation practices, care must be taken to present a barrier to termites and other pests. Moisture migration from the slab or ground must be contained with a minimum of 6-mil polyethylene sheeting. In many areas of the US, radon gas infiltration from soil is problematic and barriers with ventilation techniques are recommended (ASHRAE 2003, chapter 43, Figure 1).

Walls

The vertical members of frame walls are made from 2 × 4 in. (1½ × 3½ in.) or 2 × 6 in. (1½ × 5½ in.) studs (pine, spruce, fir). These studs are pre-cut to provide the rough wall height with the floor plate and two top plates. The studs are fastened to the plate on 16 or 24 in. centers. Door and window openings are framed as shown in Figure 7.1 with headers and trimmer studs to provide structural support over the open spans. Roof rafters (or wood trusses) are attached to the top plates on 24 in. centers (in most applications). The roof ends can be gables as shown in Figure 7.1, which are relatively simple to ventilate. Hip roofs are another alternative design that offers better wind damage resistance in coastal regions.

Figure 7.3 provides details of two typical frame wall construction practices. Conventional walls have 2 × 4 in. (1½ × 3½ in.) studs placed on 16 in. centers. The interior finish is typically 1/2 or 5/8 in. thick gypsum board panels. The 3½ × 14½ in. cavity is filled with fiberglass or cellulose insulation. External sheathing (4 ft width of plywood or OSB) is required to provide diagonal support in the corners. The remainder of the wall sheathing can also be plywood, OSB, or a rigid board with a higher insulating value. An additional layer of insulation sheathing is optional, but a felt or building wrap ("weather barrier" or "moisture barrier") is recommended to minimize wind infiltration and moisture migration. The vapor barrier is usually located on the warm side of the insulation. More on this issue is presented in later sections of this chapter.

A 2 × 6 in. (1½ × 5½ in.) frame wall offers a more energy-efficient option since an additional thickness of insulation can be placed in the cavity. The sketch shown on the right in Figure 7.3 demonstrates a more efficient use of insulation compared to conventional practice, shown on the left in Figure 7.3. Conventional corners require three 2 × 4 in. studs and spacer blocks and permit no insulation. The 2 × 6 in. corner allows the corner

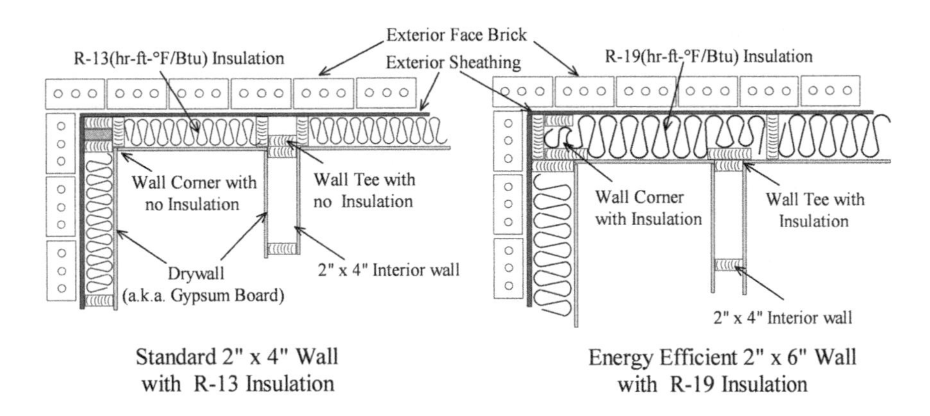

Figure 7.3 Plan view of standard and energy-efficient wood-frame wall construction.

to be insulated. The intersection of exterior walls with interior walls requires "tees" to be made from four 2 × 4 in. studs, and no insulation is possible. A more energy-efficient alternative is to fasten the 2 × 4 in. stud in the interior wall to a 2 × 6 in. stud located in the exterior wall, which is turned so the insulation can be placed behind it as shown in the figure. Finally, 2 × 6 in. frame walls also provide additional space for larger waste drains and vents.

Options of exterior finish include face brick (more expensive but very durable), wood (less expensive but requires painting and is prone to decay), hardboard (less prone to decay but needs painting), vinyl or aluminum siding (lower maintenance than wood but less durable than brick), masonry panels (more difficult to install and require painting), and stucco (less durable and requires installation care to prevent moisture problems). The insulation value of the exterior finish is typically of little consequence, as demonstrated later in "Thermal Properties of Building Materials and Insulation."

A variety of alternatives to wood frame construction are well suited to meet constrains of residential and light commercial construction. Metal studs have several advantages, including better resistance to fire damage, consistent dimensions (especially the lack of curvature and twist), and termite and rot resistance. They are well suited to the use of recycled steel and not as susceptible to the supply interruptions and price fluctuations that periodically occur as a result of environmental issues.

However, several limitations of metal studs merit consideration. Metal studs warrant careful placement of insulating sheathing to prevent thermal bridges from the interior to the exterior. Additionally, an adjustment in installation tools and work crew training is necessary. A recent limitation that is currently of growing importance is the lack of moisture-holding capacity compared to dimensional lumber. Wood will absorb a certain amount of moisture (see "Building Infiltration and Moisture Transfer") during periods of high latent loads. This moisture-holding capacity delays the onset of moisture accumulation that is a primary contributor to mold growth. Use of metal studs reduces the overall moisture-holding capacity of the structure. This lack of capacity can be offset with the use of masonry products inside the vapor barrier of the structure.

Additional wall construction options include prefabricated panels and permanent concrete forms that provide insulation. Stress skin panels consist of a rigid insulation core sandwiched between an exterior sheathing (typically OSB) and an interior sheathing (OSB, drywall, finished paneling). A common product is a 4 ft wide × 8 ft high panel with a 3.5 in. expanded polystyrene board with two 7/16 in. thick OSB panels. Special tools are required to cut openings and for alternative techniques for routing wiring and plumbing.

Another option that provides an added measure of structural and thermal mass benefits is insulated concrete forms. Interlocking extruded polystyrene forms are installed in a method resembling a block wall. (Reinforcing bar (rebar) is installed and concrete is poured into the forms. This provides the insulating benefit of polystyrene and the structural and thermal mass benefit of concrete. Cost is significantly higher than for frame wall construction.

Note: The design engineer must evaluate the actual thermal properties of these options. Marketing literature is used to promote these alternatives since the elevated cost is often difficult to justify in terms of energy savings. Such literature may contain "equivalent" thermal properties or "average reduction in energy cost" values that are intended to adjust energy calculation assumptions but that result in unrealistic economic projections. Methods discussed in this chapter and chapter 8 deter-

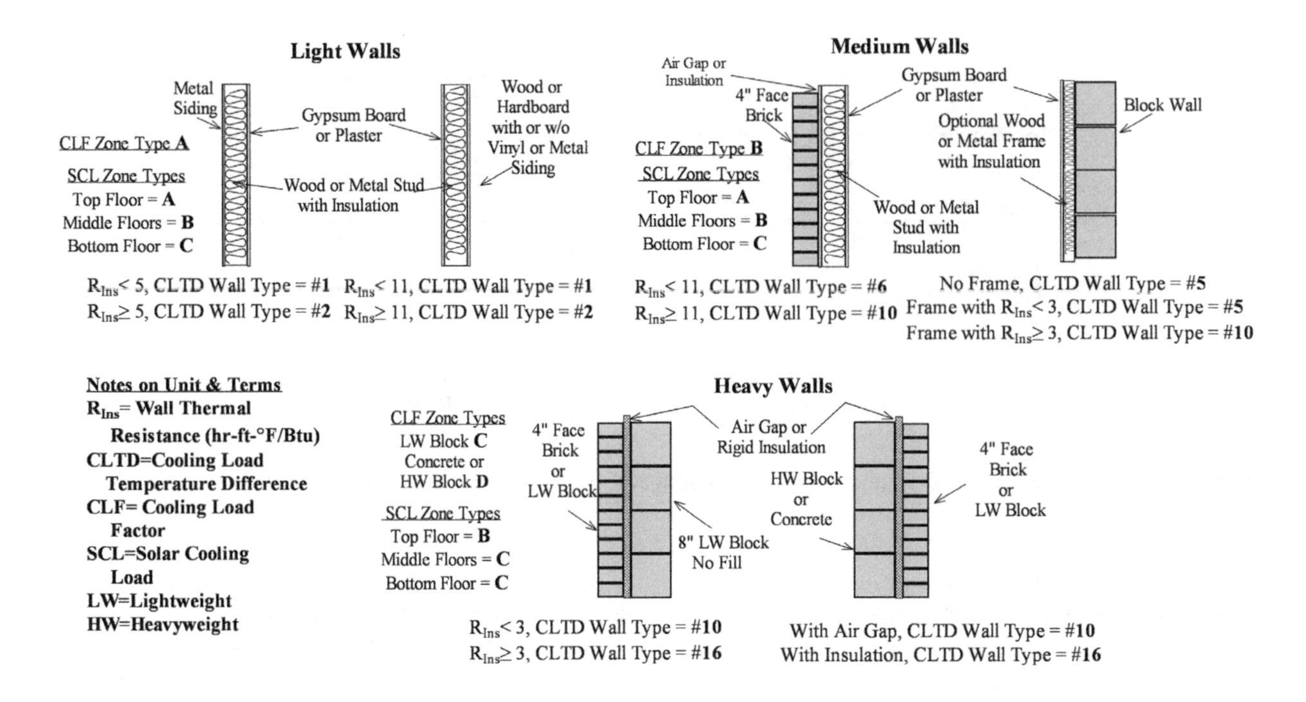

Figure 7.4 Wall construction options with CLTD, CLF, and SCL types for cooling load calculations.

mine the actual thermal resistance of wall structures and the impact of thermal mass used to offset the instantaneous heat gain during periods of peak thermal loads.

Figure 7.4 demonstrates several conventional wall options for all types of construction including light, medium, and heavy buildings. The rough categorization of walls as "light," "medium," and "heavy" is done based on thermal storage capacity. Light walls have little heat storage capacity and tend to reach thermal equilibrium quickly so that maximum heat gain into the building is near the time of peak outdoor conditions. Heavy walls more readily store heat during peak outdoor conditions, so they tend to delay the maximum heat gain into the building until later in the day (or night). This phenomenon, which will be discussed in more detail in the following chapter, can be accounted for with the use of cooling load temperature difference (CLTD), cooling load factor (CLF), and solar cooling load (SCL) values. Figure 7.4 indicates the "types" for wall CLTDs, CLFs, and SCLs, which are used to locate tabular values and factors (see chapter 8) that account for the thermal heat storage effects of building components. CLTD types are designated with a numeric value of 1 (lightweight, poorly insulated) through 16 (very heavy wall with insulation). Walls are classified by CLF type with letter designation from A (lightweight) through D (heavy) and SCL types by letters A though C. These values also account for location of the building and the direction of the wall or window exposure. Methods are presented in chapter 8 to adjust values for a number of other variables.

It is important to note that the R-values shown in the drawing indicate the value for the primary insulating material and not the net wall R-value. The actual thermal resistance should be determined using equations (and spreadsheets) discussed in the "Building Heat Transfer Characteristics" section.

Roof and Ceiling Types

Figure 7.5 shows several types of residential and light commercial roof-ceiling combinations. Pitched wood frame roofs have traditionally been fabricated on-site using 2 × 6 in. or 2 × 8 in. softwood. Plywood, OSB, or other wood-based sheathing serves as the roof decking and is attached to the rafters. The decking is covered with lightweight (15 lb/100 ft^2) felt paper and mineral-impregnated asphalt shingles. Other types of shingles are appearing on the market, but thermal performance varies little compared to asphalt shingles. Insulation is typically located between and above the ceiling joist, which creates an uninsulated attic space below the roof. Cathedral ceilings (no joist or attic) are also an option, but ventilation is typically warranted in all designs to vent attic heat and moisture. Figure 7.5 includes a common passive ventilation method of vents placed in the soffits and a baffled ridge vent at the roof peak. Care

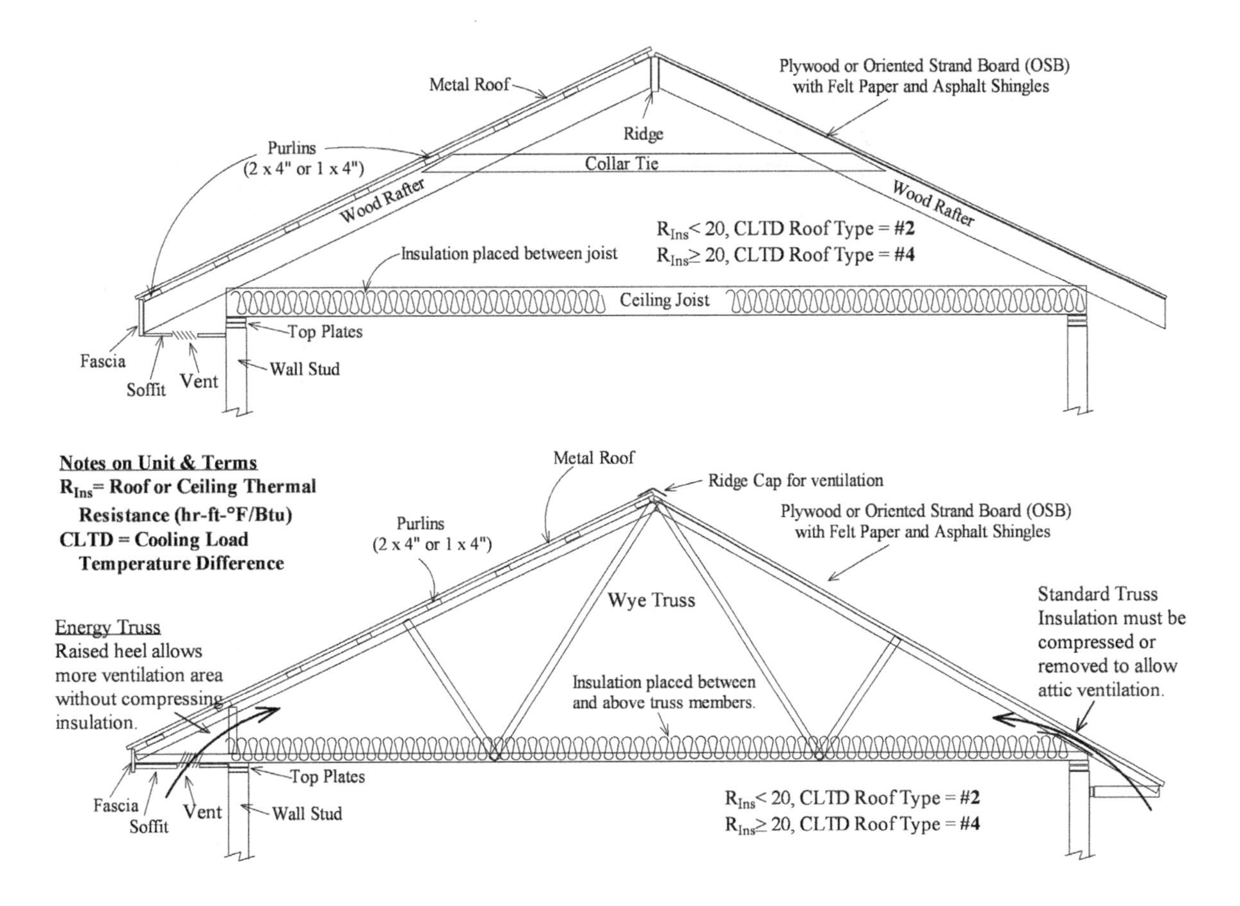

Figure 7.5 Wood frame roof-ceiling construction options.

must be taken to provide adequate vent space, especially at the attic space above the wall-roof juncture.

Lightweight metal roofs are becoming a more common alternative to shingles in the residential and light commercial market. One option is to fasten the metal panels to 1 × 4 in. or 2 × 4 in. horizontal purlins that are attached to the roof rafters. Venting methods are similar to those for wood roof decks.

Figure 7.5 also shows a prefabricated wood truss roof. This alternative reduces time and labor costs at the site and material costs are only marginally more expensive than site-built rafter-joist roofs. Care must be taken to ensure that insulation does not block the small ventilation path near the soffit vent. An alternative to the standard truss construction is a raised heel "energy truss" that provides a larger ventilation path and added head room in the attic space as shown in the lower left side of Figure 7.5. Figure 7.5 also contains the CLTD types for these roofs, which are designated with numeric values of 1 (lightweight, poorly insulated) through 13 (very heavy roof with insulation above the roof mass).

Figure 7.6 shows several types of roof and ceiling options for nonresidential construction. CLTD numbers (1 through 13) for these options are given on the figure. Flat roofs are typically unvented since the space between the roof and ceiling is often used to rout ductwork and, in some cases, used as a return air plenum. Pitched roofs with insulation in the ceiling are vented using the same methods used with residential systems. Many commercial buildings with insulation in the roof either use no ventilation or vent a small space between the roof and insulation. Asphalt shingle roofs are of special concern since some roof insulation and ventilation practices may elevate surface temperature and reduce shingle life span and/or manufacturer warranty. However, the use of roof insulation broadens the alternatives for usable space in high roof/ceiling buildings (storage, etc.). HVAC equipment and ductwork can be located in these areas without the heat loss and heat gain penalties that occur if they are located in unconditioned space or outdoors.

Windows

This section provides a summary discussion of window types, window heat transfer modes, and rating methods. Additional information can be found in the

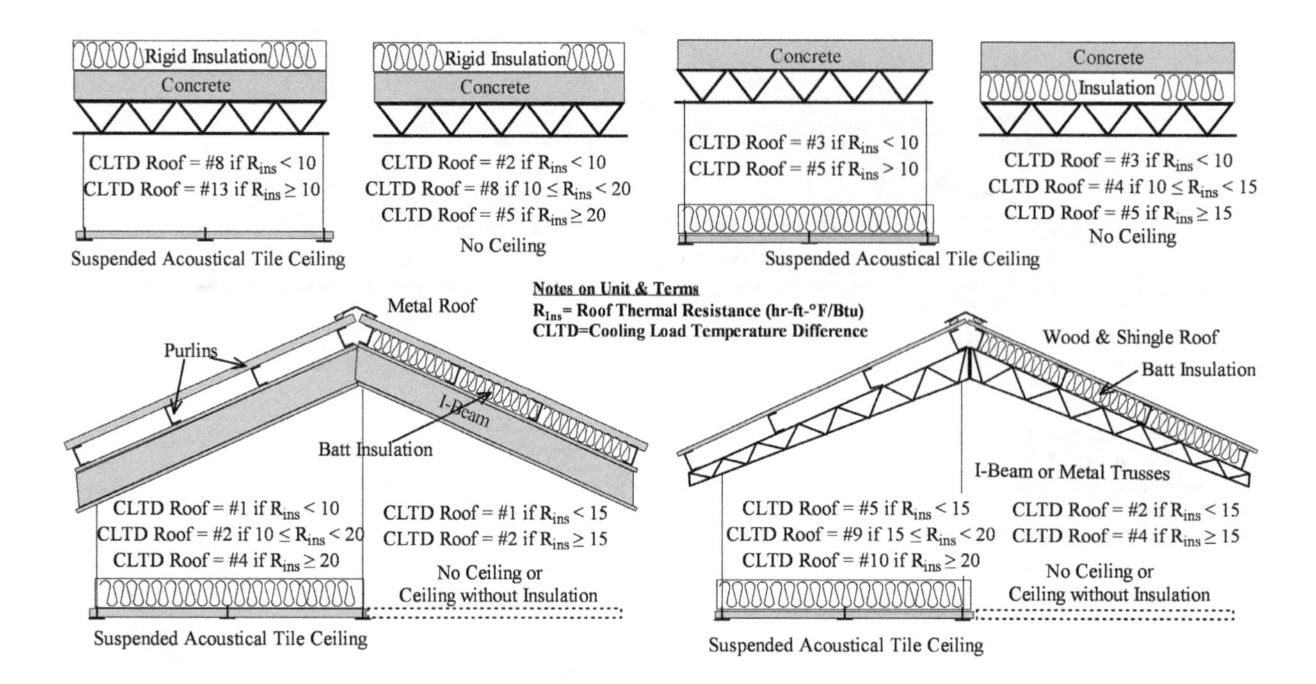

Figure 7.6 Flat and pitched roof and ceiling types for nonresidential construction.

"Fenestration" chapter of the *2005 ASHRAE Handbook—Fundamentals* (ASHRAE 2005, chapter 31) and publications of the National Fenestration Rating Council (NFRC/www.nfrc.org). ASHRAE follows architectural terms, using *fenestration* to describe window systems and *glazing* to refer to window panes.

Figure 7.7 shows several types of windows. The term *operable* is used to denote windows that can be opened and closed. A type widely used in the US is the vertical sliding window that is either single hung (one sash moves up and down) or double hung (both sashes move up and down. The term *sash* refers to the window glass and enclosed frame. Other options include horizontal sliding windows, which present a design challenge to provide ease of movement without compromising resistance to air leakage. To some degree the swing-open movement of casement and awning windows can enhance airtightness, although ease of opening may be compromised. Non-operable windows (a.k.a. picture windows) can be more easily sealed. Varieties of operable or non-operable windows include bay windows (shown) and/or greenhouse windows of similar design.

Window panes are traditionally clear glass, which transmits 80% of incident solar radiation and 75% of visible light (ASHRAE 2005, chapter 31). Tinting can alter the amount of solar radiation (desirable during cold weather, undesirable during warm periods) and the amount of visible light to offset the energy consumption and heat generation of lighting. Coatings and screens are also used to alter these characteristics. In some cases, tints, screens, and coatings are selective to maximize the transmission of heat and/or light when it is more desirable. Window glazing can also be made from plastics or laminates and can be manufactured to provide security (bulletproof), decoration (etched, stained), or privacy (obscured).

Window performance is multifaceted. Conduction-convection performance is expressed in terms of U-factor (Btu/h· ft^2· °F) or the inverse R-value (h· ft^2· °F/Btu). Single-pane windows typically have high U-factors (low thermal resistance) with the bulk of the resistance being due to the film coefficients. Double- and triple-pane windows provide additional air gap resistances. Higher-cost windows include the replacement of the air in the gap with a sealed cavity filled with low-conductivity gas such as argon. The window frame and sealing methods contribute to the overall U-factor. Metal frame windows offer rigid and durable material that provides greater integrity and life compared to wood. However, thermal breaks or spacers are required to minimize frame heat conduction. Vinyl frame windows have appeared as a compromise in many residential applications.

A second facet of window performance is solar heat gain, which is desirable when heating is required but undesirable when cooling is necessary. The solar heat gain coefficient (SHGC) is used to denote the fraction of incident solar radiation that is transmitted through the window. This has created some confusion with solar heat gain

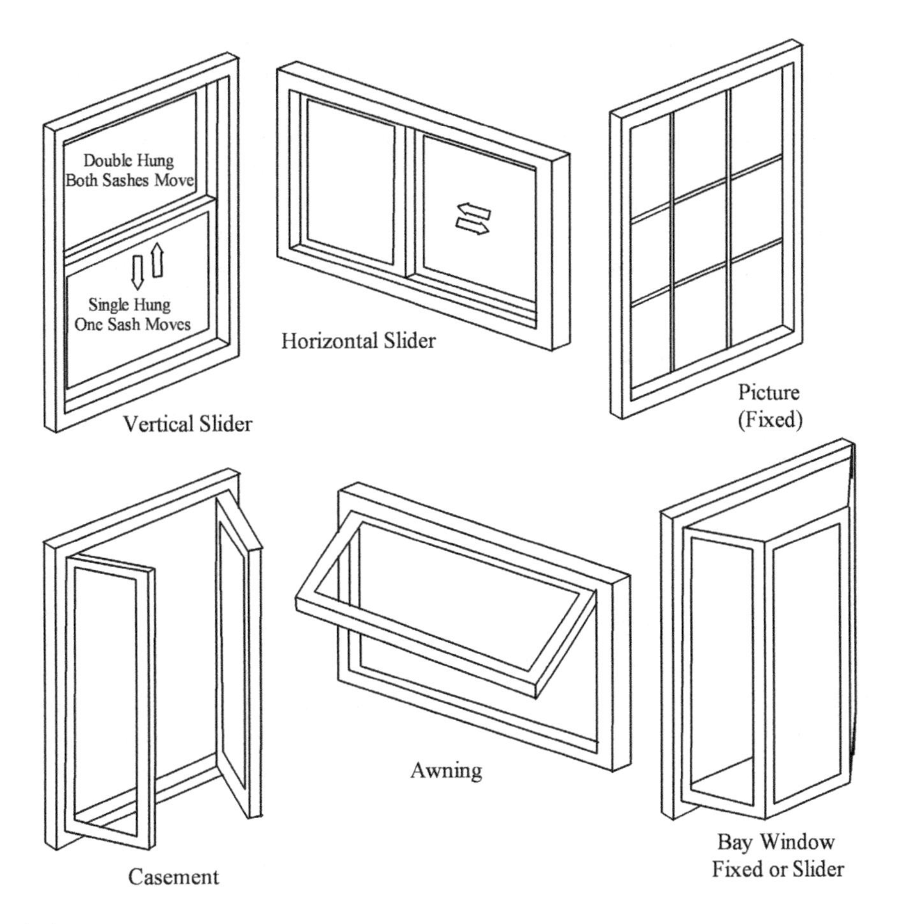

Figure 7.7 Window types.

factors (SHGFs) or solar cooling loads (SCLs), which are used to compute the heat rate (q = SCL × A_{Window}) transmitted through a clear single-pane window for various latitudes and directions. The units for SHGF and SCL are Btu/h· ft^2. The shading coefficient (SC) is used to account for the effects of tinting and shading devices (q = SC × SHGF × A_{Window}). SC is a simplified variant of SHGC in that it is used to denote the fraction of transmitted solar heat relative to clear single-pane glass, which has an SHGC of 0.87 at normal solar incidence. SC references the glazing portion of the window only and does not include frame effects such as the SHGC. Shading coefficient tables are available that include the effects of the shading devices in addition to the window properties. SC is more limited than SHGC with respect to the angular dependence of incident radiation. However, in many cases, window SC can be estimated from SHGC by

$$SC = SHGC/0.87 . \tag{7.1}$$

A related window property is the visible transmittance (VT) of a window, which is the fraction of incident visible light that is transmitted (weighted to the photopic response of the human eye). A desirable trait in some applications would be a low SHGC to minimize cooling requirements but a high VT to improve indoor visibility and psychological effects of daylighting. Advanced windows can have this trait and, in some cases, it can be varied throughout the day to further enhance performance.

A fourth measure of window performance is resistance to air leakage. The units of measurement are leakage rate per unit area of window surface (cfm/ft^2). Prior to standardization, a commonly used measure was leakage rate per unit length of crack or window sash perimeter. Unfortunately, window leakage rate is influenced by thermal, wind, and mechanical ventilation induced pressure differentials that vary widely. Therefore, a standard pressure differential of 75 Pa (0.3 in. of water) has been established.

Figure 7.8 is a representation of a window rating label that should accompany standard products. The National Fenestration Rating Council (NFRC 2004) publishes the results of an extensive rating program, which also includes test procedures and conditions. This program and rating procedure offer design engineers a method of comparison for windows and a relatively simple set of values on which to base specifications.

Shading devices can be used to minimize the negative impact and optimize the desirable properties of windows. Traditional methods include interior shades, blinds, and curtains. Although operable external shading devices can be more effective, costs, complexity, and maintenance requirements are higher. A basic understanding of the movement and position of the sun can be an important tool in optimizing the placement of windows and shading devices.

THERMAL PROPERTIES OF BUILDING MATERIALS AND INSULATION

Thermal resistance values of building components (walls, floors, roofs) can be calculated from the thermal properties of the individual materials. The *2005 ASHRAE Handbook—Fundamentals* (ASHRAE 2005, chapter 31) contains an extensive listing of these values. Often it is useful to convert these values to the transmission coefficients (U-factors), the inverse of thermal resistance. This reference also contains the transmission coefficients (U-factors) for other components, including door units. Transmission coefficients for windows and glass doors are available in the "Fenestration" chapter of the Handbook (ASHRAE 2001, chapter 30).

Table 7.1 is a short summary of the data found in these references. The data include the thermal resistance of nonreflective air-to-solid surfaces and air gaps with typical temperature differences. A short listing is provided of the thermal resistances of common exterior and interior materials of a typical thickness. Values for popular types of loose-fill wall and ceiling insulation materials are given with and without consideration for the impact of wood-framing material on the overall value. Values for common rigid insulation board and panels are included.

Values are given for both a listed thickness and for a unit thickness (per inch). Masonry and wood structural material resistance is also included. The final items in the table are both R-values and U-factors for window and door assemblies.

BUILDING HEAT TRANSFER CHARACTERISTICS

The isometric drawing in Figure 7.1 provides details of the components of a typical residential frame wall. Although 2 × 4 in. (1.5 × 3.5 in.) or 2 × 6 in. (1.5 × 5.5 in.) framing studs are typically placed on 16 in. centers, approximately 15% to 20% of the finished wall is framing since headers, trimmer studs, and plates are necessary. The remaining 80% to 85% is insulation-filled cavities. A plan view detail of a typical completed wall is shown in Figure 7.9. The thermal resistance concept is applied by visualizing two heat flow paths through the wall, one at the framing (R_{FOV}) and a second through the insulation (R_{IOV}). Since the area through each path is constant, the individual resistances can be added to find the overall thermal resistance for both paths. The overall heat transfer coefficient for each path is the inverse of the thermal resistance.

At the framing:

$$R_{Fov} = R_i + R_{2\times 6} + R_{Ply} + R_{gpg} + R_{Brk} + R_o \; ; \; U_{Fov} = 1/R_{Fov} \tag{7.2}$$

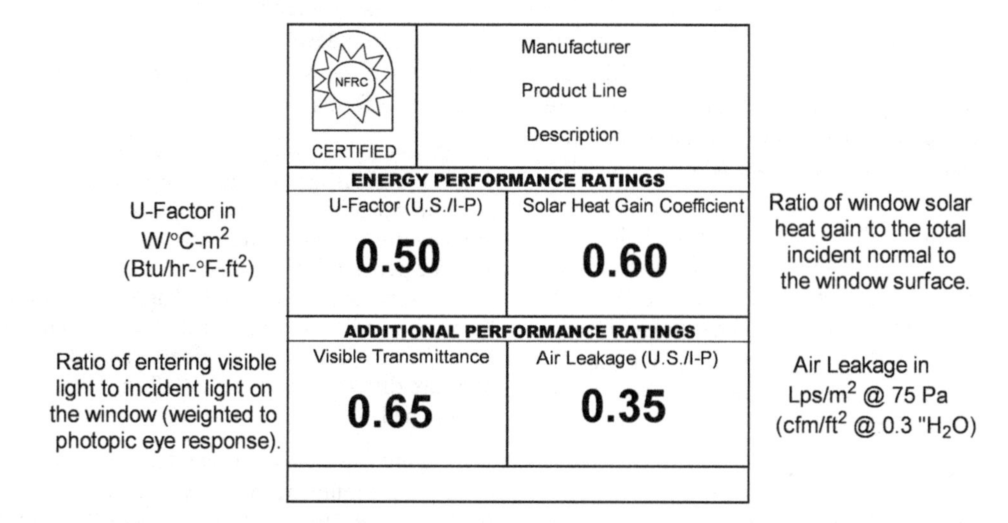

Figure 7.8 Example of National Fenestration Rating Council rating label (inch-pound units).

Table 7.1 R-Values for Building Materials and Structural Units (ASHRAE 2005, chapter 25)

Air (Heat Flow)	h· °F· ft²/Btu	Insulation	h· °F· ft²/Btu	Structural	h· °F· ft²/Btu
Ext. surface (7½ mph)	0.25	3½ in. fiberglass	11.0	8 in. LW block	2.00
Ext. surface (15 mph)	0.17	w/ 20% wood framing	9.6	with perlite cores	5.50
Inside surface	0.68	3½ in. fiberglass	13.0	12 in. LW block	2.40
1 in. vertical air gap	0.8-1.6	w/ 20% wood framing	10.5	with perlite cores	7.20
4 in. vertical air gap	0.9-1.5	5½ in. fiberglass	19.0	12 in. HW block	1.11
1 in. horizontal air gap	0.8-1.7	w/ 10% wood framing	17.3	Concrete	0.14/in.
4 in. horizontal air gap	0.8-2.2	w/ 20% wood framing	15.2	Softwood	1.25/in.
		7¼ in. fiberglass	25.0	Hardwood	0.9/in.
Exterior materials		w/ 10% wood framing	21.0	R-values in h· °F· ft²/Btu	
4 in. face brick	0.45	9¼ in. fiberglass	30.0	U-factors in Btu /h· °F· ft²	
Steel siding	0.00	w/ 10% wood framing	26.0	Windows	R / U
Vinyl siding (0.06 in.)	0.06	11¼ in. fiberglass	38.0	Sgl. pane alum. frame	0.79/1.27
1 in. stucco	0.21	w/ 10% wood framing	32.5	Dbl. pane alum. frame	1.20/0.83
3/4 in. plywood	0.93	Cellulose	3.7/in.	Dbl. vinyl/alum. frame	1.75/0.57
1/2 in. plywood	0.62	w/ 10% wood framing	3.1/in.	Dbl. wood/vinyl frame	1.82/0.55
5/8 in. hardboard	0.85	w/ 20% wood framing	2.7/in.	Triple vinyl/al. frame	2.30/0.43
3/8 in. hardboard	0.50	Panel-3½ in. beadbrd.+OSB	12.8	Doors	
1/2 in. veg. (black) board	1.32	Panel-3½ in. extr.poly+OSB	17.0	1 ¾ in. solid wood	2.50/0.40
Roof stone/slag	0.1/in.	1 in. duct liner/wrap	4.0	with storm door	3.80/0.26
Building wrap, felt	~ 0	2 in. duct liner/wrap	8.0	1 ¾ in. panel	1.85/0.54
7/16 in. OSB	0.5	Expanded polysty (beads)	3.5/in.	with storm door	2.80//0.36
Interior		Extruded polystyrene	5.0/in.	1 ¾ in. insul. metal	2.50/0.40
1/2 in. gypsum board	0.45	Polyisocyanurate	6.0/in.	with storm door	3.80/0.26
5/8 in. gypsum board	0.56	Polyurethane foam	6.0/in.	2 ¼ in. solid wood	3.7/0.27
1/2 in. acoustical tile	1.79			with storm door	5.0/0.20

At the insulation:

$$R_{Iov} = R_i + R_{ins.} + R_{Ply.} + R_{gap} + R_{brk.} + R_o \,;\; U_{Iov} = 1/R_{Iov} \quad (7.3)$$

For parallel heat flow paths, the overall UA value is the sum of the individual UA values.

The overall heat transmission coefficient (U_{ov}) of the entire wall is then found by multiplying the path heat transfer coefficients by the relative areas of each.

$$U_{ov}A = U_{Fov}A_{Frm} + U_{Iov}A_{Ins}$$

$$U_{ov} = U_{Fov}\frac{A_{Frm}}{A} + U_{Iov}\frac{A_{Ins}}{A} = U_{Fov} \times \%A_{Frm} + U_{Iov}\%A_{Ins} \quad (7.4)$$

Roof-Ceilings

Determining the overall heat transmission coefficient (U-factor) of a roof and ceiling assembly is similar to the procedure for calculating the values for walls. There are typically two heat flow paths, one through the insulation cavity and one through the wood or metal frame. However, insulation is not only installed between the floor joists but also above the top of the joist. This additional layer of insulation is accounted for by inserting a thermal resistance value for the added insulation through both paths in Equations 7.2 and 7.3. The resistance of the ceiling material, the roofing material, the roof decking, and the attic must also be considered. The equations for the framing and insulation path resistance (R_{FOV}, R_{IOV}) and U-factor of a wall are modified for the roof and ceiling.

$$R_{Fov} = R_i + R_{Ceiling} + R_{Joist} + R_{InsAdded} + R_{Attic} + R_{Deck} + R_{Roof} + R_o \,;\; U_{Fov} = 1/R_{Fov} \quad (7.5)$$

$$R_{Iov} = R_i + R_{Ceiling} + R_{ins.} + R_{InsAdded} + R_{Attic} + R_{Deck} + R_{Roof} + R_o \,;\; U_{Iov} = 1/R_{Iov} \quad (7.6)$$

The attic can be considered an air gap if it is unventilated. However, most residential and light commercial attics are ventilated and some contain reflective material. Table 7.2 is provided to account for the effects of vented attics with and without reflective surfaces (or "radiant barriers"). Reflective surfaces should be located so that a ventilation path is between the barrier and an

Example Problem 7.1

Determine the overall U-factor and resistance of the wall shown in Figure 7.9 for a wall that is 15% wood and 85% insulation.

Solution:

Equations 7.2 and 7.3 results for the wall are arranged in the table below.

$\frac{h \cdot °F \cdot ft^2}{Btu}$	% Area	Out. Surface	Out. Finish	Insul.	Wood	Gap	½ in. Ply.	Int. Finish	Inside Surface	ΣR_{Path}	$U = 1/R$
$R_{FramePath}$	15	0.25	0.45		6.875	1.2	0.62	0.45	0.68	10.5	0.095
$R_{Ins.Path}$	85	0.25	0.45	19		1.2	0.62	0.45	0.68	22.7	0.044

Equation 7.4 is applied to find the overall U-factor of the wall.

$$U_{ov} = A_{Frame} \times U_{Frame} + A_{Ins} \times U_{Ins} = 0.15 \times 0.095 + 0.85 \times 0.044 = 0.052 \text{ Btu /h} \cdot °F \cdot ft^2$$

The overall R-value is the inverse of the overall U-factor.

$$R_{ov} = 1/U_{ov} = 1/0.052 = 19.3 \text{ h} \cdot °F \cdot ft^2/ \text{ Btu}$$

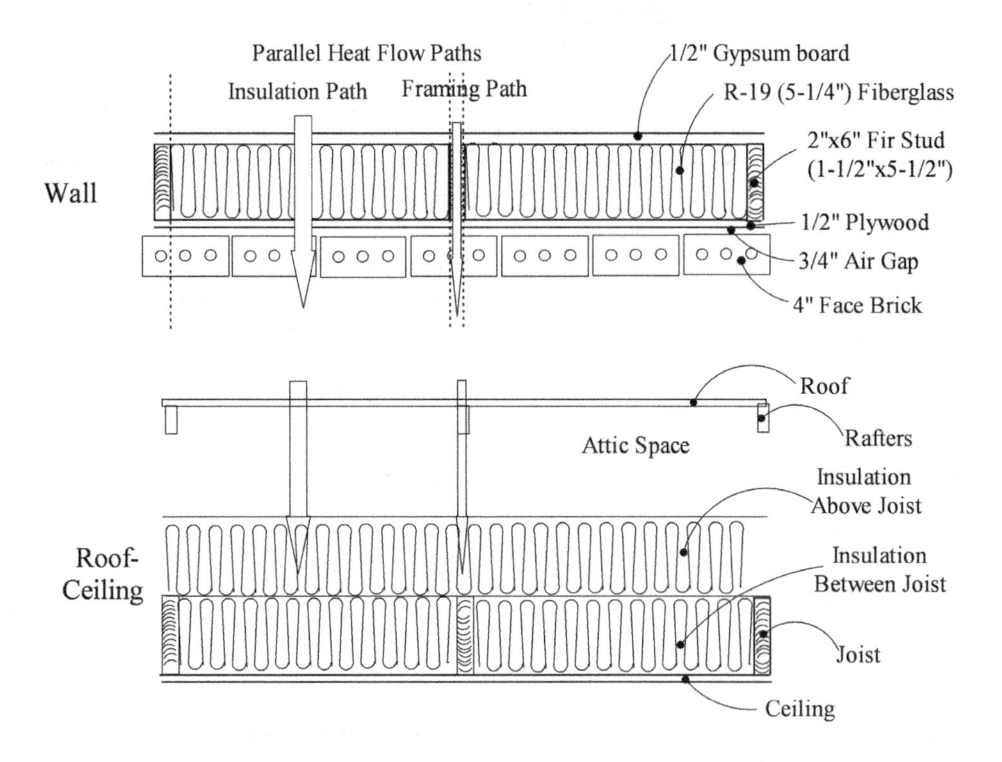

Figure 7.9 Parallel heat flow through insulation and framing paths—wall and roof/ceiling.

Table 7.2 Effective Thermal Resistance of Attics—Summer Conditions (ASHRAE 2001, chapter 25)

Sol-Air Temp.=140°F	No Ventilation		Natural Ventilation 0.1 cfm/ft^2		Power Ventilation 0.5 cfm/ft^2	
Vent. Air Temp. (°F)	$R_{Ceiling}$ (h· °F· ft^2/ Btu) 10	20	$R_{Ceiling}$ (h· °F· ft^2/ Btu) 10	20	$R_{Ceiling}$ (h· °F· ft^2/ Btu) 10	20
			Non-Reflective Surfaces			
80	1.9	1.9	2.8	3.5	6.5	10
90	1.9	1.9	2.6	3.1	5.2	7.9
100	1.9	1.9	2.4	2.7	4.2	6.1
			With Reflective Barrier (= 0.05)			
80	6.5	6.5	8.2	9.0	14	18
90	6.5	6.5	7.7	8.3	12	15
100	6.5	6.5	7.3	7.8	10	12

Table 7.3 U-Factor (Btu/h· °F· ft^2), Visible Transmission, Shading Coefficient (SC), and Solar Heat Gain Coefficient (SHGC) for Unshaded Windows (ASHRAE 1997, chapter 29)

		Glass Only				Aluminum Frame				Wood/Vinyl Frame			
Glass Type	Gap	U	VT	SC	SHGC	U	VT	SC	SHGC	U	VT	SC	SHGC
Sgl. 1/8 in. Clear		1.04	0.90	1.00	0.86	1.27	0.65	0.86	0.75	0.89	0.65	0.72	0.63
Sgl. 1/4 in. Acryl		0.88	0.89	0.93	0.81	1.21	0.65	0.82	0.71	0.83	0.65	0.69	0.60
Dbl. 1/8 in. Clear	1/4 in. air	0.55	0.81	0.87	0.75	0.87	0.59	0.76	0.66	0.55	0.59	0.63	0.55
Dbl. 1/8 in. Clear	1/2 in. argon	0.41	0.81	0.87	0.75	0.79	0.59	0.76	0.66	0.49	0.59	0.63	0.55
Dbl. 1/8 in. ε = 0.2	1/4 in. air	0.45	0.76	0.70	0.60	0.79	0.55	0.66	0.57	0.49	0.55	0.55	0.48
Dbl. 1/8 in. ε = 0.2	1/2 in. argon	0.30	0.76	0.70	0.60	0.67	0.55	0.66	0.57	0.39	0.55	0.55	0.48
Trpl. 1/8 in. Clear	1/4 in. air	0.38	0.74	0.71	0.61	0.72	0.51	0.68	0.59	0.43	0.50	0.59	0.51
Trpl. 1/8 in. Clear	1/2 in. argon	0.29	0.74	0.71	0.61	0.65	0.51	0.68	0.59	0.37	0.50	0.59	0.51

asphalt shingle roof. This will prevent elevated roof surface temperatures that can reduce the service life and warranty of these types of roofing materials. The table provides values for a sol-air temperature (the temperature of the surface adjusted to account for the effect of radiation) of 140°F. Values for 120°F and 160°F are provided in the *2001 ASHRAE Handbook—Fundamentals* (ASHRAE 2001, chapter 25).

Equation 7.4 can be applied to the results of Equations 7.5 and 7.6 to obtain the overall heat transmission coefficient (U_{ov}) for the roof-ceiling assembly.

Windows

The NFRC rating values discussed above provide values for the four primary indicators of window performance. The U-factor (overall heat transmission coefficient) provides a measure of the conduction-convection heat loss or gain of a window. Table 7.1 provides this value and the inverse R-value for several generic window types. Table 7.3 provides values for a wider variety of glazing-only arrangements and windows assemblies (glazing, frame, and sash).

Table 7.3 also lists the visible transmission of these windows, which is an indicator of how well the systems admit useful sunlight that can be used to offset artificial lighting. Windows also transmit solar heat that is a beneficial supplement during the heating season but an undesirable heat gain when cooling is required. Table 7.3 lists the two common indicators of the fraction of incident solar gain that is transmitted. These are shading coefficient (SC ≡ fraction compared to a clear single-glass window) and the solar heat gain coefficient (SHGC ≡ fraction compared to total incident on outside window surface). As noted previously, a clear single-pane glass admits a fraction of 0.87 of incident radiation, thus SC = SHGC/0.87 at normal angle of incidence.

Table 7.3 applies to unshaded windows, although some glazing types have reflective treatments designed to reduce the effects of solar heat gain. A more common method of reducing solar heat gain is operable internal shading devices. Table 7.4 list the shading coefficients of several types of blinds and drapes. It is emphasized that

Table 7.4 Typical Shading Coefficients (SC) for Windows with Interior Shades (ASHRAE 1997, chapter 29)

Blinds	Med. Blind	Lt. Blind	Med. Blind	Lt. Blind	Dark Roller	White Roller
Position	45°	45°	Closed	Closed	Closed	Closed
Single 1/8 in. Clear	0.74	0.67	0.63	0.58	0.81	0.39
Double 1/8 in. Clear	0.62	0.58	0.63	0.58	0.71	0.35
Double 1/8 in. ε = 0.2	0.39	0.36			0.4	0.22
Drapes			Drape Color / Weave			
	Lt./Open	Med./Open	Dk./Open	Lt./Closed	Med./Closed	Dk./Closed
Single 1/8 in. Clear	0.69	0.74	0.82	0.69	0.59	0.39
Double 1/8 in. Clear	0.63	0.65	0.71	0.63	0.53	0.35
Double 1/8 in. ε = 0.2	0.36	0.37	0.37	0.36	0.32	0.22

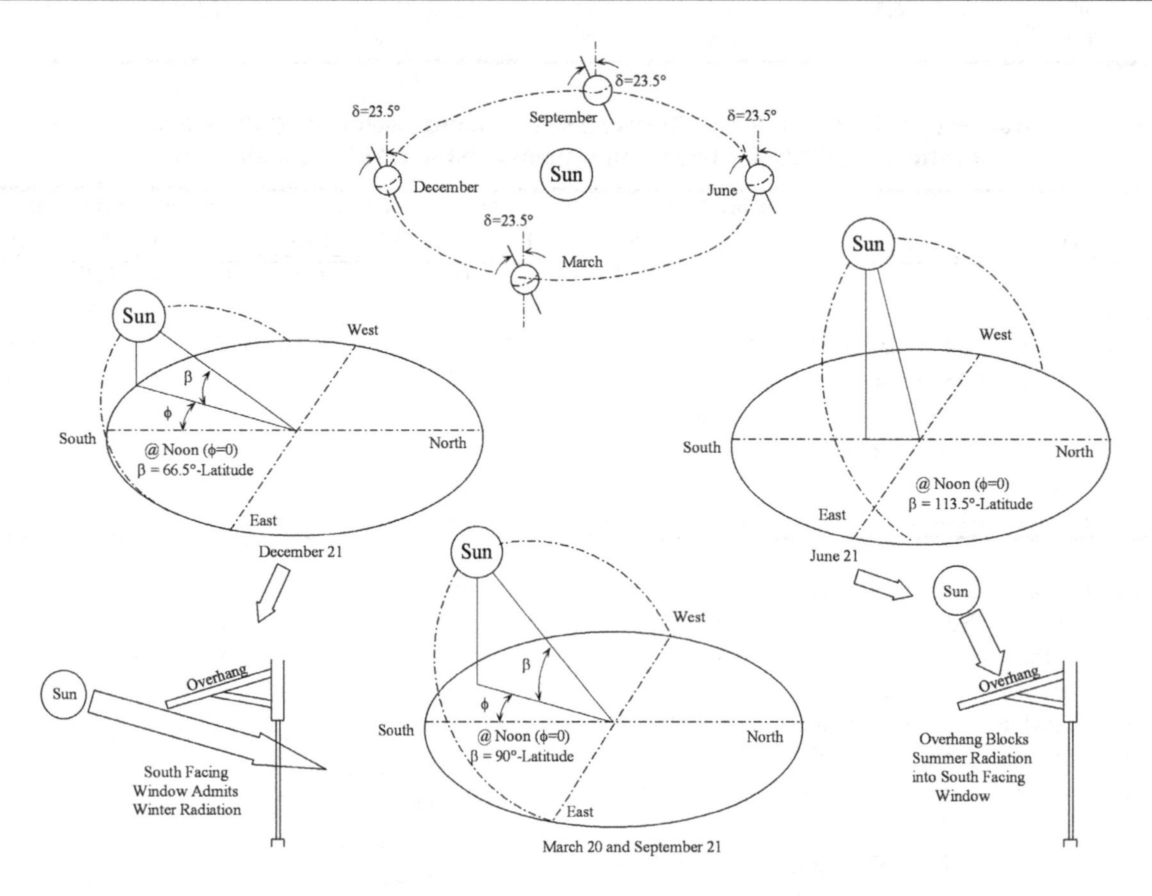

Figure 7.10 Rotation and tilt of earth and path of sun in the winter, summer, fall, and spring.

these values are typical and individual blinds and drapes may have a significant deviation from these values even though the description is similar. Thus, verifiable data from a shade manufacturer or testing laboratory would provide more accurate values.

External shading is often more effective at reducing direct solar heat gain. The arrangement and location of the shading device is critical to minimize cooling mode heat gain while optimizing visible light and solar heat gain during the heating mode. An understanding of the annual and daily movement and location of the sun relative to the earth is useful in arranging the location of windows and external shading devices.

Figure 7.10 includes a set of diagrams that demonstrate the relative movement of the sun and earth with the resulting solar angles for the northern hemisphere. The axis of daily rotation of the earth relative to the plane of the annual orbit around the sun is offset by an angle of 23.5°. This is referred to as the angle of declination (δ) (ASHRAE 2003, chapter 33). In the summer this angle or tilt results in the northern hemisphere being exposed at angles of incidence closer to normal (90°). The length of days exceeds 12 hours and increases at higher latitudes between the tropic of Capricorn (l = 23.5°) and the Artic Circle (l = 90 – 23.5°). During the spring and fall, days are shorter in the northern hemisphere and the angle of incidence is reduced compared to summer values. The length of day is 12 hours for all latitudes on the fall (September 21) and spring (March 20) equinox (equal nights). In the winter, the angles of incidence are reduced further and the days are shortest.

The resulting paths of the sun in the northern hemisphere are also shown in Figure 7.10. During the winter the sun angle remains relatively low to the horizontal surface. On December 21, this angle is only 66.5° (local latitude) even at the peak elevation, which occurs at solar noon. The angle of movement in the sky in the winter is reduced with the sun rising at an angle south of due east and setting at an angle south of due west. Days are shorter since the rate of angular movement of the sun across the horizon (15° per hour [= 360° ÷ 24 hours]) is constant. The shortest day of the year is also December 21 in the northern hemisphere. A latitude of 66.5°N represents the Artic Circle and the solar noon elevation is zero on December 21, which corresponds to no sunlight.

In the spring and fall the elevation of the sun is greater. Days are longer. In the summer, the elevation of the sun is the greatest and the days are even longer. The sun rises north of due east and sets north of due west.

Note the diagrams in the figure show the solar angle of incidence for a south-facing wall in the winter and summer. In the winter, solar rays are not blocked by a properly located window overhang on a south-facing wall. Thus, passive solar heat gain is available to assist in warming the building. However, this same overhang can block unwanted direct solar radiation in the summer when the elevation of the sun is high. Other passive solar techniques include the location of trees that shed leaves in the winter (deciduous) on the east, southeast, southwest, and west sides of a building.

Window performance is the prime beneficiary of optimum consideration of solar angles. However, walls and roof design and location to maximize winter heat collection and minimize summer solar gain are also important.

Slabs, Crawlspaces, and Basements

Heat loss through slabs, crawlspaces, and basements is complicated by the parallel heat flow into and through the surrounding soil. The location of insulation must not interfere with the structural integrity of the load-bearing perimeter footing and foundation wall. Insulation placed on the external interface between the slab, crawlspace wall, or basement wall is effective in this regard. However, precautions must be taken to prevent the insulation from becoming or creating a pathway to the building interior for pests, especially termites. Foam insulation is a concern with regard to the spread of fire, and precautions are warranted.

Figure 7.2 shows typical locations for rigid insulation for slab-on-grade applications. Table 7.5 provides a list of slab-on-grade heat transmission coefficients (F_s) that consider the location of the insulation (horizontal, vertical, or both), thermal resistance of the insulation (R_{Ins}), width (or height) of insulation, and slab temperature (heated or unheated). The reference for this table uses the term "F-factor" for the heat transmission coefficient (F_s). Note the heat loss is based on the length of the slab perimeter (P_{Bldg}) rather than on area.

$$q_{slab}(\text{Btu/h}) = F_s \times P_{Bldg}(\text{ft}) \times (t_o - t_i)°\text{F} \qquad (7.7)$$

Table 7.6 provides values for crawlspaces and basement walls that incorporate a similar heat transmission coefficient (F_{csb}). Table 7.6 was developed from the procedure presented in the *ASHRAE Handbook—Fun-*

Table 7.5 Slab Perimeter Heat Transmission Coefficients (F_s) in Btu/h· ft· °F (ASHRAE 2004)

Slab Insulation		Unheated Slabs					Heated Slabs				
Insulation Position*	h· °F· ft²/Btu	Insulation Depth or Width (in.)					Insulation Depth or Width (in.)				
		0	12	24	36	48	0	12	24	36	48
Horizontal	R_{Ins} = 5	0.73	0.72	0.7	0.68	0.67	1.35	1.31	1.28	1.24	1.20
	R_{Ins} = 10	0.73	0.71	0.7	0.66	0.64	1.35	1.30	1.26	1.20	1.13
Vertical	R_{Ins} = 5	0.73	0.61	0.58	0.56	0.54	1.35	1.06	0.99	0.95	0.91
	R_{Ins} = 10	0.73	0.58	0.54	0.51	0.48	1.35	1.00	0.90	0.84	0.78
Horz.+ Vert. (Fully Insul.)	For R_{Ins} = 5, F = 0.46			For R_{Ins} = 10, F = 0.36			For R_{Ins} = 5, F = 0.74			For R_{Ins} = 10, F = 0.55	

* See Figure 7.2.

Table 7.6 Crawlspace and Basement Heat Transmission Coefficients (F_{csb}) in Btu/h· ft· °F (ASHRAE 2005, chapter 29)

Crawlspace or Basement Heights			F_{csb} (Btu/h· ft· °F) $\{q_{csb} = F_{csb} \times P_{Bldg}$ (ft) $\times [t_i - t_o]$ (°F)$\}$			
L_{ag} (ft)	L_{bg} (ft)	L_{csb} (ft)	$R_{Ins.} = 0$ h· ft²· °F/Btu	$R_{Ins.} = 4$ h· ft²· °F/Btu	$R_{Ins.} = 8$ h· ft²· °F/Btu	$R_{Ins.} = 12$ h· ft²· °F/Btu
1	1	2	1.13	0.39	0.24	0.17
	4	5	4.42	2.08	1.41	1.07
	7	8	9.0	4.71	3.34	2.62
4	1	5	2.95	0.92	0.55	0.40
	3	7	4.96	1.91	1.22	0.90
	5	9	7.62	3.41	2.30	1.75
7	1	8	4.76	1.45	0.86	0.61
	2	9	5.66	1.87	1.14	0.82
	3	10	6.78	2.44	1.54	1.12

*Insulation can also be located on interior wall.

damentals for below-grade walls and is coupled to a calculation of the above-grade wall (Equation 7.3 for a concrete or block wall). The values are also based on the building perimeter (P_{Bldg}) and assume the wall insulation extends from the footing or basement floor to the building floor.

$$q_{csb}\ (\text{Btu/h}) = F_{csb} \times P_{Bldg}\ (\text{ft}) \times (t_o - t_i)°\text{F} \tag{7.8}$$

If the wall is partially insulated, a weighted average estimate for an uninsulated wall and a fully insulated wall can be used or the procedure outlined in the ASHRAE Handbook can be followed if greater accuracy is required. It should be noted that addressing moisture issues in crawlspaces (and basements) is a first priority. The exterior water should be directed away from foundations, and soils must be completely covered with a vapor retarder or barrier. Recent studies have indicated the practice of not venting crawlspaces is often preferred to venting (DeWitt 2003).

Ductwork

Heat losses and gains through well-sealed ductwork can be significant when it is located in hot or cold locations. Ducts that leak and are located in unconditioned spaces will likely be a primary source of high energy consumption. Attics and vented crawlspaces are popular locations since there is adequate room for installation and main duct runs do not have to penetrate walls. These locations are problematic since they have higher heat transfer potential due to high temperature difference between the supply air (t_S) in the ducts and the attic (t_A) space. Attics are often the hottest location in the cooling season and are near outdoor temperature (t_o) in the heating season. Although crawlspace ducts have lower heat gain in the cooling mode, moisture levels are high and condensation problems are possible if precautions are not taken.

Heat transfer takes place in multiple paths, and an energy balance can best handle the interconnected modes. In the case of an attic, heat is transferred in the form of convection/conduction through the duct to the supply air (q_{Duct}), through the roof to the outdoors (q_{Roof}), and through the ceiling to the conditioned space ($q_{Ceiling}$). Attics are typically ventilated (Q_{AV}) so heat is transferred in this mode and by exfiltration from leaky ductwork (Q_{DL}) and ceilings. In the winter the design conditions typically occur near sunrise when the impact of radiation on roof surface temperature is negligible. Neglecting the infiltration between the room and attic and the impact of radiation, an energy balance on the attic in the winter yields:

$$\sum q_{Attic} = 0 = q_{Ceiling} + q_{Roof} + q_{AtticVent} + q_{Duct} + q_{Ductleak} \tag{7.9}$$

$$0 = \frac{A_C}{R_C}(t_i - t_A) + \frac{A_R}{R_R}(t_o - t_A) + \frac{c_p Q_{AV}}{\upsilon_o}(t_o - t_A) + \frac{A_D}{R_D}(t_s - T_A) + \frac{c_p Q_{DL}}{\upsilon_s}(t_s - t_A) \tag{7.10}$$

When the values of standard air and English units (A [ft²], R [°F· ft²/Btu], Q [ft³/min], c_p [Btu/lb· °F], q [Btu/h], and t [°F]) are used, the temperature of the attic (t_A) is

$$t_A(°\text{F}) = \frac{(A_C/R_C)t_i + (A_R/R_R + 1.08Q_{AV})t_o + (A_D/R_D + 1.08Q_{DL})t_s}{A_C/R_C + A_R/R_R + A_D/R_D + 1.08(Q_{AV} + Q_{DL})} \tag{7.11}$$

For crawlspaces, the foundation wall is exposed to the outdoors and the floor is the barrier to heat flow into the space. Therefore, the foundation wall perimeter

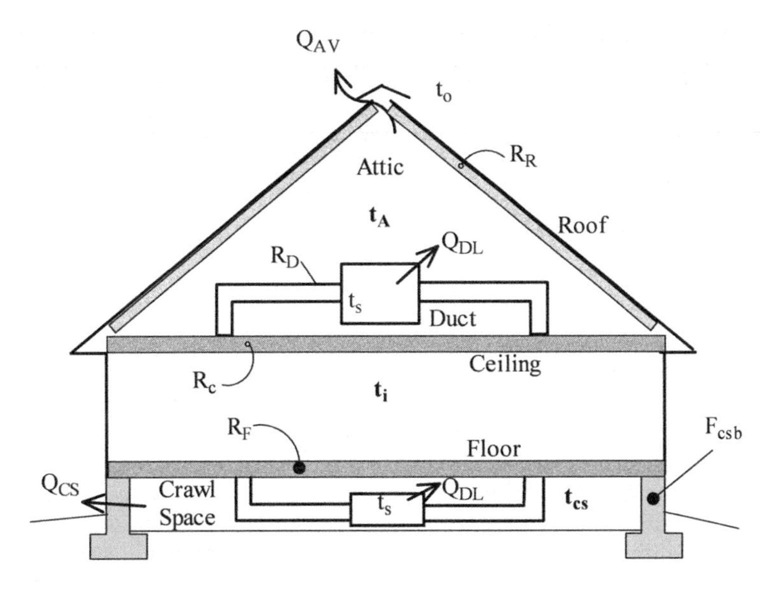

Figure 7.11 Heat loss terms for attic and crawlspace ductwork locations.

(P_{Bldg}) and transmission coefficient (F_{csb}) replace the roof terms, and the floor area (A_F) and thermal resistance (R_F) are substituted for ceiling values. Since the foundation wall transmission coefficient (F_{csb}) is a function of crawlspace height, which varies around the perimeter of the building, an average value should be used. The temperature for a crawlspace (t_{CS}) in which ductwork is present is

$$t_{CS}(°F) = \frac{(A_F/R_F)t_i + (F_{csb} \times P_{Bldg} + 1.08Q_{CSV})t_o + (A_D/R_D + 1.08Q_{DL})t_s}{A_F/R_F + F_{csb} \times P_{Bldg} + A_D/R_D + 1.08(Q_{CSV} + Q_{DL})} . \tag{7.12}$$

In the summer the value for the outdoor temperature (t_o) must be adjusted to account for elevated roof temperatures. This can be estimated by substituting an equivalent roof temperature difference (cooling load temperature difference [see Table 8.5 in this book]) added to the indoor temperature (t_i).

$$t_A(°F) = \frac{(A_C/R_C)t_i + (A_R/R_R)(CLTD_R + t_i) + 1.08Q_{AV}t_o + (A_D/R_D + 1.08Q_{DL})t_s}{A_C/R_C + A_R/R_R + A_D/R_D + 1.08(Q_{AV} + Q_{DL})} \tag{7.13}$$

In all cases, the heat transfer of concern is from the conditioned space through the ceiling or floor (by conduction/convection) and from the duct to the attic (by conduction/convection and by leakage). When ductwork is located in an attic,

$$q_{Ceiling} = \frac{A_C}{R_C}(t_i - t_A) \tag{7.14}$$

and

$$q_{Duct} = q_{DuctCond} + q_{DuctLeak} = \left(\frac{A_D}{R_D}(t_s - t_A) + 1.08Q_{DL}(t_s - t_A)\right) . \tag{7.15}$$

When the ductwork is located in a crawlspace,

$$q_{Floor} = \frac{A_F}{R_F}(t_i - t_{CS}) \tag{7.16}$$

and

$$q_{Duct} = q_{DuctCond} + q_{DuctLeak} = \frac{A_D}{R_D}(t_s - t_{CS}) + 1.08Q_{DL}(t_s - t_{CS}) . \tag{7.17}$$

Leakage can be a major component of heating and cooling loads, especially with unsealed ductwork that is located outdoors or in unconditioned attics or crawlspaces. Additionally, leakage into negative pressure return air ducts can contribute to added latent load and need for cleaning to prevent mold growth. A convenient duct leakage classification (C_L) has been developed (ASHRAE 2005, chapter 35) that provides the value of duct leakage per duct surface area (A_D) at a standard inside-to-outside differential duct pressure (Δp_s) of 1.0 in. of water.

$$C_L \equiv Q/A_D \text{ (cfm/100 ft}^2\text{)} @ \Delta p_s = 1.0 \text{ in. of water} \tag{7.18}$$

Duct leakage (Q_{DL}) for other differential pressures (Δp_s) can be estimated.

$$Q_{DL} = \frac{C_L \Delta p_s^{0.65}}{100} \times A_D \tag{7.19}$$

Table 7.7 Duct Leakage Class (C_l) in cfm/100 ft^2 @ Δp_s = 1.0 in. of Water (ASHRAE 2005, chapter 35)

Duct Material and Shape	Sealed	Unsealed		
		Low	Med.	High
Metal round and oval	3	6	30	70
Metal—Rectangular	12	12	48	110
Metal flex duct—Round	8	12	30	54
Non-metal flex duct—Round	12	4	30	54
Fiber duct board—Round	3		N/A	
Fiber duct board—Rectangular	6		N/A	

Table 7.7 lists the leakage classifications for common duct types.

Calculation of duct leakage and duct heat loss and gain presents a design challenge. Heating and cooling loads must be calculated as recommended in the following chapter. This includes an estimate of duct losses/gains. However, duct sizes (a primary input for duct loss/gain calculation) are unknown until the heating and cooling equipment is selected and airflow rates are specified. But heating and cooling equipment cannot be selected until the heat losses/gains are determined. Procedures for using single calculation estimates and iterative methods for more detailed results are discussed in chapter 8. However, Example Problem 7.2 is included to demonstrate the procedure once the ductwork dimensions have been estimated or specified.

BUILDING INFILTRATION AND MOISTURE TRANSFER

The movement of moisture into buildings impacts the cooling and heating loads. Engineers are increasingly concerned with the computation of moisture (latent) loads since indoor air quality is a strong function of relative humidity and moisture content in building materials. The primary sources of moisture transfer are plumbing leaks, sprinkler system malfunctions, roof and wall leakage, infiltration and ventilation airflow, moisture migration through the envelope, and internal generation. Ventilation rates are discussed in chapter 6 in the section titled, "Ventilation for Acceptable Indoor Air Quality."

Infiltration

The rate of moisture migration from outdoors to the interior space due to infiltration can be determined using Equation 4.20 or 4.21 and the latent heat of vaporization (h_{fg}). Modifying Equation 4.21, which assumes standard air at 70°F and 50% RH,

$$q_L\,(\text{Btu/h}) \approx 4680 \times Q_{Inf}(\text{cfm}) \times (W_o - W_i)\ \text{lb}_\text{w}/\text{lb}_\text{a} \quad (7.20)$$

$$m_w\,(\text{lb/h}) = q_L\,(\text{Btu/h}) \div h_{fg}\,(\text{Btu/lb})\,. \quad (7.21)$$

Usually infiltration rates are estimated from procedures that have a high degree of uncertainty in buildings due to the variation in building quality and designs. This uncertainty for large buildings is summarized in the *Humidity Control Design Guide* (Harriman et al. 2001):

> There is no historical evidence to suggest that the infiltration rate can be determined accurately without actual field measurements. A study of eight buildings...measured leakage rates between 0.15 and 0.55 air changes per hour (Persily and Grot 1986; Persily 1999). The best available prediction calculations... suggested air leakage should have been between 0.01 and 0.25 air changes per hour .

Infiltration is an important factor in the determination of heating and cooling load, and improved methods for estimating values in commercial structures are highly desirable. The above summary statement would indicate these methods are not accurate. Furthermore, recent studies have also shown that measured ventilation rates far exceed design values in a study of 100 office buildings (Persily et al. 2005). Designers can deal with this uncertainty by post-installation measurement (in the commissioning/test and balance process) of building ventilations rates and possibly CO_2 concentrations after occupancy. This will provide some means of minimizing excessive ventilation and infiltration.

Methods are available to predict infiltration in residential structures in terms of air changes per hour (ACH).

$$\text{ACH} = \frac{Q\,(\text{cfm}) \times 60\ \text{min/h}}{Area_{Floor}\,(\text{ft}^2) \times Height_{Ceiling}\,(\text{ft})} \quad (7.22)$$

Tables 7.8 and 7.9 provide ACH for tight, medium, and loose homes for summer and winter. It should be noted that these designations may not be appropriate for colder climates in which infiltration precautions may be more rigourous.

A more detailed procedure that can account for the impact of individual leak sites is also available (ASHRAE 2001, chapter 26) for residential structures. An equivalent building leakage area is computed by summarizing the contribution of each component. The infiltration rate is computed from this area as a function of local wind speed, temperature difference, building height, and wind shielding from nearby buildings and vegetation. The infiltration computation method in the most recent version of the *ASHRAE Handbook—Fundamentals* (ASHRAE 2005, chapter 27) requires a blower door test to compute building leakage area.

Example Problem 7.2

Ductwork is to be located in an attic of a 3000 ft^2 building with a metal roof with an 8/12 pitch and acoustical ceiling tile. The ductwork consists of a 50 ft main rectangular metal duct (16 × 24 in.) and 200 linear ft of 12 in. round metal duct. The duct is not sealed but is wrapped with insulation to provide $R = 4$ (h· ft· °F/Btu). Outdoor temperature is 10°F, the indoor temperature is 70°F, and the indoor fan of a 140 MBtu/h output furnace delivers 3200 cfm. The attic is ventilated naturally at 0.1 cfm/ft^2. The ceiling R-value is 24 h· ft· °F/Btu and the roof R-value is 3 h· ft· °F/Btu. Compute the duct losses.

Repeat the calculation if the ducts are sealed, wrapped with R+ 6 insulation, and the insulation is placed in the roof ($R = 22$ h· ft· °F/Btu) rather than the ceiling ($R = 5$ h· ft· °F/Btu).

Solution:

Areas: $A_C = 3000$ ft^2

For a roof pitch of 8 ft rise/12 ft run: $A_R = A_C\dfrac{\sqrt{\text{rise}^2 + \text{run}^2}}{\text{run}} = 3000\dfrac{\sqrt{8^2 + 12^2}}{12} = 3600\ \text{ft}^2$

$A_D = A_{D\text{-}Main} + A_{D\text{-}Round} = 2(16 \text{ in.} + 24 \text{ in.})/12 \text{ in./ft} \times 50 \text{ ft} + \pi(12 \text{ in.} / 12 \text{ in./ft}) \times 200 \text{ ft} = 333 + 628 = 961 \text{ ft}^2$

Airflows: $Q_{AV} = 0.1$ cfm/ft^2 × 3000 ft^2 = 300 cfm

Assuming medium leakage from unsealed ducts (Table 7.7):

$Q_{DL\text{-}Main} = 48 \text{ cfm}/100 \text{ ft}^2 \times (0.5 \text{ in. of water})^{0.65} \times 333 \text{ ft}^2 = 102 \text{ cfm}$

$Q_{DL\text{-}Round} = 30 \text{ cfm}/100 \text{ ft}^2 \times (0.5 \text{ in. of water})^{0.65} \times 628 \text{ ft}^2 = 120 \text{ cfm}$

$Q_{DL} = 222$ cfm

Supply Air Temperature (From Equation 4.14):

$$t_s = 70°\text{F} + (140{,}000 \text{ Btu/h} \div [1.08 \times 3200 \text{ cfm}]) = 110.5°\text{F}$$

Applying Equation 7.11:

$$t_A(°\text{F}) = \frac{(3000/24) \times 70 + (3600/3 + 1.08 \times 300) \times 10 + (961/4 + 1.08 \times 222) \times 110.5}{3000/24 + 3600/3 + 961/4 + 1.08(300 + 222)} = 36.2°\text{F}$$

Duct loss is determined from Equation 7.15:

$$q_{Duct} = q_{DuctCond} + q_{DuctLeak} = \frac{961}{4}(110.5 - 36.2) + 1.08 \times 222(110.5 - 36.2)$$

$$= 17{,}900 + 17{,}800 = 35{,}700 \text{ Btu/h}$$

For the alternate condition of sealed ducts, R6 duct insulation, and roof insulation:

Leakage from sealed ducts (Table 7.7):

$Q_{DL\text{-}Main} = 12 \text{ cfm}/100 \text{ ft}^2 \times (0.5 \text{ in. of water})^{0.65} \times 333 \text{ ft}^2 = 26 \text{ cfm}$

$Q_{DL\text{-}Round} = 3 \text{ cfm}/100 \text{ ft}^2 \times (0.5 \text{ in. of water})^{0.65} \times 628 \text{ ft}^2 = 12 \text{ cfm}$

$Q_{DL} = 36$ cfm

$$t_A(°\text{F}) = \frac{(3000/5) \times 70 + (3600/22 + 1.08 \times 300) \times 10 + (961/6 + 1.08 \times 36) \times 110.5}{3000/5 + 3600/22 + 961/6 + 1.08(300 + 36)} = 93.5°\text{F}$$

$$q_{Duct} = q_{DuctCond} + q_{DuctLeak} = \frac{961}{6}(110.5 - 93.5) + 1.08 \times 36(110 - 93.5)$$

$$= 2720 + 660 = 3380 \text{ Btu/h}$$

This example illustrates the importance of proper duct location and care in duct sealing. The heat loss (35,700 Btu/h) from the poorly sealed ductwork located in a vented attic was approximately 25% of the furnace capacity (140,000 Btu/h). When the duct was placed inside the insulation envelope, sealed, and better insulated, this loss declined to 2.4% of the furnace capacity even though the attic space was ventilated.

Chapter 9 contains procedures for sizing and routing the ductwork that are to be used in conjunction with the heat loss and gain calculations discussed in this section. The procedure for optimum design is iterative since load is dependent on duct size and location and duct size is dependent on load. However, spreadsheet design tools are available for both load calculation and duct design that can reduce computation time.

Table 7.8 Winter Air Exchange Rates (ACH) for US Residential Buildings (ASHRAE 2001, chapter 28)

	Outdoor Temperature (°F)									
Class	50	40	30	20	10	0	–10	–20	–30	–40
Tight	0.41	0.43	0.45	0.47	0.49	0.51	0.53	0.55	0.57	0.59
Med	0.69	0.73	0.77	0.81	0.85	0.89	0.93	0.97	1.00	1.05
Loose	1.11	1.15	1.2	1.23	1.27	1.30	1.35	1.40	1.43	1.47

Based on 15 mph wind and 68°F indoor temperature.
Tight – Weather stripped, close fitting doors/windows, vapor retarder, well caulked, no fireplace, one story, < 1500 ft^2.
Medium – New two-story or older one-story homes, 1500 ft^2+, average fit windows and doors, fireplace with damper.
Loose – Old poorly caulked/sealed structures, window and door cracks, no damper fireplaces, ventilated appliances.

Table 7.9 Summer Air Exchange Rates (ACH) for US Residential Buildings (ASHRAE 2001, chapter 28)

	Outdoor Temperature (°F)					
Class	85	90	95	100	105	110
Tight	0.33	0.34	0.35	0.36	0.37	0.38
Med	0.46	0.48	0.50	0.52	0.54	0.56
Loose	0.68	0.70	0.72	0.74	0.76	0.78

Based on 7.5 mph wind and 75°F indoor temperature.

Moisture Control and Migration

Another potential large source of internal moisture is by water leakage in poorly assembled structures and by moisture migration through the envelope materials. Moisture problems in buildings are frequently caused by leaking roofs, poorly sealed penetrations, or window and wall flashings that do not direct water away from the building. Failure to clean gutters and downspouts or provide drainage away from exterior walls may result in entry of water through walls and foundations, which are fertile sites for mold growth.

Another common moisture problem is overflows that result from a failure to properly slope and clean cooling coil drain pans. ASHRAE Standard 62.1 specifically addresses drain pan design, carryover from cooling coils, access for cleaning and inspection, and proper application of spray devices (ASHRAE 2004). Drain pans must be sloped at least 1/8 in. per ft to the outlet. Friction loss across individual coils cannot exceed 0.75 in. of water for a 500 fpm face velocity. The drain outlet must be located at the lowest point of the pan, be of sufficient size to ensure no overflow, and must contain a P-trap to maintain a seal of water to prevent reverse airflow from the exterior. Some building codes require a second drain outlet at a slightly higher elevation. This outlet is routed to a conspicuous location (exterior entryway) to alert occupants that the primary outlet is plugged and the system needs service.

Moisture migration through the structure is unavoidable but can be minimized to prevent unwanted buildup that will result in mold growth and wood rot. Wood frame and masonry construction offer a buffer to unwanted moisture buildup because of the hygric buffer capacity. A wood frame-and-sheathing building structure needs to exceed 16% moisture content by weight to support mold growth. In most climates, the equilibrium content is 5% to 6%. Therefore, a moisture storage capacity equal to 10% by weight of the wood is available. An average (2000 ft^2) wood frame-and-sheathing structure contains 4,000 to 5,000 lb of wood, so 400 to 500 lb of moisture storage capacity is available (Lstiburek 2002). This amount of storage capacity can accommodate normal migration rates without exceeding the 16% moisture limit above which mold is likely to grow.

This buffer capacity is significantly enhanced with masonry exterior walls and interior masonry cladding, which yields a moisture storage capacity of ten times the value for the wood frame-sheathing home. However, the buffer capacity of structures that have steel frame and gypsum sheathing construction is lower. Steel has no water storage capacity and gypsum can store only one percent by weight. An average-sized US home can therefore only store 40 to 50 lb of moisture (Lstiburek 2002).

Energy flows (heat, air) through the building speed the drying time of the envelope. A well-insulated, tight building with moisture retarders or barriers will minimize these flows and therefore extend drying time. Wood frame-sheathing and masonry buildings usually provide adequate moisture buffer capacity in these types of buildings. However, the combination of low buffer capacity with slow drying times can be highly problematic for well-insulated and tight steel frame-gypsum construction (Lstiburek 2002).

The control of moisture movement is critical. However, in some applications the movement of moisture is also advantageous. In cold climates in the winter, the exterior air is dryer with a lower vapor pressure (humidity ratio lower). Moisture in the walls, floors, and ceilings must be allowed to flow outward. Thus, the use of vapor impermeable materials (vapor barriers < 1 perm) should not be used near exterior surfaces. The recommended practice is to determine the placement of vapor barriers using dew-point temperature calculations so that building materials will be above the dew point so that condensation does not occur.

The location of the vapor barrier in hot climates is reversed since a cooled interior space has a lower humidity ratio and vapor pressure than the warm exterior. The moisture flows to the drier interior that is created by the removal of the moisture by the cooling unit. It is especially critical in humid climates to select a cooling unit with a high latent heat capacity to create interior conditions that are conducive to moisture removal from the space and structure. The interior dew point should be maintained below 55°F in the summer, but overcooling may result in unwanted condensation in and around air delivery systems.

In mixed climates, location of a vapor barrier in the wall is problematic since both modes of moisture migration to the exterior (winter) and interior (summer) are required. Permeable or semi-permeable material should be used; care should be taken to ensure drainage can occur if moisture does condense near the interior or exterior surfaces. Air barriers should be used on both the interior and exterior, and the interior dew point should be maintained between 40°F in the winter and 55°F in the summer (Lstiburek 2002). However, if these barriers are also vapor barriers, moisture could be trapped inside the assembly.

The rate of water vapor migration through building materials is a function of the vapor pressure difference and the "permeance" of the material. The unit of permeance is the "perm," which has the units of grains of moisture flow per inch of mercury vapor pressure differential per hour per ft^2 of cross section (gr/in. hg· h· ft^2).

$$m_w \text{ (gr/h)} = A \text{ (ft}^2) \times \text{Perm} \times (p_o - p_i) \text{ in. hg} \qquad (7.23)$$

Converting Equation 7.23 to the units of lb and psi yields

$$m_w \text{ (lb/h)} = \frac{A \text{ (ft}^2) \times \text{Perm} \times (p_o - p_i) \text{ psi}}{14.260}. \qquad (7.23a)$$

Table 7.10 provides values of perms for a variety of building materials. The ability of a material to transmit moisture may also be expressed in permeability, which is a permeance per unit thickness. In the English system, the units are perms/in. However, this term is also used in the hydro-geological community with the units of ft/min (ft^3/min· ft^2· area· ft of water head/ft of thickness)

Internal moisture generation from occupants and appliances constitute another source. Moisture rates and loads for occupants, their activities, and appliances are provided in chapter 8 in conjunction with the discussion of cooling load calculation.

CHAPTER 7 PROBLEMS

7.1 Find the overall U-factor and R-value for a 2 × 4 in. wall with R-13 fiberglass batts (approximately 20% of the wall is framing). The exterior wall is 5/8 in. hardboard (standard tempered) over 1/2 in. vegetable fiberboard with no air gap. The interior finish is 1/2 in. gypsum board.

7.2 Repeat Problem 7.1 if the exterior finish is 110 lb/ft^3 4 in. face brick.

7.3 Find the overall U-factors (Btu/h· °F· ft^2) and R-values (h· °F· ft^2/Btu) of a 2.25 in. thick solid wood door with and without a metal storm door.

Table 7.10 Permeance Values of Building Products (ASHRAE 2005, chapter 25)

Material	Thickness		Permeance	
	in.	mm	gr/h· ft^2· in. hg	ng/s· m^2· Pa
Concrete	8	203	0.40	23
Hollow block	8	203	2.4	137
Gypsum board	1/2	13	38	2150
Ext. plywood	1/2	13	0.35	20
Int. plywood	1/2	13	0.95	55
Tempered hardboard	1/2	13	1.25	72
6 mil polyethylene	0.006	0.15	0.06	3.4
Aluminum foil	0.001	0.03	0	0
Aluminum foil	0.0004	0.01	0.05	2.9
Exterior trim paint			5.5	313
Oil base + primer			1.6–3.0	91–172
Semi-gloss enamel			6.6	380
15lb/100ft^2 felt paper			5.6	320
Asphalt kraft paper			1.8	103
Kraft paper on insulation			0.6–4.2	34–240

7.4 Find the overall U-factor (Btu/h· °F· ft^2) and R-value (h· °F· ft^2/Btu) of the wall in the building shown below.

7.5 Find the overall U-factor (Btu/h· °F· ft^2) and R-value (h· °F· ft^2/Btu) of the roof/ceiling in the building shown below if the insulation is polyisocyanuarate.

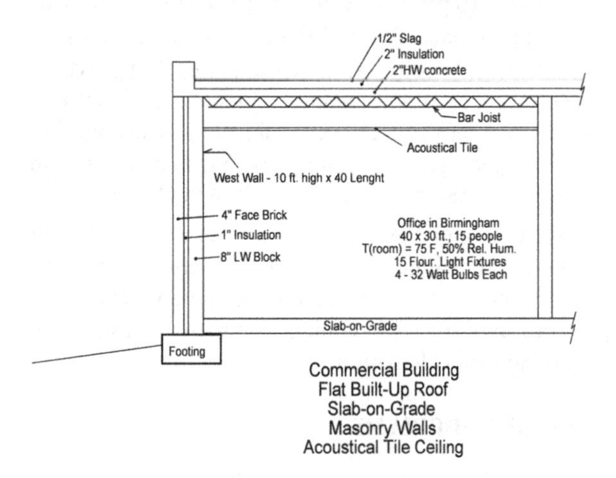

7.6 A roof-ceiling system consists of a metal roof, 6 in. fiberglass insulation, 24 in. attic space, and ½ in. acoustical ceiling tile. Find the R-value and the most appropriate roof number.

7.7 Find the overall U-factor and R-value for a wall with 2 × 4 in. fir studs on 16 in. O.C. with R13 fiberglass batt insulation (approximately 15% of the wall is framing). The wall exterior is 4 in. face brick (110 lb/ft^3) and 1/2 in. extruded polystyrene with a 1/2 in. air gap (no reflective foil). The interior finish is 1/2 in. gypsum board.

7.8 Repeat problem 7.7 for wall with 2 × 6 in. studs 24 in. O.C. with R19 batts.

7.9 Find the overall U-factors (Btu/h· °F· ft^2) and R-values (h· °F· ft^2/Btu) of 1.75 in. thick steel door with a urethane core (no thermal break) with and without a metal storm door.

7.10 Find the overall U-factor (Btu/h· °F· ft^2) and R-value (h· °F· ft^2/Btu) of a vinyl frame, double-glass window with a ½ in. air gap and no thermal break.

7.11 Find the R-value of a wall that is 4 in. face brick, 2 in. insulation, and 8 in. heavyweight concrete walls. What wall number most closely matches this wall?

7.12 Find the overall U-factor (Btu/h· °F· ft^2) and R-value (h· °F· ft^2/Btu) of an aluminum frame, double-glass window with a ½ in. air gap and no thermal break.

7.13 Find the overall U-factors (Btu/h· °F· ft^2) and R-values (h· °F· ft^2/Btu) of a 1.75 in. thick wood door with and without a metal storm door.

7.14 Find the overall U-factor (Btu/h· °F· ft^2) and R-value (h· °F· ft^2/Btu) of a vinyl frame, double-glass window with a ½ in. argon gap.

7.15 Find the CLF and SCL zone type letters for the top floor of a building made with walls like the room of Problems 7.4 and 7.5.

7.16 Find the shade coefficient for the window of Problem 7.12 with (a) no interior shade, (b) closed, medium-colored blinds, and (c) dark-colored drapes with a closed weave fabric.

7.17 Find the shade coefficient for the window of Problem 7.10 with (a) no interior shade, (b) dark roller shades, and (c) light-colored drapes with an open weave fabric.

7.18 Ductwork is to be located in an attic of a 5000 ft^2 building with a roof of $R = 3$ h· ft^2· °F/Btu and an $R = 20$ h· ft^2· °F/Btu ceiling. The ductwork consists of a 100 ft main rectangular metal duct (20 × 28 in.) and 150 linear ft of 10 in. round metal duct. The duct is not sealed but is wrapped with insulation to provide $R = 6$ (h· ft· °F/Btu). The average main duct ESP is 1.3 in. of water and the round duct is at 0.75 in. of water. The outdoor temperature is –5°F, the indoor temperature is 68°F, and the indoor fan of a 200 MBtu/h output furnace delivers 5200 cfm. The attic is ventilated naturally at 0.1 cfm/ft^2. Compute the duct losses.

7.19 Compute the losses through the roof and the attic ventilation air in Problem 7.18.

7.20 Repeat Problem 7.18 if the duct is located in a 5 ft high crawlspace that has a 1 ft exterior exposure with no insulation. The building is 50 × 100 ft and has an $R = 5$ h· ft^2· °F/Btu floor. Crawlspace ventilation is 0.05 cfm/ft^2.

REFERENCES

ASHRAE. 1991. *ASHRAE Terminology of Heating, Ventilation, Air Conditioning, & Refrigeration*. Atlanta: American Society of Heating, Refrigerating and Air-Conditioning Engineers, Inc.

ASHRAE. 1997. *1997 ASHRAE Handbook—Fundamentals,* chapter 29, Fenestration. Atlanta: American Society of Heating, Refrigerating and Air-Conditioning Engineers, Inc.

ASHRAE. 2001. *2001 Handbook—ASHRAE Fundamentals*, chapter 25, Thermal and water vapor transmission data; chapter 26, Ventilation and infiltration; chapter 28, Residential heating and cooling load calculations. Atlanta: American Society of Heating, Refrigerating and Air-Conditioning Engineers, Inc.

ASHRAE. 2003. *2003 ASHRAE Handbook—HVAC Applications*, chapter 33, Solar energy use; chapter 43, Building envelopes. Atlanta: American Society of Heating, Refrigerating and Air-Conditioning Engineers, Inc.

ASHRAE. 2004. *ANSI/ASHRAE Standard 62.1-2004, Ventilation for Acceptable Indoor Air Quality.* Atlanta: American Society of Heating, Refrigerating and Air-Conditioning Engineers, Inc.

ASHRAE. 2004. *ANSI/ASHRAE Standard 90.1-2004, Energy Standard for Buildings Except Low-Rise Residential Buildings*. Atlanta: American Society of Heating, Refrigerating and Air-Conditioning Engineers, Inc.

ASHRAE. 2005. *2005 ASHRAE Handbook—Fundamentals*, chapter 25, Thermal and water vapor transmission data; chapter 27, Ventilation and infiltration; chapter 29, Residential heating and cooling load calculations; chapter 31, Fenestration; chapter 35, Duct design. Atlanta: American Society of Heating, Refrigerating and Air-Conditioning Engineers, Inc.

DeWitt, C. 2003. Crawlspace myths. *ASHRAE Journal*. 45(11).

Harriman, L., G.W. Brundrett, and R. Kittler. 2001. *Humidity Control Design Guide*. Atlanta: American Society of Heating, Refrigerating and Air-Conditioning Engineers, Inc.

ICC. 2003. *International Building Code*. Falls Church, VA: International Code Council, Inc.

Lstiburek, J. 2002. Moisture control for buildings. *ASHRAE Journal* 44(2).

NFRC. 2004. National Fenestration Rating Council, Silver Springs, MD. www.nfrc.org.

Perisly, A.K., and R.A. Grot. 1986. Measured air leakage rate in eight large office buildings. STP 904, American Society of Mechanical Engineers, Philadelphia, PA.

Perisly, A.K. 1999. Myths about building envelopes. *ASHRAE Journal* 41(3).

Perisly, A.K., J. Gorfain, and G. Brunner. 2005. Ventilation design and performance of U.S. office buildings. *ASHRAE Journal* 47(4).

8 Cooling Load and Heating Loss Calculations and Analysis

FACTORS AFFECTING BUILDING COOLING AND HEATING LOADS

The calculation of the cooling and heating loads on a building or zone is the most important step in determining the size and type of cooling and heating equipment required to maintain comfortable indoor air conditions. Building heat and moisture transfer mechanisms are complex and as unpredictable as the weather and human behavior, both of which strongly influence load calculation results. The variation in outdoor temperature is accompanied by a variation in internal heat generation from occupants, equipment, and lights. Buildings made of heavier components will more effectively delay the rate at which heat is converted to building cooling load. The designer must not only compute heat transfer but also account for thermal mass effects.

A summary of the primary factors that influence loads are:

- Conduction and convection of heat through walls, roofs, floors, doors, and windows
- Radiation through windows and heating effects on wall and roof surface temperatures
- Thermal properties of materials (insulation, glass transmittance, surface absorbtivity)
- Building thermal mass and corresponding delay of indoor temperature change
- Construction quality in preventing air, heat, and moisture leakage (infiltration)
- Heat and moisture added and/or lost, with ventilation air needed to maintain air quality
- Heat generated by lights, people, appliances, and equipment
- Heat added and/or lost by air, water, and refrigeration distribution systems
- Moisture movement through the building envelope
- Moisture generated by occupants and equipment
- Activity level, occupancy patterns, and makeup (male, female, child) of people
- Acceptable comfort and air quality requirements of occupants
- Weather conditions (temperature, moisture, wind speed, latitude, elevation, solar radiation, etc.)

These many factors combine to force engineers to develop procedures that minimize the load calculation complexity without compromising accuracy. A combination of measured data and detailed simulations has generated techniques that can be done either with a pocket calculator and a one-page form or with more complex numerical simulations that take hours to complete using modern computers. However, many assumptions and simplifications must be made for all methods.

The many approaches to calculating cooling and heating loads are attempts to consider the fundamental principle that heat flow rates are not instantaneously converted to cooling or heat loads primarily because of the thermal and moisture capacitance of the buildings. Thus, building heat inflows or outflows do not immediately result in a change in temperature. Thermally heavy buildings can effectively delay the cooling or heating load for several hours. The space temperature in an unconditioned, thermally light building will rise more quickly on a hot, clear day but will also decline more rapidly as the outdoor temperature falls after sunset. The heat balance (HB) method accounts for these transient effects but also considers the interaction between all surfaces and loads. A previous simplification of the HB method is the transfer function method (TFM) as presented in the *1997 ASHRAE Handbook—Fundamentals* (ASHRAE 1997). This procedure was used to develop an even simpler method by generating cooling load temperature difference (CLTD) and cooling load factor (CLF) data for a one-step load calculation method. The accuracy of the method was improved with the recent development of solar cooling load (SCL) factors for computing the impact of solar loads through glass. CLTDs, CLFs, and SCLs all include the effect of time lag in conductive heat gain through opaque surfaces and time delay by thermal storage in converting radiant gain to cooling load. Although data are not available for all building types, the current information provides a wide array of building types so that accuracy is consistent with the TFM and the presentation is suited to instruction (ASHRAE 1997, chapter 28).

LOAD CALCULATION METHOD OPTIONS

Chapter 30 of the *2005 ASHRAE Handbook—Fundamentals*, "Nonresidential Heating and Cooling Load Calculations," promotes a revised version of the heat balance (HB) method for nonresidential buildings (ASHRAE 2005). The cooling load temperature difference/cooling load factor (CLTD/CLF) method is no longer discussed in either the residential (ASHRAE 2005, chapter 29) or nonresidential load calculation procedures. The heat balance method has superseded (not invalidated) other methods, including CLTD/CLF. As the name implies, this method employs a series of heat balance calculations on the outside surfaces, the surface conduction, the internal surfaces, and the indoor air. Numerical methods are required and modern computers are essential. A separate publication is available from ASHRAE that includes a basic software package that will calculate loads for a single-zone building (Pedersen et al. 1998).

Unfortunately, a detailed presentation of this method is well beyond the scope and resources of a traditional single HVAC course or seminar. This would leave little time for other important HVAC concepts. Likewise, the cost and time to learn a computerized commercial version of the program is beyond the resources of most education institutions. A simplified version of the HB method is referred to as the radiant time series (RTS) method. However, the *2005 ASHRAE Handbook—Fundamentals* (chapter 30) does not provide sufficient information to use the RTS method in a wide variety of applications at this time, and users must be prepared to generate the required coefficients (24) for each component.

Heat Balance Method and Input Uncertainties

Although the heat balance is a more scientific and rigorous computation, there are limitations that result from the high degree of uncertainty for several input factors. In many cases, the added detail required of this method is not warranted given the variability and uncertainty of several significant heat gain components. Several of these components are listed and discussed here.

The new methods focus on improvements primarily in conduction and solar transmission. However, in modern buildings where most zones have zero, one, or, occasionally, two exposed surfaces, these items are typically less than 20% of the total cooling load. Often the heat gain from internal sources exceeds the heat gain of the building envelope. The 1997 and 2001 versions of the *ASHRAE Handbook—Fundamentals* contained improved internal gain values for office equipment. However, limited data are available for many types of equipment, and the suggested value is 25% to 50% of the nameplate power consumption, which could result in appreciable error. The improved office load information suggests variations between 0.5 W/ft^2 to 1.5 W/ft^2. The value used is left to the interpretation of the designer. Consider a laser printer. The load is a strong function of the frequency of use, so the engineer must be able to predict this frequency in order to "accurately" predict cooling load. This value will vary significantly from office to office. Thus, the primary source of uncertainty with regard to internal loads is not the choice of method (complex HB vs. simple CLTD/CLF) but the ability to predict the behavior of the occupants and equipment use factors.

Duct losses can be a significant component, but they are not considered as part of the HB correlation. The *2001 ASHRAE Handbook—Fundamentals* suggests supply duct losses should be "estimated as a percentage of space sensible cooling load (usually about 1%)." It suggests return duct losses are usually insignificant. The engineer is referred to another chapter (chapter 34 in the 2001 Handbook) to calculate heat gain/loss in the supply duct. An example problem is provided, but little information is given as to how to determine the assumptions made in the problem. Furthermore, the duct loss is computed to be 39,400 Btu/h for a flow of 17,200 cfm with a 122°F supply temperature. Assuming a 70°F room temperature, the total heating capacity would be 984,000 Btu/h. Duct loss is, therefore, 4% of the total, or four times the amount suggested for the HB method.

The Air-Conditioning Contractors of America *Commercial Load Calculation, Manual N* (ACCA 1988) suggests much higher supply and return duct gains and losses compared to the HB procedure. Table 16 indicates that supply gains are 10% of the sensible total, and losses are 10% to 15% with insulated ducts (1 in. thickness) placed in unconditioned spaces. Return gains and losses would be approximately half the supply duct values. Percentages vary with the amount of insulation and the location of the duct. Additionally, 10% should be added for ducts that are not sealed from leakage. These discrepancies suggest the HB method will significantly underpredict losses unless ducts are located inside the conditioned space. There are no references provided for the assertion that duct losses are "about 1%."

The HB method adds the sensible contribution of the ventilation and infiltration in the final air heat balance. These contributions are added to the convective transfer from the surfaces and the heat transfer from the

HVAC system and directly to the convective part of the internal loads. It is stated in the HB method discussion that this "violates the tenets of the heat balance approach." Additionally, "the amount of infiltration air is uncertain … however it is determined" (ASHRAE 2001, p. 29.23). Ventilation and infiltration also contribute significantly to the latent loads as discussed in the next section.

The rate at which moisture levels change in the indoor space is a significant factor in cooling load computation. Relative humidity levels must be maintained by the HVAC system to ensure comfort and health. The primary sources are ventilation air and infiltration, occupant perspiration, internal sources, and migration through walls without vapor barriers or retarders. The current heat balance method offers no improvement compared to any previous method. All methods use the same equations to compute both ventilation and infiltration. Ventilation rates are dictated by *ANSI/ASHRAE Standard 62.1, Ventilation for Acceptable Indoor Air Quality* (ASHRAE 2004a). Little direction is provided for computation of infiltration in nonresidential buildings in chapter 30 of the *2005 ASHRAE Handbook—Fundamentals* (ASHRAE 2005). Chapter 27, "Ventilation and Infiltration," provides an overview of the procedure, but research indicates there is a large discrepancy between computed values and measured values (Persily and Grot 1986; Persily 1999). Infiltration in residential buildings can be handled in greater detail by summarizing the component-by-component equivalent building leakage area (Table 1, chapter 26, ASHRAE 2001a) and applying a series of equations. Chapter 28 (ASHRAE 2001a) provides a less rigorous computation of residential infiltration by classifying buildings as tight, medium, or loose and correcting for outdoor air temperature. However, all methods appear to have a high degree of uncertainty.

Occupant latent loads are also significant and uncertainty is also high given the difficulty in predicting both numbers and activity level. Table 1 in chapter 30 of *2005 ASHRAE Handbook—Fundamentals* provides sensible and latent contributions for 14 different activities (ASHRAE 2005). The latent load can vary from a value of 105 Btu/h per person for an inactive occupant to 1090 Btu/h per person for highly active occupants (assuming a normal percentage of men, women, and children). Intuition and experience appear to have an important influence upon accuracy.

Data for latent loads for cooking applications are abundant in the *2005 ASHRAE Handbook—Fundamentals*. However, data for other applications are limited. Additional research appears to be warranted in this area. The Handbook offers limited detail with regard to the computation of latent load due to moisture migration. However, the *Humidity Control Design Guide* (Harriman et al. 2001) offers greater detail, which suggests that this component is only a factor when the building envelope does not include recommended moisture control precautions.

There remains a high degree of uncertainty in the inputs required to determine cooling and heating loads. Much of this is due to the unpredictability of occupancy, human behavior, complex and varied building practices, ductwork location and leakage rates, lack of and variation in heat gain data for modern equipment, and the introduction of new building products and HVAC equipment with unknown thermal characteristics. These generate uncertainties that far exceed the increased error generated by simple methods, such as CLTD/CLF, compared to the more complex HB method. Therefore, for most applications the current added time and expense required for the HB method is likely to be unwarranted for most buildings until additional research is conducted in these areas to reduce these input uncertainties. It is recognized that the heat balance and radiant time series methods are likely to provide greater accuracy when they have been more fully developed, independently verified, and corrected. A detailed presentation of the HB and RTS method can be found in the most recent version of the *ASHRAE Handbook—Fundamentals* (ASHRAE 2005, chapter 30).

THE CLTD/CLF/SCL METHOD

Many engineers use some form of the cooling load temperature difference/cooling load factor/solar cooling load (CLTD/CLF/SCL) method. This method has the same limitations as the heat balance method. However, it is well suited to spreadsheet computations, and individual components are readily identified so that load analysis is simple to designers who wish to identify and quantify the impact of the most critical components.

The combined effects of convection, conduction, radiation, and thermal lag for opaque surfaces are combined into a modification of the conduction equation:

$$q = UA \times \mathrm{CLTD}_{Cor} \tag{8.1}$$

An array of CLTD tables is used to account for thermal mass, insulation levels, latitude, time of day, direction, temperature swing, and other variables. The values in the tables must be corrected since they are based on an indoor temperature of 78°F and a mean outdoor temperature of 85°F. CLF factors are used to account for the fact that building thermal mass creates a time lag between heat generation from internal sources (lighting, people, appliances, etc.) and the resulting cooling load. CLF factors are presented in a set of tables that account for the number of hours the heat source has been on, thermal mass, type of floor covering and window shading, number of walls, and the presence of ventilation hoods. A CLF represents the fraction of the heat gain that is converted to cooling load at a particular time.

$$q = q_{IntLoad} \times \text{CLF} \quad (8.2)$$

Solar gains through glass are computed in a similar manner with solar cooling load (SCL) factors, with the units of heat rate per unit area tabularized by facing direction (N, E, S, W, horizontal) and latitude. The fraction of solar gain that is transmitted is accounted for with a shading coefficient (SC) to correct for glass transmittance and shading devices.

$$q = A \times \text{SC} \times \text{SCL} \quad (8.3)$$

All of these factors are summed and added to some estimate of the latent (dehumidification) load to arrive at the cooling load. Latent load estimating is a critical shortcoming of all load calculation methods and will be treated in greater detail in this chapter as each of the three major sources are discussed (infiltration/ventilation, occupants, and internal latent sources).

Recent publications have devoted less attention to the heating loads in larger buildings, since they are often small even in colder climates due to the internal heat generation of equipment. The most recent version of the *ASHRAE Handbook—Fundamentals* (ASHRAE 2005) contains a half-page discussion of heating load. More detailed discussions are provided for residential buildings in an earlier version of the Handbook (ASHRAE 2001a, chapter 28) and the *Residential Load Calculation, Manual J* (ACCA 1986). However, increased attention to heating load calculations is warranted due to the growing awareness of the need for adequate ventilation air at all times to maintain indoor air quality (IAQ). The heat losses due to the recommended ventilation rates in high-occupancy buildings often exceed the heat losses from all other components combined.

A Modified CLTD/CLF Method for Cooling Load Calculation

The intention here is to discuss the CLTD/CLF procedure and related items in a single text that is integrated with a computer-based spreadsheet with psychometric functions (*TideLoad.xls* on the CD accompanying this book). The most recent versions of the *ASHRAE Handbook—Fundamentals* (1997, 2001, 2005) will serve as references and provide more detailed discussion of many facets of load calculation. Greater emphasis will be placed on latent load components in order to address the increasing awareness of the contribution of relative humidity on indoor air quality. Additionally, the newer information on duct leakage and gains/losses will be incorporated into an optional computation.

Table 8.1 is a reproduction of a portion of the main sheet for the CLTD/CLF/SCL cooling and heating program *TideLoad.xls* on the CD accompanying this book. The explanation of the CLTD/CLF/SCL cooling and heating method will follow the order of this table, which is for a single zone. Multiple zones are accommodated by reproducing columns F through I for rows 7 through 64. The program has additional worksheets that include CLTD, CLF, SC, and SCL tables; drawings of walls and roof types; heat gains of lighting, appliances, and people; ACCA *Manual N* values for duct losses; ASHRAE Standard 62.1-2004 ventilation air requirements; a table of thermal properties of building materials with calculators to determine overall R-values and U-factors; and weather data.

Location and Climatic Data

Information regarding the outdoor design conditions and desired indoor conditions are the starting point for the load calculation program. Row 1 of Table 8.1 requires the location, elevation, and latitude of the building site. The elevation is required to develop psychrometric relationships and the latitude is required to determine CLTDs and SCLs. This information can be found in Table 8.2, columns C and B, respectively. Table 8.2 is a condensed version of Table 6.5 and is reproduced in this section to aid the reader. Table 6.5 is a condensed version of Tables 1A and 1B of the "Climatic Design Information" chapter of the *2001 ASHRAE Handbook—Fundamentals* (ASHRAE 2001a, chapter 27). The second row in Table 8.2 references the columns of Tables 1A and 1B of the Handbook. These tables provide a reference for a large number of locations in the US and many worldwide sites. These data are expanded in the *2005 ASHRAE Handbook—Fundamentals* with the introduction of an electronic format.

The outdoor air design temperature is entered in cell C3 and values can be located in column D of Tables 6.5 or 8.2. The 99.6% subheading represents the temperature whose annual frequency of occurrence is exceeded 99.6% of the time (or the value that the outdoor temperature is equal to or lower than 0.4% of the time). The Handbook also contains the values for the 99.0% temperature.

The cooling design conditions are entered in cells C4 for the outdoor air dry-bulb (DB) temperature and C5 for the mean coincident wet-bulb (MWB) temperature. These values can be found in column E in Tables 6.5 or 8.2 under the heading "Cooling DB/MWB." Both values are needed to determine the sensible and latent (dehumidification) loads in the cooling mode. The 0.4% subheading represents the temperature that is exceeded 0.4% of the time on an annual basis. The values for 1% and 2% can be found in the Handbook. Tables 6.5 and 8.2 list only the 0.4% values, which are 95°F and 77°F for Tuscaloosa, Alabama, for 0.4%.

Table 8.1 CLTD/CLF/SCL Spreadsheet for a Single Zone

	A	B	C	D	E	F	G	H	I
1	City =		State =		Elev. (ft)=		Lat. °N=		
2	Temps/ Hum.	Design Points	Outdoor		Indoor				
3	Heating	tdb (99.6%) =		tdb =					
4	Cooling (DB)	tdb (0.4%) =		tdb =		°F DR =		°F T (h)=	
5	Cooling (WB)	twb (0.4%) =		twb =		°FΔT (c) =		°F T (m) =	
6	Hum.Rat. (lb)	W (Outdoor) =		W (Indoor)=		ΔW		CLTD (Cor)=	
7	Rel. Hum (%)	RH (Outdoor) =		RH (Indoor)=			Zone 1		
8		*The spreadsheet version of this table converts column G, H, and I values to MBtu/h				Area (ft^2)	qc (a.m.)	qc (p.m.)	qh
9	Solar	Shade Coeff.	SCL (a.m.)	SCL (p.m.)					
10	Windows (N)								
11	Windows (E)								
12	Windows (S)								
13	Windows (W)								
14	Other								
15	Conduction	U (Btu/h· ft^2· °F)	CLTD (a.m.)	CLTD (p.m.)	ΔT	Area (ft^2)			
16	Windows (N)								
17	Windows (E)								
18	Windows (S)								
19	Windows (W)								
20	Other								
21	Conduction	U (Btu/h· ft^2· °F)	CLTD (a.m.)	CLTD (p.m.)	ΔT	Area (ft^2)			
22	Walls (N)								
23	Walls (E)								
24	Walls (S)								
25	Walls (W)								
26	Other								
27	Conduction	U (Btu/h· ft^2· °F)	CLTD (a.m.)	CLTD (p.m.)	ΔT	Area (ft^2)			
28	Doors (N)								
29	Doors (E)								
30	Doors (S)								
31	Doors (W)								
32	Roof/Ceiling	U (Btu/h· ft^2· °F)	CLTD (a.m.)	CLTD (p.m.)	ΔT	Area (ft^2)			
33	Type A								
34	Type B								
35	Floor	U (Btu/h·ft^2·F)			ΔT (flr)	Area (ft^2)			
37	Slab	UP (Btu/h·ft·F)			ΔT (slab)	Per (ft)			
38									
39	Ventilation		ΔT (a.m.)	ΔT (p.m.)	ΔT	cfm			
40	Sensible	1.1							
41	HRU Sen Eff.		ΔW	ΔW					
42	Latent	4840							
43	HRU Lat Eff.					People			
44	People	Btu/person	CLF (a.m.)	CLF (p.m.)					
45	Sensible								
46	Latent		1	1					
47	Internal		CLF (a.m.)	CLF (p.m.)		watts			
48	Sensible	3.412							
49	Latent		1	1	Btu/h→				
50			CLF (a.m.)	CLF (p.m.)	F (ballast)	watts			
51	Lighting	3.412							
52	Net Sen.								
53		U(duct)	ΔT (a.m.)	ΔT (p.m.)	ΔT (htg)	Area (ft^2)			
54	Duct Cond.								
55	Alternate or								
56	Duct Cond.	DuctGain Fac.=		DuctLossFac					
57	Duct Leaks		ΔT (a.m.)	ΔT (p.m.)	ΔT-duct	cfm			
58	Sensible	1.1							
59	Latent	4840							
60		Entire Building Totals					Zone 1		
61	Total Sensible	(MBtu/h)				Sensible			
62	Total Latent	(MBtu/h)				Latent			
63	Total Gain	(MBtu/h)				Tot. Gain			
64	Total Loss	(MBtu/h)		Net Loss		Tot. Loss		Net Loss	

Table 8.2 Climatic Data for Heating and Cooling Calculations

See Table 6.5 for additional locations.

A	B	C	D	E		F		G				H
Location	Lat.	Elev.	Heating	Cooling[1]		Evaporation[2]		Dehumidification[3]				Range
Handbook Column #	1c	1e	2a	2a	2b	3a	3b	4a	4b	4c		5
			(99.6%)	DB	MWB	WB	MDB	DP	HR	DB	WB	DB
City/State	°N	ft	°F	°F	°F	°F	°F	°F	Gr	°F	°F	°F
Boston, MA	42	30	7	91	73	75	87	72	119	80	74	15
Denver, CO	40	5330	–3	93	60	65	81	60	96	69	63	27
Miami, FL	26	13	46	91	77	80	87	78	144	83	79	11
Minneapolis, MN	45	837	–16	91	73	76	88	73	124	83	76	19
Sacramento, CA	39	23	31	100	69	72	96	62	84	82	69	33
St. Louis, MO	39	564	2	95	76	79	90	76	138	85	78	18
Tuscaloosa, AL	33	171	20	95	77	80	90	77	142	84	79	20

[1] These design conditions typically result in the highest sensible cooling load and highest total load when ventilation requirements and infiltration are low.
[2] These design conditions represent the maximum outdoor air enthalpy; therefore, they will result in higher total cooling loads when the ventilation air requirements and infiltration are high in humid and moderate climates.
[3] These design conditions typically result in cooling loads with the lowest sensible heat ratio (SHR_{Load}). The sensible heat ratios (SHR_{Unit}) of the HVAC equipment should be less than or equal to SHR_{Load} to maintain indoor air relative humidity.

In moderate and humid climates, for buildings that have significant ventilation air requirements (> 10% of supply), the maximum load often occurs during periods of maximum wet-bulb (WB) conditions due to the high enthalpy of the outdoor air (which corresponds closely to high wet-bulb temperatures). Column F of Tables 6.5 and 8.2 presents these values and the corresponding coincident dry-bulb (MDB) temperature under the heading "Evaporation WB/MDB." The 0.4% values are 80°F (WB) and 90°F (MDB) for Tuscaloosa. These two values should be used to replace the "Cooling DB/MWB" in cells C4 and C5 of Table 8.1 in order to compute the load at the maximum WB conditions.

Often the maximum dehumidification load will occur during the periods when the outdoor air dew-point temperature (DP) is greatest. Data are presented in the columns under G that contain the maximum DP values with the corresponding humidity ratio (HR), coincident dry-bulb temperature, and the corresponding wet-bulb temperature (WB). Tables 6.5 and 8.2 list only the 0.4% values under the heading "Dehumidification." These conditions will typically result in elevated indoor relative humidity if the HVAC system does not have adequate latent capacity. Thus, sensible and latent loads at these conditions should be analyzed by entering the "dehumidification" DB and WB in cells C4 and C5 in Table 8.1.

Note: *Complete analysis dictates that the designer use both the DB/MWB and WB/MDB conditions to check for maximum total cooling load and repeat the computation for the DP/MDB and HR conditions to determine total load, sensible load, and sensible heat ratio (SHR_{Load}). The HVAC equipment should meet the total load at both DB/MWB and WB/MDB conditions and have a SHR_{Unit} less than or equal to SHR_{Load}.*

The final required input from Table 6.5 or 8.2 is the design-day daily range (DR) of the dry-bulb temperature on the design day. This value and the indoor temperature are used to correct the tabular CLTD values that assume an indoor temperature of 78°F and average temperature of 85°F for a location with a 95°F design temperature and a 20°F daily range DR.

$$\begin{aligned} \text{CLTD}_{Cor} &= \text{CLTD}_{Table} + (78 - t_i) + (t_m - 85) \\ &= \text{CLTD}_{Table} + (78 - t_i) + \left[\left(t_o - \frac{DR}{2}\right) - 85\right] \end{aligned} \tag{8.4}$$

Once the design dry-bulb and wet-bulb temperatures are entered into cells C4 and C5 in Table 8.1, the outdoor humidity ratio and relative humidity are computed using functions (macros) developed from psychrometric correlations given in Figure 4.2. The humidity ratio is needed to calculate latent outdoor air loads, and the relative humidity is provided for information purposes. These simple functions also consider the impact of elevation, which design engineers often neglect when using sea level psychometric charts. Considerable error will result at higher elevations if sea level charts are used.

Indoor Conditions

Figure 6.1 is a modification of a portion of the psychrometric chart that outlines the summer and winter comfort zones (ASHRAE 2005, chapter 8, Figure 5). Recommended relative humidity levels are between 30% and 65%. The winter indoor temperature comfort range is between 68°F and 75°F. Occupants are typically dressed in lighter clothes in the summer and are more conditioned to warm weather, so the recommended range is between 73°F and 80°F for the dry-bulb temperature. The suggested indoor design temperature is 70°F for the winter. The suggested summer temperature is 75°F with a wet-bulb temperature of 63°F. This corresponds to a relative humidity near 50%. Values are entered in cells E3, E4, and E5. The indoor humidity ratio and relative humidity are also computed and appear in cells E6 and E7 of Table 8.1.

Load Calculation Psychrometrics

The need to provide ventilation for indoor air quality (ASHRAE Standard 62.1-2004) and comfort conditions (ASHRAE Standard 55-2004) has resulted in greater attention being devoted to both indoor temperature and humidity levels. Engineers must now calculate both the sensible and latent cooling loads since the ventilation air requirements are being enforced and IAQ litigation is increasing (especially mold-related litigation). The primary sources of latent load are from the ventilation air, occupants, and internal sources (coffee pots, cooking, dishwashing, showers, etc.). The psychometric functions (macros) also compute the humidity ratio and relative humidity for the indoor air. The elevation and dry-bulb and wet-bulb temperature values are used to find the indoor humidity ratio and the relative humidity. The indoor relative humidity is now a critical design parameter with the recent revision of ASHRAE Standard 62.1, which prescribes a maximum value of 65% (ASHRAE 2004a).

Several intermediate output values appear in Table 8.1 to the right of the input cells for indoor and outdoor conditions. These include:

- Indoor-outdoor cooling design temperature difference (Δt_c) in cell G5
- Indoor-outdoor cooling design humidity ratio difference (ΔW) in cell G6
- Indoor-outdoor heating design temperature difference (Δt_h) in cell I4
- Outdoor cooling design-day mean temperature difference ($t_m = t_o - \mathrm{DR}/2$) in cell I5
- CLTD correction for indoor and mean temperatures ($\Delta\mathrm{CLTD} = 78 - t_o + t_m - 85$) in cell I6

CLTDs for Walls, Roof, and Doors

Equation 8.2 ($q = UA \times \mathrm{CLTD}_{Cor}$) is used to calculate the conduction-convection heat gain through walls, roof, doors, and windows. The arrangement of Table 8.1 takes advantage of the fact that the value of U is the same for almost all walls of a building. The value for U is computed using Equations 7.2, 7.3, and 7.4. Results are entered in column B, rows 22 through 25, for north-, east-, south-, and west-facing walls.

Table 8.3 contains a set of uncorrected CLTDs for six different general types of walls. Almost all conventional construction variations can be categorized under these types. Figure 7.4 is consulted to determine the wall type number that most closely matches the actual construction. The insulation values for various types of building material and insulation, which are often needed to determine wall type, are given in Table 7.1. CLTDs for walls that do not fall within these categories can be found in the *1997 ASHRAE Handbook—Fundamentals* (chapter 28). Table 8.3 contains only the CLTDs for hours 10 (10 a.m.) and 15 (3 p.m.). This permits loads to be calculated in the morning and afternoon in order to better determine when the peak load occurs. Values for other hours can be found in the 1997 Handbook. The values are also given in 6° latitude increments, and the user can use interpolated values to match other latitudes. However, it should be noted that this typically only affects the south-facing wall to any extent. Additionally, direct interpolation for walls facing other directions is also possible, or the Handbook can be consulted for CLTDs for walls facing northwest, northeast, southwest, and southeast.

The values in Table 8.3 assume a mean outdoor temperature of 85°F (t_m), an indoor temperature (t_i) of 78°F, a daily range (DR) of 21°F, a dark surface, and a clear sky on July 21. When conditions are different, CLTD values from the table must be corrected before being used. No correction is suggested for outdoor color since most light-colored surfaces decay with time.

The arrangement of Tables 8.1 and 8.3 allows the uncorrected CLTD values for N, E, S, and W at 10 a.m. and 3 p.m. to be copied and pasted into columns C and D, rows 22 through 25. Table 8.3 values are also located in the "CLTDs" worksheet (*TideLoad.xls* on the CD accompanying this book), and Equation 8.4 is automatically applied. The user can also access the "Wall & Roofs" type worksheet (*TideLoad.xls* on the CD accompanying this book) to view Figure 7.4. All heat gain/loss values in *TideLoad* are given in MBtu/h (Btu/h × 1000).

Values for Zone 1 wall areas ($A_{Wall} = h \times w - A_{Windows} - A_{Doors}$) are entered in Column F, rows 22 through 25, in Table 8.1. If the zone does not have walls facing all four directions, the area is left blank ($A_{Wall} = 0$). Cell G22 is the 10 a.m. heat gain for the north wall, which is computed by

$$q_{N\text{-}Wall@10am} = U\,(\text{cell B22}) \times A\,(\text{cell F22}) \times \mathrm{CLTD}_{Cor}\,(\text{cell C22 with Equation 8.4})\,. \quad (8.5)$$

The 3 p.m. heat gain for the wall is

$$q_{N\text{-}Wall@3pm} = U\,(\text{cell B22}) \times A\,(\text{cell F22}) \times \mathrm{CLTD}_{Cor}\,(\text{cell D22 with Equation 8.4})\,. \quad (8.6)$$

The process is repeated for the remaining walls. The actual spreadsheet (*TideLoad.xls* on the CD accompanying this book) has additional columns for multiple zones. Only the last four columns (F, G, H, I) are repeated for additional zones since the values in the other columns are not likely to change if the wall con-

Table 8.3 Uncorrected CLTDs for Walls (°F) for Morning (Hr 10) and Afternoon (Hr 15)

See Figure 7.4 for wall type descriptions.

Wall Type 1										
Lat.	**24°N**	**24°N**	**30°N**	**30°N**	**36°N**	**36°N**	**42°N**	**42°N**	**48°N**	**48°N**
Dir.	**Hr 10**	**Hr 15**	**Hr 10**	**Hr 15**	**Hr 10**	**Hr 15**	**Hr 10**	**Hr 15**	**Hr 10**	**Hr 15**
N	18	30	16	30	14	29	14	29	13	28
E	63	31	64	31	64	31	64	31	64	30
S	12	31	15	39	18	46	23	53	28	59
W	13	59	13	59	13	59	13	59	13	58
Wall Type 2										
Lat.	**24°N**	**24°N**	**30°N**	**30°N**	**36°N**	**36°N**	**42°N**	**42°N**	**48°N**	**48°N**
Dir.	**Hr 10**	**Hr 15**	**Hr 10**	**Hr 15**	**Hr 10**	**Hr 15**	**Hr 10**	**Hr 15**	**Hr 10**	**Hr 15**
N	12	25	11	25	9	24	9	24	9	24
E	42	38	44	38	46	38	47	38	49	38
S	4	27	5	35	6	42	8	49	9	55
W	5	33	5	33	5	33	6	33	6	32
Wall Type 5										
Lat.	**24°N**	**24°N**	**30°N**	**30°N**	**36°N**	**36°N**	**42°N**	**42°N**	**48°N**	**48°N**
Dir.	**Hr 10**	**Hr 15**	**Hr 10**	**Hr 15**	**Hr 10**	**Hr 15**	**Hr 10**	**Hr 15**	**Hr 10**	**Hr 15**
N	7	19	7	18	6	17	6	17	6	17
E	22	37	24	37	25	38	27	39	28	39
S	3	18	4	23	4	27	5	32	6	37
W	4	20	5	20	5	20	6	20	6	20
Wall Type 6										
Lat.	**24°N**	**24°N**	**30°N**	**30°N**	**36°N**	**36°N**	**42°N**	**42°N**	**48°N**	**48°N**
Dir.	**Hr 10**	**Hr 15**	**Hr 10**	**Hr 15**	**Hr 10**	**Hr 15**	**Hr 10**	**Hr 15**	**Hr 10**	**Hr 15**
N	8	17	8	17	7	16	7	16	7	16
E	22	34	24	35	25	35	31	36	27	36
S	4	16	5	21	5	25	7	30	8	34
W	7	20	7	20	7	20	8	20	8	20
Wall Type 10										
Lat.	**24°N**	**24°N**	**30°N**	**30°N**	**36°N**	**36°N**	**42°N**	**42°N**	**48°N**	**48°N**
Dir.	**Hr 10**	**Hr 15**	**Hr 10**	**Hr 15**	**Hr 10**	**Hr 15**	**Hr 10**	**Hr 15**	**Hr 10**	**Hr 15**
N	8	14	7	13	5	12	6	12	6	12
E	15	30	15	33	14	35	15	36	16	36
S	5	12	5	15	4	18	5	21	5	24
W	11	15	9	14	7	13	8	13	8	13
Wall Type 16										
Lat.	**24°N**	**24°N**	**30°N**	**30°N**	**36°N**	**36°N**	**42°N**	**42°N**	**48°N**	**48°N**
Dir.	**Hr 10**	**Hr 15**	**Hr 10**	**Hr 15**	**Hr 10**	**Hr 15**	**Hr 10**	**Hr 15**	**Hr 10**	**Hr 15**
N	8	11	8	11	7	10	8	10	8	10
E	11	26	12	27	12	28	13	29	14	29
S	6	8	7	10	7	12	10	15	10	17
W	12	11	13	12	13	12	14	12	14	12

struction is common throughout the entire building. In the event the construction is different, additional rows (labeled "other") can be used to describe different walls.

Table 8.4 contains CLTD values for glass doors and windows while Table 8.5 is for roof/ceiling combinations. Light wall CLTDs are appropriate for opaque doors. Figure 7.5 provides a representation of various residential and light commercial roof/ceiling combinations that correspond with roof/ceiling numbers to be used to determine the most the most appropriate CLTD

Table 8.4 Uncorrected Glass Door and Window Cooling Load Temperature Differences (°F)

All Directions	Hr 10	Hr 15
	4	14

Table 8.5 Uncorrected Roof/Ceiling Cooling Load Temperature Differences (°F) for Morning (Hr 10) and Afternoon (Hr 15)

Roof Num.	24°N		30°N		36°N		42°N		48°N	
	Hr 10	Hr 15	Hr 10	Hr 15	Hr 10	Hr 15	Hr 10	Hr 15	Hr 10	Hr 15
1	44	92	45	91	45	90	45	87	44	83
2	30	90	31	89	32	88	32	85	32	82
3	22	74	23	74	24	73	24	71	24	68
4	5	67	6	67	7	66	8	64	8	62
5	10	61	11	61	12	61	13	59	13	57
8	14	59	15	54	15	49	16	48	16	46
9	5	46	6	46	7	46	8	45	8	43
10	8	37	9	48	9	38	10	37	10	36
13	16	38	17	38	17	38	18	38	18	37
14	19	36	20	36	20	36	20	36	20	35

Example 8.1

Find the $CLTD_{cor}$ and the heat gain through a west wall that is 10 ft high by 40 ft wide with two 3 × 5 ft windows in Tuscaloosa, Alabama (lat. = 33.2°N), with a face brick exterior (4 in.), airspace, 1/2 in. blackboard sheathing, 2 × 4 in. wood studs on 16 in. centers, with R13 (h· ft²· °F/Btu) insulation, and 1/2 in. gypsum board (dry wall) on a single-story building at 3 p.m.

Solution:

This wall is best described as a medium wall (see Figure 7.4) with a CLTD designation of Type 10 since the wall insulation is R > 11. Also note that it carries a CLF Zone Type B designation and a SCL Zone Type A designation since it is the top floor. Table 8.3 is consulted and the uncorrected CLTD value for the Type 10 wall at the closest latitude (36°N) is corrected using Equation 8.4.

$$CLTD_{Cor} = CLTD_{Table} + (78 - t_i) + \left(\left(t_o - \frac{DR}{2}\right) - 85\right) = 13 + (78 - 75) + \left[\left(95 - \frac{19.6}{2}\right) - 85\right] = 16°F$$

Equations 7.2 and 7.3 are applied (in tabular form) to compute the wall transmission coefficient for the heat flow paths through the wood frame and through the insulation.

h·°F·ft²/Btu	% Area	Out. Sur.	Brick	Ins.	Wood	Gap	Black board	Gyp. Board	Int. Sur.	ΣR_{Path}	$U = 1/R$
$R_{FramePath}$	15	0.25	0.45		3.5 × 1.25 in.	1.2	1.32	0.45	0.68	8.7	0.115
$R_{Ins.Path}$	85	0.25	0.45	13		1.2	1.32	0.45	0.68	17.4	0.057

Equation 7.4 is applied to find the overall U-factor of the wall.

$$U_{ov} = A_{Frame} \times U_{Frame} + A_{Ins} \times U_{Ins} = 0.15 \times 0.115 + 0.85 \times 0.057 = 0.066 \text{ Btu/h·°F·ft}^2$$

The heat gain is

$$q_{W\text{-}Wall@3pm} = U \times A \times CLTD_{Cor} = 0.066 \text{ Btu/h·°F·ft}^2 \times (10 \times 40 - 2 \times 3 \times 5 \text{ ft}) \times 16°F = 390 \text{ Btu/h.}$$

at hours 10 (10 a.m.) and 15 (3 p.m.). Figure 7.6 provides the CLTD types for roof and ceiling construction more commonly used in larger buildings. The value for U is computed using Equations 7.4, 7.5, and 7.6. Results are entered in column B, row 33. If there are two different roof types in the building, row 34 is used for the second roof. Heat gains for roofs are computed with Equation 8.1. The area where the greatest amount of insulation is located should be the value used for pitched roof/ceilings when the roof area is not equal to the ceiling area.

Partition and Shaded Walls

Partitions have one surface facing a conditioned space and another surface facing an enclosed but unconditioned space. An example would be a wall separating an office and an attached garage or warehouse. Solar effects on the surface would have limited impact upon heat gain; thus, the basic conduction-convection equation (Equation 3.10 in this book) applies if the space is ventilated with outdoor air and the wall type is light or medium (wall types 1 though 6).

$$q = U_{ov}A(t_o - t_i) = \frac{A(t_o - t_i)}{R_{ov}} \tag{3.10}$$

However, if the space is not ventilated, an alternative would be to determine the combined overall resistance of the partition, the unconditioned space (treated as an air gap), and the wall or surface between the space and exterior. It is suggested that the $CLTD_{Cor}$ for the north wall be applied for exterior walls that are continuously shaded and for heavy partitions or floors (wall types 10 though 16, heavy floors above parking areas) exposed to well-ventilated unconditioned spaces.

Windows

Windows must be considered for both conduction and convection using CLTDs and radiation (solar) using SCL factors (Btu/h· ft^2) and SCs (dimensionless). The conduction-convection contribution follows the standard CLTD calculation (Equation 8.1) using the same values for doors found in Table 8.4. U-factors are given for common windows in Table 7.3 and the conduction gain is found for windows in the same manner as walls (Equations 8.5 and 8.6). Solar gain is calculated using Equation 8.3. Values for SCL factors appear in Table 8.6, but they do not have to be corrected for outdoor and indoor temperatures. Table 7.3 provides a set of SC values for unshaded windows, and Table 7.4 lists values for windows with shades. If windows are NFRC rated, the recommendation is to compute SC from the rated SHGF (SC = SHGF/0.87). A large set of data can be found in the "Fenestration" chapter of the *ASHRAE Handbook—Fundamentals* (ASHRAE 2005, chapter 31), but the user must match the description to the actual window, which is difficult given the large number of variants. SC and SCL values are also found on the "SCs & SCLs" worksheet in the spreadsheet program *TideLoad* on the CD accompanying this book. As an example, application of Equation 8.3 for north-facing windows is:

$$q(Solar)_{N\text{-}Window1@10am} = \text{SC (cell B10)} \times \text{SCL}_{10am} \text{ (cell C10)} \times A \text{ (cell F10)} \tag{8.7}$$

$$q(Solar)_{N\text{-}Window1@3am} = \text{SC (cell B10)} \times \text{SCL}_{3am} \text{ (cell D10)} \times A \text{ (cell F10)} \tag{8.8}$$

Floor and Slabs-on-Grade

Heat gain from crawlspace floors and slabs-on-grade are neglected when computing cooling load. Heat gain through floors above open spaces (i.e., a second-floor apartment building with an open air parking garage) can be handled by using Equation 8.1 with the north wall CLTD. Since north walls are shaded, floors above open spaces will have nearly equal thermal characteristics.

Ventilation Air and Infiltration

As stated in chapter 7, existing methods of calculating air infiltration rates in commercial buildings are inaccurate. Ventilation air is assumed to be the dominant mode of outside air entry. Validity is added to this assumption if CO_2 concentration is used as a control variable. ASHRAE Standard 62.1-2004 suggests that a concentration of 700 ppm above the outdoor air concentration indicates ventilation airflows that are equivalent to those specified in the standard. Ventilation and infiltration of outdoor air are lumped together in Table 8.1 and in *TideLoad*. The sensible component and latent component are considered independently. Equation 4.14 is used to compute the sensible component.

$$q_{sen} = \rho Q \times c_{pa} \times (t_o - t_i) \quad \text{or}$$
$$q_{sen}\text{ (Btu/h)} \approx 1.08 \times Q\text{ (cfm)} \times \Delta t(°\text{F}) \tag{8.9}$$

Values for density (ρ) and specific heat (c_p) can be computed using the psychrometric chart or the spreadsheet functions from Figure 4.2. Values for air at standard conditions are assumed with the use of the 1.08 coefficient in Equation 8.9. The outdoor-to-indoor humidity ratio difference ($\Delta W = W_o - W_i$) is necessary to compute the latent load due to the ventilation and infiltration. Again, the psychrometric chart or the spreadsheet functions are necessary to find these values, and Equation 4.21 is applied.

$$q_{lat} = \rho Q h_{fg} \times (W_o - W_i) \quad \text{or}$$
$$q_{lat}\text{ (Btu/h)} \approx 4680 \times Q\text{(cfm)} \times \Delta W\text{ (lb}_w/\text{lb}_a) \tag{8.10}$$

Table 8.6 Window Solar Cooling Load (SCL) Factors (Btu/h· ft²) for Morning (Hr 10) and Afternoon (Hr 15)

See Figure 7.4 for SCL zone types.

Zone Type	24°N	24°N	30°N	30°N	36°N	36°N	42°N	42°N	48°N	48°N
A	**Hr 10**	**Hr 15**	**Hr 10**	**Hr 15**	**Hr 10**	**Hr 15**	**Hr 10**	**Hr 15**	**Hr 10**	**Hr 15**
N	38	38	36	37	37	36	35	36	34	35
E	154	40	155	40	155	39	154	39	153	38
S	35	37	44	45	53	52	72	70	90	88
W	35	160	35	160	35	159	35	159	34	156
Zone Type	**24°N**	**24°N**	**30°N**	**30°N**	**36°N**	**36°N**	**42°N**	**42°N**	**48°N**	**48°N**
B	**Hr 10**	**Hr 15**	**Hr 10**	**Hr 15**	**Hr 10**	**Hr 15**	**Hr 10**	**Hr 15**	**Hr 10**	**Hr 15**
N	35	37	34	35	36	35	34	35	32	34
E	141	48	142	48	143	48	143	48	143	47
S	31	36	39	44	47	52	63	81	78	109
W	32	141	32	140	32	140	32	139	31	137
Zone Type	**24°N**	**24°N**	**30°N**	**30°N**	**36°N**	**36°N**	**42°N**	**42°N**	**48°N**	**48°N**
C	**Hr 10**	**Hr 15**	**Hr 10**	**Hr 15**	**Hr 10**	**Hr 15**	**Hr 10**	**Hr 15**	**Hr 10**	**Hr 15**
N	32	34	32	33	31	32	31	32	30	31
E	123	47	124	47	124	47	124	47	124	47
S	29	32	37	39	45	45	60	61	75	76
W	34	132	34	132	34	132	34	131	33	130
Zone Type	**24°N**	**24°N**	**30°N**	**30°N**	**36°N**	**36°N**	**42°N**	**42°N**	**48°N**	**48°N**
D	**Hr 10**	**Hr 15**	**Hr 10**	**Hr 15**	**Hr 10**	**Hr 15**	**Hr 10**	**Hr 15**	**Hr 10**	**Hr 15**
N	29	33	28	32	27	31	27	31	27	30
E	108	52	109	53	110	53	111	53	111	53
S	25	30	32	37	39	43	52	58	64	72
W	32	112	33	112	33	112	33	111	33	110

Required ventilation airflow rates (cfm) for both commercial and residential structures are discussed under the "Ventilation for Acceptable Air Quality" section in chapter 6 of this text. Ventilation rates for buildings designed before the publication of ASHRAE Standard 62.1-2004 and ASHRAE Standard 62.2 (residential) should refer to Table 1 of ASHRAE Standard 62-2001.

The ventilation load can be reduced with the use of heat recovery equipment by an amount equal to sensible effectiveness and latent effectiveness (Equations 4.25 and 4.26). The *TideLoad* program on the CD accompanying this book can compute the reduced loads (or the loads without HRUs by setting the effectiveness to zero). Equations 8.9 and 8.10 appear in Table 8.1 in spreadsheet format as shown in Equations 8.11 and 8.12, which represent the sensible and latent gains for 3 p.m. (hour 15).

$$q_{S@3pm} = 1.08 \times (t_o - t_i) \text{ (cell G5)} \times Q \text{ (cell F40)} \times [1 - \varepsilon_S \text{(cell B41)}] \quad (8.11)$$

$$q_{L@3pm} = 4680 \times (W_o - W_i) \text{ (cell G6)} \times Q \text{ (cell F40)} \times [1 - \varepsilon_L \text{ (cell B43)}] \quad (8.12)$$

However, all heat gain/loss values in *TideLoad* are converted to MBtu/h (Btu/h × 1000).

Heat Gain from People

Occupants generate both sensible and latent heat components according to activity level. The sensible heat rate increases slightly with higher activity, but latent heat increases dramatically because of greater perspiration rates needed to maintain body temperature. Table 8.7 lists the sensible, latent, and total heat gain from an average size population in Btu/h per person. The entire sensible heat rate from people is not immediately converted into cooling load in a building that is intermittently occupied because of thermal mass effects. The sensible heat gain is corrected with cooling load factors (CLFs), which are functions of the building thermal mass and the hour relative to when the zone is first occupied. Equation 8.13 is the general equation and Equation 8.14 provides an example for the 10 a.m. (hour 10) sensible heat gain (with Table 8.1 cell designations). CLF values for people and equipment (without ventilation hoods) are given in Table 8.8. These values are functions of the length of occupancy for a typical day and the amount of time after the start of occupancy. For example, if a type A zone is occupied for 12 hours per day beginning at 8 a.m., the CLF for people and unhooded equipment would be 0.98 at 3 p.m. (hour 15).

$$Q_S = q_S/\text{person (Table 8.7)} \times \text{CLF (Table 8.8)} \times N_{\text{People}} \quad (8.13)$$

$$q_{S@10\text{am}} = q_S/\text{person (cell B45)} \times CLF_{10am} \text{ (cell C45)} \times N_{People} \text{ (cell F45)} \quad (8.14)$$

The latent heat gain component for people is immediately converted to cooling load so no CLF correction is needed, as shown in Equations 8.15 (general) and 8.16 (Table 8.1 format).

$$Q_L = q_L/\text{person (Table 8.7)} \times N_{\text{People}} \quad (8.15)$$

$$q_{L@10am} = q_L/\text{person (cell B46)} \times N_{People} \text{ (cell F45)} \quad (8.16)$$

Internal Loads and Lighting

Heat gain from equipment (office equipment, motors, video equipment, vending machines, process equipment, kitchen equipment, etc.) is handled in a similar manner. Table 8.10 contains a summary of office equipment heat gain values found in the *2005 ASHRAE Handbook—Fundamentals* (ASHRAE 2005, chapter 30). Extensive values for kitchen equipment in Btu/h (Table 5), medical equipment in watts (Table 6), and laboratory equipment in watts (Table 7) also can be found in the Handbook (ASHRAE 2005, chapter 30). A CLF must be applied to all sensible loads, which are typically expressed in watts. The factor of 3.412 (Btu/W·h) is used to convert to Btu/h (which can also be used to convert kW to MBtu/h using 3.412 MBtu/kW·h).

$$q_S \text{ (Btu/h)} = 3.412 \text{ Btu/W·h*} \times \text{CLF (Table 8.8 or 8.9)} \times w_S \text{ (Table 8.10 or equal)} \quad (8.17)$$

$$q_{S@10am} = 3.412^* \times \text{CLF}_{10am} \text{ (cell C48)} \times w_S \text{ (cell F48)} \quad (8.18)$$

* Omit this factor when using Table 5 from the *2005 ASHRAE Handbook—Fundamentals*, chapter 30. Heat gain values are given in Btu/h.

Latent internal loads in many tables are often provided in Btu/h, so diligence is required to ensure proper unit conversions are applied.

Table 8.7 Heat Gain from People (ASHRAE 2005, chapter 30)

All values in Btu/h per person Activity	Level	Sensible Heat Gain	Latent Heat Gain	Total Heat Gain
Seated	At rest	245	105	350
Seated	Light work	245	155	400
Office work	Moderate	250	200	450
Standing	Walking	250	250	500
Light	Bench work	275	475	750
Nightclub	Dancing	305	545	850
Heavy	Work	580	870	1450

Table 8.8 Cooling Load Factors for People and Unhooded Equipment for Morning (Hr 10) and Afternoon (Hr 15) (ASHRAE 1997, chapter 28)

	Zone Type A*		Zone Type B*		Zone Type C*		Zone Type D*	
Occupancy	Hr 10	Hr 15	Hr 10	Hr 15	Hr 10	Hr 15	Hr 10	Hr 15
8 h/day	0.9	0.98	0.78	0.94	0.72	0.89	0.71	0.85
12 h/day	0.9	0.98	0.78	0.94	0.73	0.89	0.73	0.87
16 h/day	0.91	0.98	0.8	0.94	0.76	0.91	0.78	0.89

* See Figure 7.4 for zone types.

Table 8.9 Cooling Load Factors for Hooded Equipment for Morning (Hr 10) and Afternoon (Hr 15) (ASHRAE 1997, chapter 28)

	Zone Type A*		Zone Type B*		Zone Type C*		Zone Type D*	
Occupancy	Hr 10	Hr 15	Hr 10	Hr 15	Hr 10	Hr 15	Hr 10	Hr 15
8 h/day	0.86	0.97	0.68	0.91	0.6	0.82	0.6	0.77
12 h/day	0.86	0.97	0.69	0.92	0.6	0.82	0.62	0.82
16 h/day	0.87	0.97	0.72	0.92	0.66	0.88	0.68	0.85

* See Figure 7.4 for zone types.

Example Problem 8.2

Estimate the 3 p.m. heat gain due to 10 computers with 19 in. monitors and one interconnected laser printer for an office occupied from 8 a.m. to 6 p.m. The construction is described in Example Problem 8.1.

Solution:

It is assumed all 10 computers are operating at 3 p.m. From Table 8.10 the corresponding heat gain is 135 watts per computer and the laser printer is large (550 watts) since it can handle ten computers. From Example 8.1 it is noted that the wall has a Type B designation for CLF. Consulting Table 8.8 for a building occupied for 8 hours per day, the 3 p.m. CLF is 0.94. The value for a 12-hour-occupied building is also 0.94. Thus, the value for the 10-hour building is also 0.94. Applying Equation 8.17,

$$q_S\text{(Btu/h)} = 3.412\text{ Btu/W·h} \times \text{CLF} \times w_S\text{(W)} = 3.412 \times 0.94 \times (10 \times 135 + 500) = 5930\text{ Btu/h} = 5.93\text{ MBtu/h}.$$

Table 8.10 Heat Gain Rates for Equipment (Watts) (ASHRAE 2005, chapter 30)

Latent heat values in Btu/h.

	Continuous	Average	Idle
Computer —15 in. monitor	110	-	20
—17 in. monitor	125	-	25
—19 in. monitor	135	-	30
Laser printer —Desktop	130	100	10
—Small office	320	160	70
—Large office	550	275	125
Fax machine		30	
Other office equipment		25% nameplate (watts ≈ volts × amps)	
Coffee maker —10 cup		1050 W + 1540 Btu/h latent	
Microwave oven —1 ft³		400	
Refrigerator —15 ft³		300	
Water cooler —8 gal/hr		350	
1/4 hp motor	270		
3/4 hp motor	750		
1 hp motor	930		
10 hp motor	8500		
($W = 746 \times hp/\eta_{Motor}$)			

$$q_L = q_L \text{ (Table 8.10 or Table 5, chapter 30, } \textit{2005 ASHRAE Handbook—Fundamentals}\text{)} \quad (8.19)$$

$$q_{L@10am} = q_L \text{ (cell F49)} \quad (8.20)$$

Input values in Btu/h in F49 in *TideLoad* will convert to MBtu/h in cells G49 and H49.

Heat gains due to lighting can be the largest single component in many commercial and institutional buildings. Cooling load analysis should include a careful review of this component to ensure lighting power per unit area limits (W/ft²) imposed by ASHRAE 90.1-2004 are followed (see chapter 11 for details).

Lighting loads are sensible only, heat gain values in watts are provided in Tables 8.11 and 8.12, and CLFs are given in Table 8.13. Incandescent bulbs convert electrical energy to light at the line voltage supplied to the fixture. Other types of lighting require a ballast to convert energy to usable voltages. Recent developments of electronic ballast have substantially reduced the energy loss of this component. Ballast heat gains are accounted for by multiplying a ballast factor (F_{bal}) by the bulb rated wattage. However, some bulb heat gains are expressed in net watts, which include the ballast factor. In these cases F_{bal} = 1.0 as it is for incandescent bulbs.

$$q_S\text{(Btu/h)} = 3.412 \times \text{CLF (Table 8.13)} \times F_{bal} \times w_S/\text{bulb (Tables 8.10, 8.11)} \times N_{bulbs} \quad (8.21)$$

$$q_{S@10am} = 3.412\ CLF_{10am} \text{ (cell C51)} \times F_{bal} \text{ (cell E51)} \times [w_S/\text{bulb} \times N_{bulbs}] \text{ (cell F51)} \quad (8.22)$$

Ductwork

Duct heat losses or heat gains can be significant if ductwork is located in uninsulated attics or is poorly insulated and sealed. The chapter 7 section on "Building Heat Transfer Characteristics" discusses a method of calculating attic and crawlspace duct heat losses due to conduction-convection-radiation and duct leakage. This method should be used when losses are expected to be significant (attics, poorly sealed ducts) or analysis is needed to evaluate the economic value of locating ductwork inside conditioned spaces. These methods also consider the latent heat gains that can be important if duct leakage is significant since the supply air is relatively dry compared to the outdoor air.

A common popular method of estimating duct heat gains and losses is to multiply the total of all other sensible heat gains by sensible duct gain fractions (F_{DG}) and heat loss fractions (F_{DL}) that consider duct location, insulation, sealing, and heating equipment supply temperature (ACCA 1988). A summary of the values is provided in Table 8.14. The factor for the supply and

Table 8.11 Power Consumption of Selected Lighting Types (ASHRAE 2005, chapter 30)

Fluorescent Bulbs and Ballasts						
			Magnetic Ballast		Electronic Ballast	
Length (in.)	Dia.	Lamp Wattage	Ballast Factor	Net Wattage	Ballast Factor	Net Wattage
18	T8	15	1.27	19	–	–
18	T12	15	1.27	19	–	–
24	T8	17	1.41	24	0.94	16
24	T12	20	1.4	28	–	–
36	T8	25	–	–	0.96	24
36	T12	30	1.35	40	–	–
48	T8	32	1.1	35	0.97	31
48	T12	34	1.18	40	0.92	31
60	T8	40	–	–	0.90	36
60	T12	50	1.26	63	0.90	45
72	T12	55	1.2	66	1.07	59
96	T8	59	–	–	0.94	55
96	T12	60	1.15	70	0.92	55

Other Non-Incandescent Lighting							
Compact Fluor.		HP Sodium		Mercury Vapor		Metal Halide	
Rated watts	Actual (W/bal)	Rated watts	Actual (W/bal)	Rated watts	Actual (W/bal)	Rated watts	Actual (W/bal)
13	17	50	66	50	74	50	72
18	23	70	95	75	93	100	128
22	24	100	138	100	125	175	215
26	33	200	250	250	290	250	295
		400	465	400	455	400	458

Table 8.12 Incandescent Lighting Performance (Grainger 2004)

F_{bal} = 1.0 for incandescent bulbs.

W	Type	Lumens	Life (h)
40	Standard	500	1000
40	PAR-halogen	510	2500
60	Standard	865	1000
60	PAR-halogen	800	3000
75	Standard	1190	750
75	PAR-halogen	1030	2500
100	Standard	1710	750
100	PAR-halogen	1400	2000
150	Standard	2850	750
150	PAR-halogen	1690	2000
250	PAR-halogen	3600	4200

Table 8.13 Cooling Load Factors for Lights for Morning (Hr 10) and Afternoon (Hr 15) (ASHRAE 1997, chapter 28)

Occup.	Zone Type A*		Zone Type B*		Zone Type C*		Zone Type D*	
	Hr 10	Hr 15	Hr 10	Hr 15	Hr 10	Hr 15	Hr 10	Hr 15
8 h/day	0.93	0.97	0.87	0.95	0.82	0.9	0.74	0.85
12 h/day	0.94	0.98	0.88	0.96	0.84	0.92	0.77	0.87
16 h/day	0.95	0.98	0.9	0.97	0.97	0.93	0.81	0.89

* See Figure 7.4 for zone types.

return duct should be averaged and entered into the cooling heat gain factor (cell C56) and heating heat loss factor (cell E56) in Table 8.1. Note that if there is no return duct the factor is zero, and both factors are zero if the duct is located inside the conditioned space, a practice that is *highly recommended for high-efficiency buildings.*

$$q_{S\text{-}Duct} = q_{S\text{-}Building} \times [(\text{Supply Fraction} + \text{Return Fraction})/2] \text{ (Table 8.14)} \quad (8.23)$$

$$q_{S@10am} = (\text{cells G10} \rightarrow \text{G35} + \text{G40} + \text{G45} + \text{G48} + \text{G51}) \times [(F_{DG\text{-}Sup} + F_{DG\text{-}Ret})/2] \text{ (cellC56)} \quad (8.24)$$

It should be noted that this method is an estimate with a high degree of uncertainty. The design engineer

Table 8.14 Duct Sensible Heat Gain (F_{DG}) and Loss Fractions (F_{DL}) (ACCA 1988)

Multiply average of supply and return fractions by building sensible gains or losses to obtain duct gains/losses.

Duct Location*	Duct Insulation (h· ft· °F/Btu)	Cooling Gains Supply	Cooling Gains Return	Heating Losses Supply t_S < 120°F	Heating Losses Supply t_S > 120°F	Heating Losses Return
Outdoors, open to well-vented space, uninsulated, unconditioned space	None	0.25	0.15	0.25	0.3	0.15
	R2	0.15	0.08	0.15	0.2	0.08
	R4	0.1	0.06	0.1	0.15	0.06
	R6	0.05	0.03	0.05	0.1	0.03
Located in enclosed, insulated, unvented, unconditioned spaces	None	0.2	0.12	0.2	0.25	0.12
	R2	0.1	0.06	0.1	0.15	0.06
	R4	0.05	0.03	0.05	0.1	0.03
	R6	0.03	0	0.03	0.05	0
Located in return air ceiling plenum	None	0.1	0	0.1	0.15	0
	R2	0.05	0	0.05	0.1	0
	R4	0.03	0	0.03	0.08	0
	R6	0	0	0	0.03	0
Buried under or in slab	None	0.1	0.06	0.1	0.15	0.06
	R2	0.05	0.03	0.05	0.1	0.03
	R4	0.03	0	0.03	0.08	0
	R6	0	0	0	0.03	0
Located in conditioned space	Any	0	0	0	0	0
		0	0	0	0	0

* Increase factors by 0.10 if ducts are not taped or sealed.

Table 8.15 Typical Cooling Loads for Various Buildings (ft²/ton)

Building Type	ft²/ton	Building Type	ft²/ton
Apartment	450	Hospital	270
Bank	240	Hotel room	275
Bar/tavern	90	Ext. office	320
Church	330	Int. office	360
Computer	85	New home	600
Dental office	230	Old home	400
Department. store	350	Restaurant	200
Drugstore	150	School	260

can minimize this uncertainty by ensuring the ductwork complies with the insulation levels and sealing practices prescribed by ASHRAE Standard 90.1-2004 (ASHRAE 2004c).

Summary and Analysis

Table 8.1 presents a summary of the cooling (and heating) loads by individual component contribution. This overview format allows designers to identify the components that are the most significant contributors to building energy loss or gain. Designers can consider the likely demand or energy impact of design alternatives when attempting to optimize the value of installation cost vs. operating cost trade-offs. Figure 8.1 is included as an additional resource to help ensure that all heat gain sources are considered.

Table 8.15 is included as a set of reference points that indicate the floor area per unit cooling load for several typical building types in North America. Calculated cooling loads that fall well outside these typical values should be analyzed for possible errors, unusual loads, or building practices.

As stated previously, analysis should be performed at all three design conditions:

1. Maximum dry-bulb and coincident wet-bulb (DB/MWB) temperatures
2. Maximum wet-bulb and coincident dry-bulb (WB/MDB) temperatures
3. Maximum dew-point (humidity ratio) and coincident dry-bulb (DP/MDB and HR) temperatures

The total load, sensible load, and sensible heat ratio (SHR_{Load}) should be computed for all three conditions so the HVAC equipment can be selected to meet the sensible load without failing to meet the coincident latent load (i.e., SHR_{Unit} must be less than or equal to SHR_{Load}).

HEAT LOSS

In many aspects, the computation of building design heat loss is less complex than cooling load calculation because maximum heat loss values typically

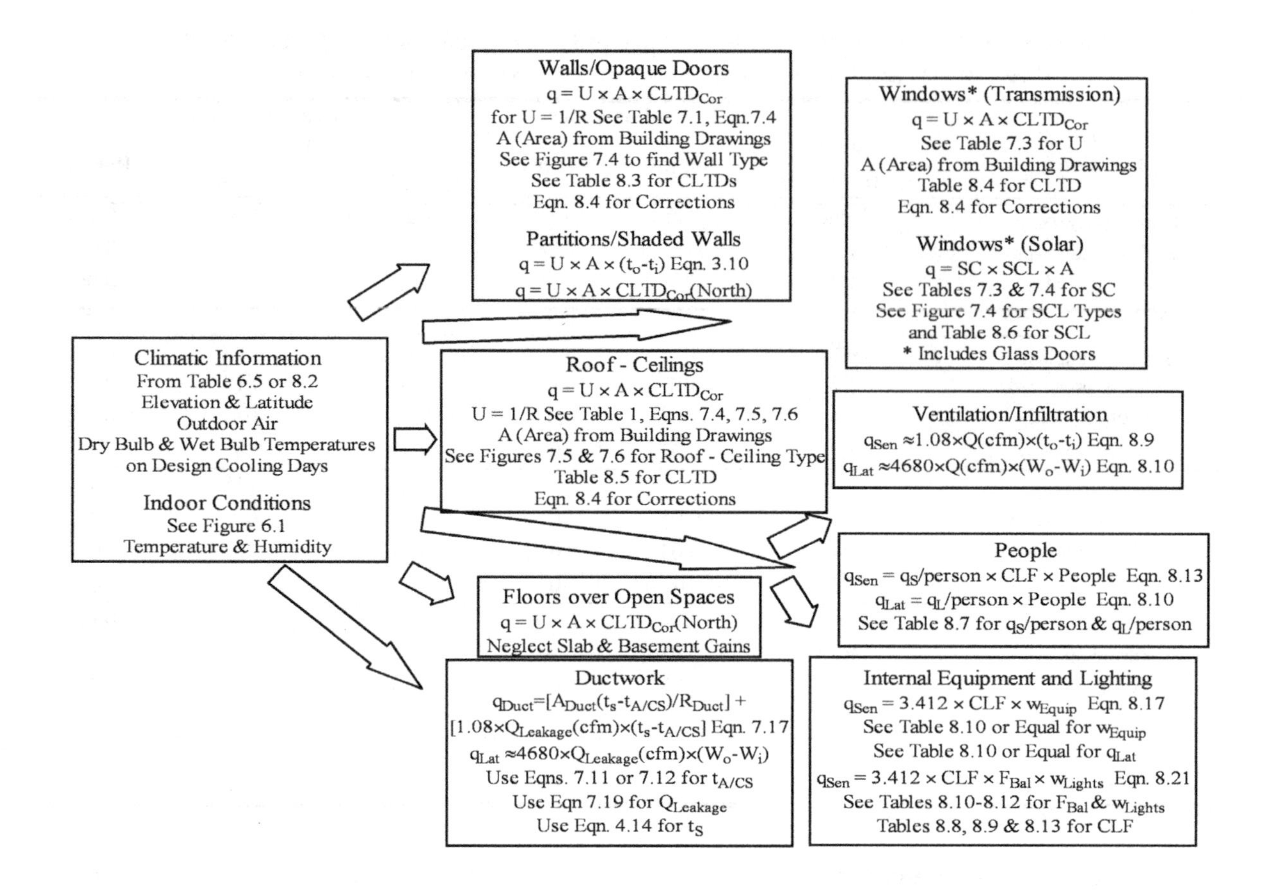

Figure 8.1 Cooling load sources.

occur in the morning when the impact of solar radiation is small. Outdoor temperatures are often relatively constant for several hours preceding the morning peak load, so thermal mass effects are also at a minimum. Therefore, the error introduced by summing the contributions of individual components using the product of the overall heat transfer coefficient, cross-sectional area, and indoor-to-outdoor temperature difference ($q_h = \Sigma U_o \times A \times [t_i - t_o]$) is small. The component overall heat transfer coefficient and area values are also found in the computation of cooling load. For most components, the indoor-to-outdoor temperature difference is used to compute heat loss in lieu of the CLTD value used for cooling load. Internal heat and solar gains, which offset heat losses, are typically set to zero at the critical morning hour when net heating requirements are at a maximum. Table 8.1 demonstrates the convenience of calculating building heat loss using the same spreadsheet used to compute cooling loads. Note the value of indoor-to-outdoor temperature difference ($\Delta T = t_i - t_o$) in column E replaces the values of morning and afternoon CLTD (columns C and D) so that the value of heat loss appears in column I.

However, complications do arise given the change in activity level, internal loads, ventilation rates, and, in some cases, thermostat setpoint during the onset of building occupancy in the morning. This is especially true if energy analysis is performed and the net heating load does not include the positive contribution of internal and solar gains. Therefore, the following caveats are included in order to recognize the limitations of computing the heating requirement based on the summation of the heat losses of individual components to the outdoors ($q_h = \Sigma U_o \times A \times [t_i - t_o]$).

- The contributions of internal and solar gains toward reducing heating requirement are neglected.
- The heat losses to unconditioned spaces will likely be overpredicted if the outdoor temperature is used in lieu of the actual space temperature, which is likely to be warmer.
- Conduction-convection and leakage losses in ducts located in unconditioned spaces can be estimated using Equation 8.23 and Table 8.14 or calculated using Equations 7.11 or 7.12.

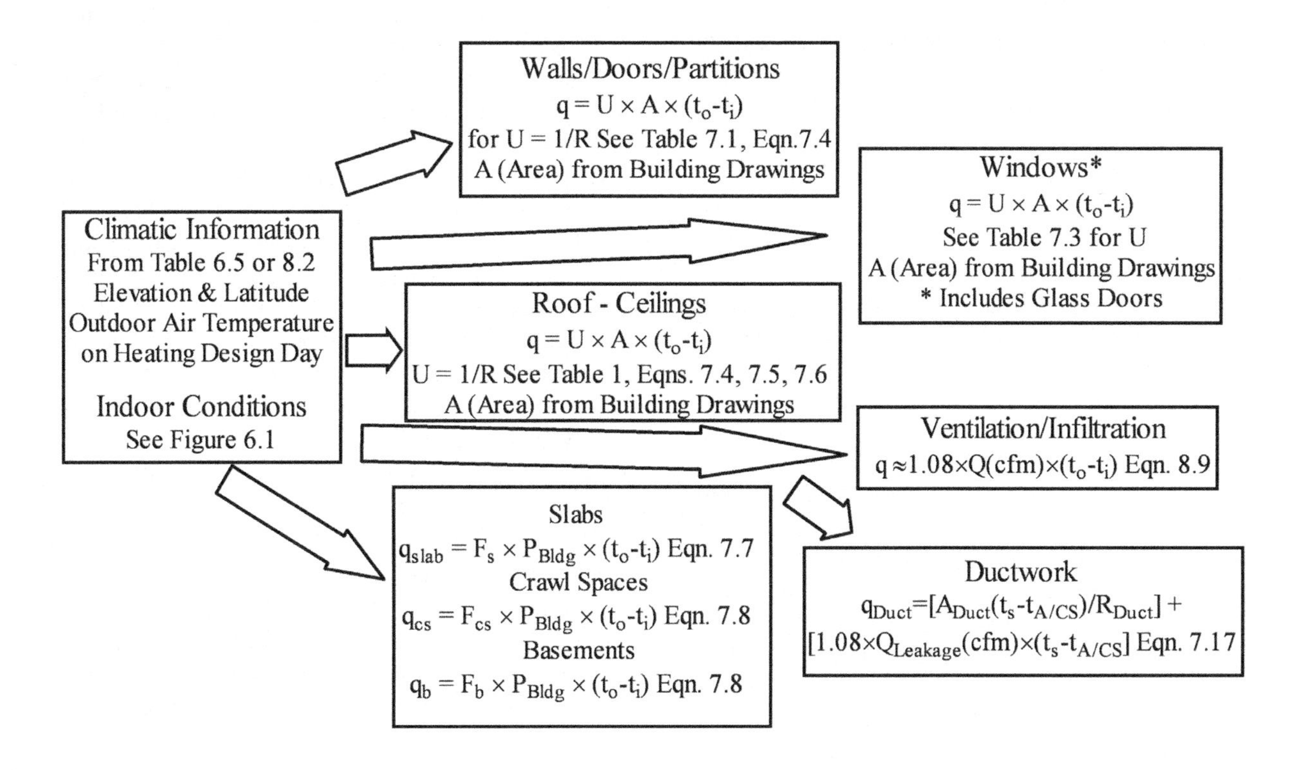

Figure 8.2 Heat loss sources.

- Slab-on-grade heat losses are computed using Equation 7.7 and Table 7.5.
- Crawlspace and basement wall heat losses are computed using Equation 7.8 and Table 7.6.
- This heat computation method does not include the impact of reduced moisture level due to the ventilation and infiltration of cold, dry outdoor air. Humidification to improve indoor comfort will result in a cooling effect and a corresponding increase in heating requirement. This cooling effect is $q_{Hum} = m_w \times (h_g@t_{room} - h_w)$ for a humidifier that injects liquid in the supply air or $q_{Hum} = m_w \times (h_g@t_{room} - h_s)$ for a steam humidifier.

Another caution in computation of heat loss is warranted due to the high degree of uncertainty in determining infiltration (Persily and Grot 1986) and ventilation (Persily et al. 2005). Infiltration and ventilation are often the largest contributors to building heat loss, and care is necessary to ensure building quality and ventilation rates correspond to values estimated for the computation. Figure 8.2 provides a graphic summary of building heat loss components.

CHAPTER 8 PROBLEMS

8.1 Compute the 3 p.m. heat gain of the roof/ceiling of the single office described in Figure C.4 in Appendix C if it is located in St. Louis, Missouri, and room conditions are 75°F/63°F (db/wb).

8.2 Compute the 3 p.m. heat gain of the 10 ft high walls of the single office described in Figure C.4 in Appendix C if it is located in St. Louis.

8.3 Compute the 3 p.m. heat gain of the windows of the single office described in Figure C.4 in Appendix C if it is located in St. Louis.

8.4 Compute the 3 p.m. internal heat gain due to lighting and office equipment of the single office described in Figure C.4 in Appendix C if it is located in St. Louis and is occupied for 12 hours per day.

8.5 Compute the 3 p.m. heat gain (sensible and latent) due to ventilation air of the single office described in Figure C.4 in Appendix C if it is located in St. Louis. Assume it is a single zone.

8.6 Compute the total cooling load, sensible cooling load, latent cooling load, and sensible heat ratio of the single office described in Figure C.4 in Appendix C if it is located in St. Louis. Provide results for all three design conditions and compare with estimates given in Table 8.15.

8.7 Compute the heat loss of the single office described in Figure C.4 in Appendix C if it is located in St. Louis and room temperature is 70°F (db).

REFERENCES

ACCA. 1986. *Residential Load Calculation, Manual J*, 7th ed. Washington, DC: Air-Conditioning Contractors of America.

ACCA. 1988. *Commercial Load Calculation, Manual N*, 4th ed. Washington, DC: Air-Conditioning Contractors of America.

ASHRAE. 1997. *1997 ASHRAE Handbook—Fundamentals,* chapter 28, Nonresidential cooling and heating load calculations. Atlanta: American Society of Heating, Refrigerating and Air-Conditioning Engineers, Inc.

ASHRAE. 2001a. *2001 ASHRAE Handbook—Fundamentals,* chapter 26, Ventilation and infiltration; chapter 27, Climatic design information; chapter 28, Residential cooling and heating load calculations. Atlanta: American Society of Heating, Refrigerating and Air-Conditioning Engineers, Inc.

ASHRAE. 2001b. *ANSI/ASHRAE Standard 62-2001, Ventilation for Acceptable Indoor Air Quality.* Atlanta: American Society of Heating, Refrigerating and Air-Conditioning Engineers, Inc.

ASHRAE. 2004a. *ANSI/ASHRAE Standard 62.1, Ventilation for Acceptable Indoor Air Quality.* Atlanta: American Society of Heating, Refrigerating and Air-Conditioning Engineers, Inc.

ASHRAE. 2004b. *ANSI/ASHRAE Standard 55-2004, Thermal Environmental Conditions for Human Occupancy.* Atlanta: American Society of Heating, Refrigerating and Air-Conditioning Engineers, Inc.

ASHRAE. 2004c. *ANSI/ASHRAE Standard 90.1-2004, Energy Standard for Buildings Except Low-Rise Residential Buildings.* Atlanta: American Society of Heating, Refrigerating and Air-Conditioning Engineers, Inc.

ASHRAE. 2005. *2005 ASHRAE Handbook—Fundamentals,* chapter 8, Thermal comfort; chapter 29, Residential cooling and heating load calculations; chapter 30, Nonresidential cooling and heating load calculations; chapter 32, Fenestration. Atlanta: American Society of Heating, Refrigerating and Air-Conditioning Engineers, Inc.

Grainger, W.W. 2004. Catalog 395. Lake Forest, IL: W.W. Grainger.

Harriman, L.G., G.W. Brundrett, and R. Kittler. 2001. *Humidity Control Design Guide.* Atlanta: American Society of Heating, Refrigerating and Air-Conditioning Engineers, Inc.

Pedersen, C.O., D.E. Fisher, J.D. Spitler, and R.J. Liesen. 1998. *Cooling and Heating Load Calculation Principles.* Atlanta: American Society of Heating, Refrigerating and Air-Conditioning Engineers, Inc.

Persily, A.K. 1999. Myths about building envelopes. *ASHRAE Journal* 41(3):39–47.

Persily, A.K., J. Gorfain, and G. Brunner. 2005. Ventilation design and performance of U.S. office buildings. *ASHRAE Journal* 47(4):30–35.

Persily, A.K., and R.A. Grot. 1986. *Measured Air Leakage Rate in Eight Large Office Buildings*, STP 904. Philadelphia: American Society of Mechanical Engineers.

9 Air Distribution System Design

Air distribution systems vary in complexity, from those for simple room air conditioners as shown in Figure 5.1 to more elaborate, central, variable air volume (VAV) systems with primary supply and return duct systems and secondary distribution systems in each zone as shown in Figure 5.12. The possible primary components and subsystems are:

- *Supply fan* that draws or blows through the heating and/or cooling coil.
- *Air volume control* to modulate comfort or system performance characteristics by altering the fan speed and/or adjustment of flow control dampers.
- *Supply duct* to deliver the heated and cooled air from the equipment room to the occupied zones without excessive noise or friction loss.
- *Air diffusers* to deliver the supply air to the space at a sufficient velocity to evenly distribute the air to all occupants without discomfort, excessive noise, or pressure loss.
- *Return air grille and ductwork* to allow return of the room air to the equipment without excessive noise or friction loss.
- *Return air fan* to assist the supply fan in providing adequate circulation and/or to adjust the room air pressure relative to outdoor or adjacent spaces.
- *Air filters* to remove a portion of air contaminants with a minimum of friction loss and noise (filters may be located near the equipment or at the return air grille).
- *Secondary room air terminals* to supplement the air distribution and temperature control in individual zones. These terminals may include additional dampers, fans, supply and return ducts, and coils.
- *Ventilation air system* to permit the introduction of outdoor air in compliance with ASHRAE Standards 62.1-2004 and 62.2-2004. The system may be integrated with the main air distribution system or be a stand-alone dedicated system with fans, ducting, and other related components.
- *Exhaust air system* to remove local contaminants (e.g., kitchen and restroom vent hoods) and to balance the building ventilation air system requirements.

Design of the air distribution system is often a challenging and time-consuming task. An engineer must optimize the three-sided tug of war between installation cost, energy-efficient design (i.e., large ducts, filters, and coils), and available space to route the often sizable ductwork through areas that also require electrical conduit, piping, structural support, and fire protection measures.

A variety of air distribution system design procedures are employed in the HVAC industry. The steps below are not universal, but they are a common method that is effective in small- and medium-sized buildings (< 50,000 ft^2).

1. Determine the total building and individual zone airflow rates.
2. Locate and select the fan coils/air handlers.
3. Locate and select the supply air diffusers or registers.
4. Locate and select the return air registers.
5. Sketch a routing of optimum supply duct mains and take-offs to outlets.
6. If necessary, locate and select the zone terminal boxes and ductwork.
7. Sketch a routing of optimum return duct to fan coils/air handlers.
8. Locate and select the filters.
9. Size all duct runs and note friction loss per unit length.
10. Determine critical routes that are likely to have the highest friction loss.
11. Provide a means of balancing airflow through non-critical runs (i.e., dampers).
12. Compute equivalent length and head loss ($\Delta h = [\Delta h/L] \times [L + \Sigma L_{eqv}]$) for each section of the critical routes.
13. Sum head loss through each section of critical route for supply and return (include individual register losses).
14. Find total loss by adding critical route supply and return duct losses to filter loss.

15. Select a fan and/or fan speed that will provide the required airflow rate (*Q*) with a total static pressure (TSP) greater than or equal to the total loss.
16. Identify and consider redesign of any components with high losses if the fan power is excessive.

ROOM AIR DIFFUSION

Room air diffusion is critical to occupant comfort and productivity. High outlet diffuser velocity is necessary to distribute and mix the supply air, but it must be directed away from occupants until it is near room temperature and the velocity is low (especially in the winter). Excessively high velocity in supply outlets or return grilles will result in high pressure losses and distracting noise levels. The design engineer must balance the need for affordable and effective distribution of heated and cooled air with low velocity in occupant zones, minimal pressure losses, and acceptable acoustics. Selection and location of outlet air devices is made even more challenging with the use of variable air volume (VAV) systems. In these systems, temperature is controlled by adjusting the flow of primary air, which can result in "dumping" of too much air near diffusers at low loads or over-blowing of air at full load to counteract dumping.

Air distribution methods are classified as mixing, displacement ventilation, underfloor, or some interrelated variation. Greatest emphasis will be placed on mixing systems in this text since the information in the current *ASHRAE Handbook—Fundamentals* (2005, chapter 33) focuses on this method. However, much more information is being developed for other methods.

Selection of supply air diffusers must be done in order to maximize comfort while minimizing noise, pressure loss, installation difficulty, and cost. Too many diffusers will be difficult and expensive to install. However, if too few diffusers are installed, high outlet velocities are required to distribute the air across greater distances. This will likely result in high noise levels, greater pressure losses, and discomfort near the diffusers. There are also limitations on the outlet locations. For example, the height between the roof and ceiling may be too limited to install the diffusers in the ceiling, so the outlets must be placed in the sidewall. The duct system may be under the floor and routing the duct up through the wall may cause high duct losses, so the registers are placed in the wall near the floor or in the floor itself. The location of the return register relative to the supply diffusers is also critical. For example, when warm air is delivered through ceiling registers, the return should be near the floor to overcome the natural buoyancy effects of the warm air and ensure delivery to the occupants.

Figure 9.1 shows typical airflow patterns from a ceiling diffuser as well as the nomenclature used in the selection and location of mixing-type air distribution. Air is introduced at velocities greater than those acceptable for comfort in the occupied zone. Room air is entrained into the jet stream by reducing the velocity to acceptable values (< 70 fpm). The entrained air also equalizes the mixed air to near ambient temperatures. The occupied areas are ventilated by the decayed air jet or by the reverse flow. This example also shows the tendency of warmer air to remain above the occupied zone.

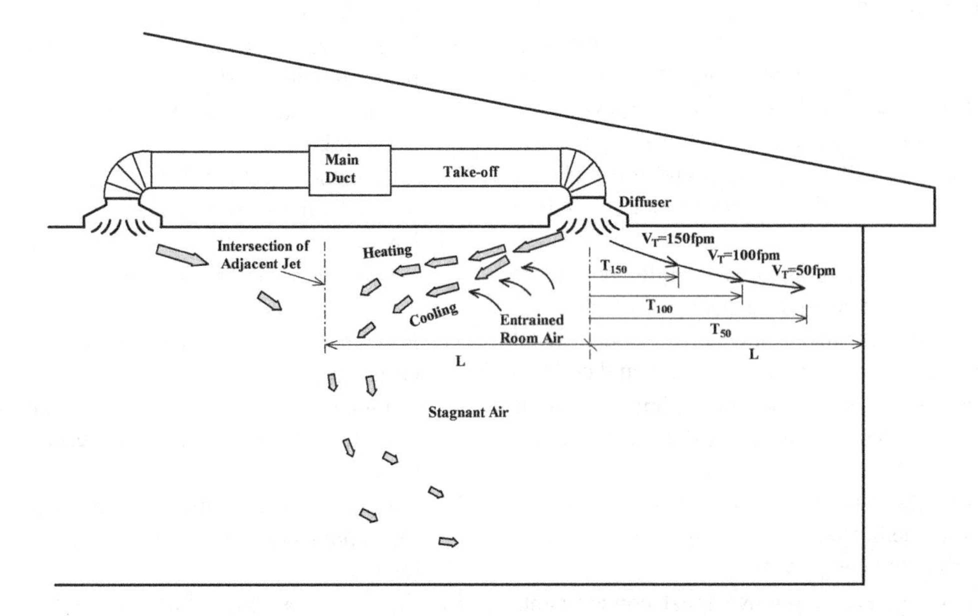

Figure 9.1 Air movement from ceiling diffuser, terminal velocities (V_T), throws (T), and characteristic lengths (L).

Thus, a low return duct location is necessary to ensure adequate distribution. The return air duct location is less critical in this example, although it is still important to avoid locating them near the supply register to prevent short-circuiting.

Figure 9.1 lists some of the nomenclature used to select and locate diffusers. At some point as the air exits the diffuser, it will reach a velocity of 150 fpm. The distance from the diffuser to this point is jet throw (T_{150}) for the terminal velocity (V_T) of 150 fpm. Other standard terminal velocities and throws are given for 100 fpm and 50 fpm. Manufacturers provide throws for diffusers at one or more of these terminal velocities. Diffusers are selected by considering characteristic lengths (L) from the diffuser to either the nearest wall or distance to the intersection of other air diffusers.

Figure 9.2 shows several types of diffusers and return air grilles. Ceiling diffusers are available to meet a variety of architectural and location applications. Round diffusers provide even distribution in all directions and are easily configured to the round duct that is commonly used for take-off runs. Lay-in T-bar ceiling diffusers provide convenient installation since they fit into the grid of a conventional ceiling tile. They are available in designs that provide one-, two-, three-, and four-way air distribution. Figure 9.2 shows both the exposed vane styles and a perforated design.

The ceiling tile grids also accommodate ceiling slot diffusers with minor modifications. These designs are typically longer, have adjustable vanes, and accommodate a wide variety of needs. Figure 9.2 also shows two types of high sidewall diffusers. The horizontal vanes, which can be adjusted to regulate flow, are typically set to have a slight upward slope to offset the tendency of cooler air to drop before reaching the areas farther away from the outlet. Vertical vanes can be either fixed or adjustable if a variation in spread angle is required. Setting the adjustable vanes parallel to the direction of flow (0°) provides the greatest throw, while setting them at a greater angle will widen the angle of spread and reduce throw. Floor registers are also available with both fixed and adjustable vanes. Near-floor outlets are preferred for heating-mode-dominant applications or when diffusers cannot be located in the ceiling for cooling-mode-dominant applications.

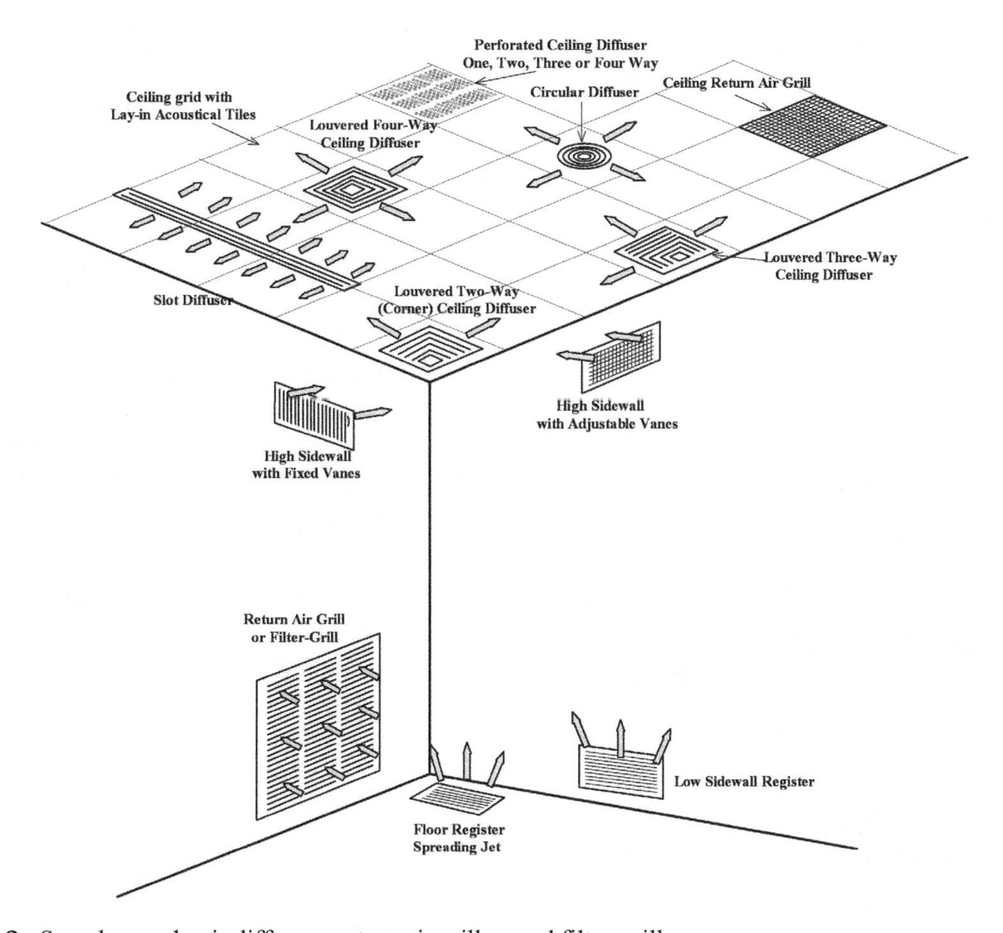

Figure 9.2 Sample supply air diffusers, return air grilles, and filter-grilles.

Sidewall and ceiling return air registers are also shown in Figure 9.2. Near-floor return air grilles are preferred in both the cooling and heating modes. If ceiling return is used in heating-dominant-mode applications, the supply diffusers must distribute air at high velocities to overcome the tendency of warm air to remain above the occupant zone. A variation is the filter-grille, which has a detachable face to permit a filter to be installed behind the grille. This design has the advantage of cleaning the air before it enters the return air ductwork to minimize dust and dirt buildup. Ease of filter replacement may be improved compared to applications where filters are located near difficult-to-access equipment. However, the designer must also make provisions to filter the outdoor air when locating the filters in the return duct rather than in the air-handling unit.

As stated previously, high outlet velocity and/or flow redirection can generate noise. This noise accompanies noise from other sources in the air distribution system (duct rumble, fan and compressor vibration) so simply choosing proper diffusers and grilles will not always ensure acceptable noise levels. However, selection of diffusers and registers must consider acoustics. Manufacturers provide information in the form of noise criteria (NC). Although many acoustics specialists prefer other measures that consider a broader range of frequencies and noise sources (Schaffer 1991; ASHRAE 2003, chapter 47), NC remains the most common single-number acoustic reference for air diffusion design (Rock and Zhu 2002).

Figure 9.3 is the representation of the pre-defined NC curves (NC = 20 through 45). The vertical axis is the sound pressure level (level of sound at a particular location) in decibels [L_p (dB) = 20 $\log_{10}(p/20\mu\text{Pa})$], and the horizontal axis is the center in octave bands (63 to 8000 Hz) in the frequency range of most audible noise. A noise source will have different sound pressure levels in different frequencies. For example, air outlets often peak in the fifth octave band (1000 Hz). The NC curves allow higher levels at lower frequencies in an attempt to adjust for perceived loudness, as noted by the downward slope of the curves with increasing frequency. The NC rating is found by plotting the sound power level (the level at which a source produces sound) in dB for each octave band. The rating is the band beneath which all points fall (Rock and Zhu 2002). For the example data shown in Figure 9.3, the level for the 1000 Hz band falls on the NC = 35 band, while all others are below 35. Thus, the NC rating for this source is 35.

A preferred alternative to the NC method is the use of the room criteria (RC), which accounts for both the sound level and the shape of the spectrum. It also considers data in the lower-frequency 31.5 and 16 Hz octave bands but not in the 8000 Hz band. Ratings also include letter descriptors of N (neutral), LF (low frequency), MF (medium frequency), or HF (high frequency). The numerical value is calculated by averaging the sound pressure levels in the 500, 1000, and 2000 Hz octave bands. A reference line is drawn through the measured level at 1000 Hz at a slope of –5 dB per octave band. A procedure (ASHRAE 2003, chapter 47) is applied to determine which (if any) of the three frequency ranges contains high deviations in sound pressure levels. Table 9.1 lists the target NC and RC levels for a variety of applications.

In addition to meeting acoustic constraints, diffusers must be located and selected so that air is evenly distributed to occupants while minimizing the sensation of draftiness. Outlets should distribute high-velocity exit air above or away from critical areas so that air velocity

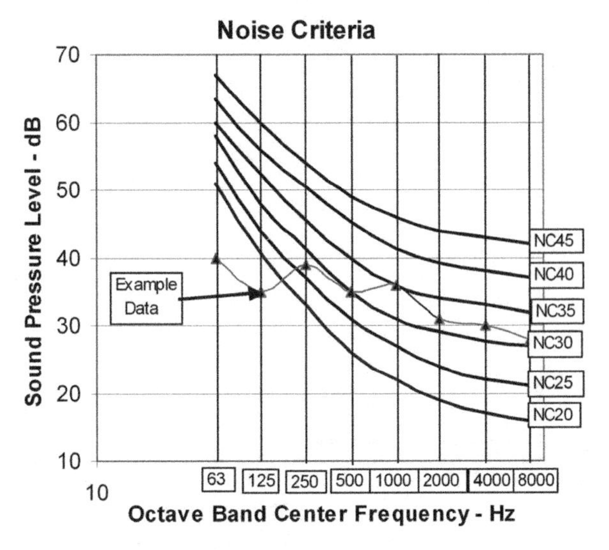

Figure 9.3 NC rating curves with data plotted on octave band center frequencies.

Table 9.1 Room Criteria (RC) and Noise Criteria (NC) Design Guides for HVAC-Related Background Sound*

Space Type	RC	NC
Residence	25–35 (N)	25–30
Apartment	25–35 (N)	30–35
Hotel room	25–35 (N)	30–35
Restaurant	–	40–45
Private office	25–35 (N)	30–35
Conference room	25–35 (N)	25–30
Hospital room	25–35 (N)	25–30
Public area	30–40 (N)	30–35
Church	25–35 (N)	30–35
Classroom	25–30 (N)	25–30
Library	30–40 (N)	35–40
Concert hall	25 (N)	15–20

* ASHRAE 2003, chapter 47; ASHRAE 1991, chapter 42.

is less than 70 fpm when it reaches the occupants. Table 9.2 is used to select diffusers to achieve a maximum or acceptable Air Diffusion Performance Index (ADPI). The ADPI is the percentage of locations within a room that are within –3°F to +2°F of the effective draft temperature and the local air velocity is below 70 fpm. Diffusers are grouped by type (see Figure 9.2). Maximum ADPI is achieved by selecting a diffuser to provide an optimum T_{50}/L (throw at 50 fpm terminal velocity/characteristic length) ratio. Table 9.2 provides this ratio for various types of diffusers at various cooling load densities. However, selecting a diffuser to provide an exact value is difficult; therefore, ranges of T_{50}/L ratios to achieve a slightly lower ADPI are provided.

Tables 9.3, 9.4, and 9.5 provide performance data for three common types of diffusers. A large number of buildings incorporate ceiling systems that consist of inverted T-bar grids on 24-inch centers. Ceiling tiles, lights, diffusers, and registers can be laid into the square grids. Table 9.3 has the performance data for this type of diffuser for four typical round duct connecting collar sizes (8-, 10-, 12-, and 14-inch diameter). Data include airflow rate (cfm), noise criteria (NC), throws at terminal velocities of 150, 100, and 50 fpm (T_{150}, T_{100}, T_{50}), and total pressure loss in inches of water (TP). These diffusers project air equally in four directions (four-way) when properly connected. Note that manufacturers choose to express the friction loss across diffusers, registers, and grilles in terms of "pressure" but use head loss units (in. of water). Correction for nonrated flows can be accomplished with Equation 9.1.

$$\Delta h \approx \Delta h_{Rated}(Q/Q_{Rated})^2$$

$$\text{or} \quad TP \approx TP_{Table}(\text{cfm}/\text{cfm}_{Table})^2 \qquad (9.1)$$

Table 9.4 provides similar information for another style of ceiling diffuser, which includes four-, three-, and two-way diffuser data. This permits flexibility in location if conditions dictate the diffusers must be near a wall (three-way) or corner (two-way). Table 9.5 is for high sidewall diffusers with adjustable vanes that permit flexibility when long throws or wide spread angles are required. It should be noted that Table 9.2 data suggest the maximum achievable ADPI is much lower with this design compared to ceiling diffusers.

Table 9.2 Outlet Jet Throw (T_{50}/L) Values for Air Diffusion Performance Index (ASHRAE 2005, chapter 33)

Terminal Device	Room Load Btu/h·ft²	Room Load ft²/ton	T_{50}/L @ Max ADPI	Max ADPI	For ADPI Greater Than	Range of T_{50}/L
High sidewall grilles	60	200	1.8	72	70	1.5–2.2
	40	300	1.6	78	70	1.2–2.3
	20	600	1.5	85	80	1.0–1.9
Circular ceiling diffusers	80	150	0.8	76	70	0.7–1.3
	60	200	0.8	83	80	0.7–1.2
	40	300	0.8	88	80	0.5–1.5
	20	600	0.8	93	90	0.7–1.3
Sill grille with spread vanes	80	150	0.7	94	90	0.6–1.5
	60	200	0.7	94	80	0.6–1.7
	40	300	0.7	94	–	–
	20	600	0.7	94	–	–
Light troffer diffusers	60	200	2.8	86	80	< 3.8
	40	300	1.0	92	90	< 3.0
	20	600	1.0	95	90	< 4.5
Louvered and perforated ceiling	11–51	235–1100	2.0	96	90	1.4–2.7
					80	1.0–3.4
	Room Load Btu/h·ft²	**Room Load ft²/ton**	**T_{100}/L @ Max ADPI**	**Max ADPI**	**For ADPI greater than**	**Range of T_{100}/L**
Ceiling slot diffusers	80	150	0.3	85	80	0.3–0.7
	60	200	0.3	88	80	0.3–0.8
	40	300	0.3	91	80	0.3–1.1
	20	600	0.3	92	80	0.3–1.5

Table 9.3 Ceiling Grid Lay-In Diffuser Performance Indicators (Titus 1996)

Collar Size	Neck Velocity (fpm)	400	500	500	700	800	1000	1200
		24 × 24 in. Lay-in Diffuser Terminal Velocities = 150-100-50 fpm						
8 in. diameter	cfm	140	175	209	244	279	349	419
	TP (in.)	0.016	0.025	0.035	0.048	0.063	0.099	0.142
	NC	–	–	–	12	16	22	28
	Throw (ft)	1-2-4	1-2-4	2-3-5	2-3-6	2-4-7	3-4-9	4-5-11
10 in. diameter	cfm	218	273	327	382	436	545	654
	TP (in.)	0.017	0.026	0.037	0.051	0.066	0.103	0.149
	NC	–	–	14	18	22	29	34
	Throw (ft)	2-4-7	3-5-9	4-5-11	4-6-13	5-7-14	6-9-17	7-11-18
12 in. diameter	cfm	314	393	471	550	628	785	942
	TP (in.)	0.017	0.027	0.038	0.052	0.068	0.106	0.153
	NC	–	11	16	21	24	31	36
	Throw (ft)	3-5-9	4-6-11	5-7-14	5-8-16	6-9-18	8-11-20	9-14-22
14 in. diameter	cfm	428	535	641	748	855	1069	1283
	TP (in.)	0.018	0.028	0.040	0.054	0.070	0.110	0.159
	NC	–	13	18	22	26	33	38
	Throw (ft)	4-5-11	5-7-14	5-8-16	6-10-19	7-11-21	9-14-23	11-16-25

Table 9.4 Ceiling Diffuser Performance Indicators (Nailor 2003)

Size (in. × in.)	Core Vel. (fpm) / TP (in.)	300 / 0.032	400 / 0.058	500 / 0.094	600 / 0.136	700 / 0.182	800 / 0.234	900 / 0.302
	Curved Blade Register 4-, 3-, and 2-way Throws (ft) for Terminal Velocities = 150-100-50 fpm							
6 × 6 10 × 4	cfm	60	80	100	120	140	160	180
	NC	–	15	21	26	29	32	35
	4-way	4-6-9	5-8-12	6-9-15	8-11-17	9-13-20	10-15-23	11-16-26
	3-way	4-6-10	6-9-14	7-10-16	8-12-19	10-14-22	11-16-25	12-17-28
	2-way	5-7-11	6-9-15	8-11-17	9-14-21	10-16-24	12-17-28	13-19-31
8 × 8 10 × 6 16 × 14	cfm	105	140	175	210	245	280	315
	NC	10	18	23	28	31	35	37
	4-way	5-7-11	6-9-15	8-11-17	9-13-20	10-16-24	12-17-27	13-18-30
	3-way	5-7-12	6-9-15	8-12-18	10-14-22	11-16-25	14-18-29	14-20-32
	2-way	5-8-13	7-11-16	9-13-20	10-16-24	12-17-28	14-20-32	15-22-36
10 × 10 18 × 6	cfm	185	250	310	370	435	495	560
	NC	12	20	26	30	34	37	40
	4-way	5-8-13	7-11-16	9-13-20	10-16-24	12-17-27	13-19-31	15-21-35
	3-way	6-9-14	8-11-17	9-14-21	11-16-26	13-18-30	14-20-33	16-24-38
	2-way	6-9-15	8-12-19	10-16-24	12-17-28	14-20-33	16-23-37	17-26-42
12 × 12 18 × 8	cfm	260	350	435	520	610	695	785
	NC	15	21	27	31	35	38	41
	4-way	6-9-14	8-12-18	10-14-22	11-16-26	13-18-30	15-21-34	16-23-38
	3-way	6-9-15	8-12-19	10-16-24	12-17-28	14-20-32	16-23-37	17-25-41
	2-way	7-11-16	9-14-21	11-16-26	13-19-31	15-22-36	17-25-41	19-28-46
14 × 14 20 × 10 36 × 6	cfm	375	500	625	750	875	1000	1120
	NC	16	23	29	33	37	40	43
	4-way	6-9-15	8-12-19	10-16-24	12-17-28	14-20-33	16-23-37	17-26-42
	3-way	7-10-16	9-14-21	11-16-26	13-19-31	15-22-36	16-25-40	18-28-45
	2-way	8-11-17	10-15-23	12-18-29	15-21-34	16-25-40	18-28-45	20-31-49
16 × 16 18 × 14 20 × 12 24 × 10 30 × 8	cfm	460	610	765	920	1070	1220	1380
	NC	17	24	29	34	38	41	44
	4-way	7-10-16	9-13-20	11-16-25	13-18-30	15-21-35	16-24-39	18-27-44
	3-way	7-11-16	10-14-22	12-17-27	14-20-32	16-23-38	17-26-43	19-30-48
	2-way	8-12-18	10-16-24	13-19-31	15-22-36	17-26-42	19-29-46	22-33-56

Table 9.5 Adjustable Vane Sidewall Diffuser Performance Indicators (Nailor 2003)

Size in. × in.	Core Vel. fpm / TP (in.) @ 0°–45°	300 / 0.015–0.026	400 / 0.026–0.046	500 / 0.041–0.072	600 / 0.059–0.103	800 / 0.106–0.186
	Angle of spread @ 50 fpm for vanes settings @ 0°±8°; @ 22.5°±15°, @ 45°±35°.					
	Throws (ft) for Terminal Velocities = 150-100-50 fpm for Vanes @ 0°, 22.5°, and 45°					
	cfm	60	80	100	120	160
6 × 6	NC	–	–	–	14	23
8 × 4	0°	5-7-13	7-9-16	8-12-18	10-14-20	12-16-23
10 × 4	22.5°	4-6-10	6-7-13	6-10-14	8-11-16	10-13-18
	45°	3-4-7	4-5-8	4-6-9	5-7-10	6-8-12
	cfm	105	140	175	210	280
8 × 8	NC	–	–	11	16	25
10 × 6	0°	6-9-18	9-13-21	10-16-24	12-19-26	17-21-30
16 × 14	22.5°	5-7-14	7-10-17	8-13-19	10-15-21	14-17-24
	45°	3-5-9	5-7-11	5-8-12	6-10-13	9-11-15
	cfm	150	200	250	300	400
	NC	–	–	13	18	27
14 × 6	0°	6-11-20	10-15-23	12-18-25	15-20-28	19-23-33
10 × 8	22.5°	5-9-16	8-12-18	10-14-20	12-16-22	15-18-26
	45°	3-6-10	5-8-12	6-9-13	8-10-14	10-12-17

NC values for vanes 0° setting: Corrections for 22.5° = +2, 45° = +7.

Table 9.6 Perforated Return Grille and Filter-Grille Performance

Size (in. × in.)	Face Vel. fpm / TP (in.)	300 / 0.024	400 / 0.042	500 / 0.067	600 / 0.095
24 × 12	cfm	555	740	925	1110
20 × 14	NC	–	–	17	24
24 × 20	cfm	951	1268	1585	1902
30 × 16	NC	–	11	18	26
24 × 24	cfm	1137	1506	1895	2274
28 × 20	NC	–	12	19	27
24 × 30	cfm	1431	1908	2385	2862
36 × 20	NC	–	13	21	28
36 × 24	cfm	1722	2296	2870	3444
30 × 28	NC	–	14	22	29
48 × 24	cfm	2307	3076	3845	4614
	NC	–	16	24	30
48 × 48	cfm	4677	6236	7795	9354
	NC	11	19	26	33

The steps below provide a suggested procedure for the selection of supply air diffusers.

1. *Determine the required supply airflow, the floor area of the zone, and the room load (Btu/h·ft² or ft²/ton).* Typical values are ~0.5 cfm/ft² and ~20 Btu/h·ft² (low load applications), ~1.5 cfm/ft² and ~60 Btu/h·ft² (high load applications), and ~1.0 cfm/ft² and 30 to 40 Btu/h·ft² being typical.
2. *Evaluate the geometry of the space.* Can diffusers be placed in the ceiling? Are there areas that are remote from where registers are located? Can the space be served by one diffuser? Can the space be divided into near-square subzone areas that can be served by a central diffuser such as a round or four-way diffuser with near-equal throw in all directions? Can the space be subdivided into areas that can be served by three-way or two-way diffusers? Divide the space into these subzones and locate the diffusers.
3. *Select diffusers for each area of the zone.* Find the range of T_{50}/L that will provide an acceptable ADPI. Find the characteristic lengths (L) in all directions of diffuser throw (distance to nearest wall or half the distance to nearest similar diffuser). Compute the acceptable range of throws in these directions [$T_{50} = (T_{50}/L) \times L$]. Using a manufacturer's chart, find a diffuser that provides the required throw (T_{50}) in all directions while meeting the noise criteria in Table 9.1 with acceptable total pressure (TP) losses (< 0.05 in. of water for residential, < 0.10 in. of water for commercial). If the number of diffusers selected cannot meet the required flow, NC limits, and pressure loss constraints, more diffusers may be warranted, other types may be considered, or alternative locations chosen.
4. *Provide a sketch of the diffuser locations with specifications and/or model numbers.*

Table 9.6 provides friction loss (TP) and NC data for return grilles and filter-grilles. Equation 9.1 can be used to correct friction losses for other flow rates. It should be noted that friction losses in filters (discussed later in this chapter) often dictate the size of filter-grilles. Note also that 24 inches is a frequent dimension, which allows convenient use of these grilles in ceiling grid applications.

Example Problem 9.1

Select louvered ceiling diffusers for a 24 × 40 ft classroom that has a cooling load of 32 MBtu/h and requires 1000 cfm of supply air.

Solution:

Step 1: The supply airflow of 1000 cfm is provided, the area is 960 ft^2, and the room load is 33 $Btu/h \cdot ft^2$ (= 32,000 ÷ 960).

Step 2: The space will divide evenly into two near-square subareas, as shown in Figure 9.4. Four-way diffusers located in the center of each area must provide 500 cfm.

Step 3: Table 9.2 indicates a T_{50}/L range of 1.4 to 2.7 will provide an ADPI of 90 for loads between 11 and 51 $Btu/h \cdot ft^2$. The required throws in each direction are:

For N and S directions:

$$T_{50} = (T_{50}/L)L = 1.4 \times 12 \text{ ft to } 2.7 \times 12 \text{ ft} = 17-32 \text{ ft}$$

For E and W directions:

$$T_{50} = (T_{50}/L)L = 1.4 \times 10 \text{ ft to } 2.7 \times 10 \text{ ft} = 14-27 \text{ ft}$$

Therefore, a four-way diffuser that will provide a throw (T_{50}) between 17 and 27 ft will meet the requirement in all directions. For a classroom, the NC must be less than 30 and preferably below 25 (see Table 9.1), and the friction loss should less than 0.1 in. of water. Consulting Table 9.4:

- A 10 × 10 in. diffuser @ 495 cfm has a throw (T_{50}) of 31 ft, which is too high.
- A 12 × 12 in. diffuser (interpolating between 435 and 520 cfm) has a throw (T_{50}) of 26 ft, which is good, and an NC of ~30, which is marginally acceptable.
- A 14 × 14 in. diffuser @ 500 cfm has a throw (T_{50}) of 19 ft, which is good, an NC of 23, which is very good, and a total pressure loss of 0.058 in., which is very good.

The 14 × 14 in. diffuser is a good choice for a constant volume airflow application because of the low NC, but the 12 × 12 in. diffuser might be a better choice for a variable air volume system since the throw is near the upper limit at design conditions and will continue to be within the acceptable range when the air is throttled at part load.

Step 4: The locations and specifications are noted on the drawing in Figure 9.4.

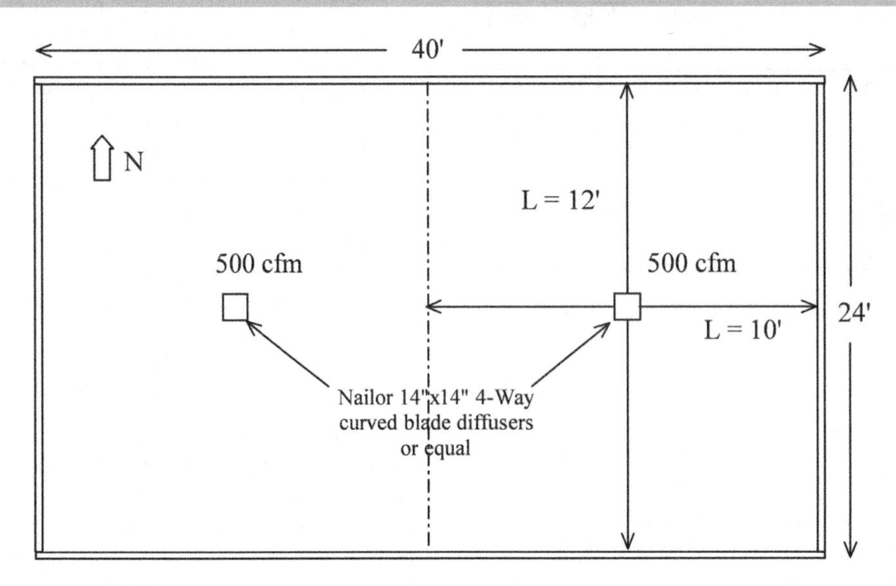

Figure 9.4 Example classroom with ceiling diffusers.

FILTRATION

The increased emphasis on indoor air quality (IAQ) has heightened the importance of proper air filtration. Filter selection must consider the efficiency requirements, the types and dimensions of particulate to be filtered, resistance to airflow, cost, physical size of the filter cabinet, and the location to permit service and replacement. ASHRAE Standard 52.2-1999 rates filters for arrestance (weight fraction of dust removed), dust spot efficiency (ability of filter to reduce soiling of fabrics and interior surfaces), fractional efficiency (percent of uniformly sized particles removed), particle size efficiency (percent of dust removed with a typical particle size distribution), and dust-holding capacity (ASHRAE 1999).

Filters of higher efficiency typically have higher friction losses. These losses ($\Delta h \equiv$ in. of water) are provided at initial (clean) conditions for typical face velocities ($V_{Face} \equiv$ fpm) and final losses (dirty and in need of replacement or cleaning). Filter losses are critical since they are responsible for a considerable portion of the total air-side fan power requirements, and dirty filters are a primary cause of equipment malfunction.

Design engineers must provide easy-to-access and obvious filter locations in the ductwork, fan-coil cabinet, or air-handling unit. They must also minimize the impact of excessive losses. Losses can be minimized by several design practices. A common method is to reduce face velocity below values that formerly were common (i.e., 500 fpm). Equation 9.2 demonstrates the impact of reducing face velocity (or flow rate) below the rated values.

$$\Delta h_{Actual} \approx \Delta h_{Rated} \times \left(\frac{V_{Actual}}{V_{Rated}}\right)^2$$

$$\text{or} \quad \Delta h_{Actual} \approx \Delta h_{Rated} \times \left(\frac{Q_{Actual}}{Q_{Rated}}\right)^2 \qquad (9.2)$$

A downside to this approach is the increased filter size, which could increase coil cabinet size (if the filter is located in the unit) or filter-grille size.

Another effective method of lowering face velocity is to choose pleated or bag filters, as shown in Figure 9.5. In a conventional filter, such as the disposable fiberglass shown on the left of Figure 9.5, the face area and the filter cross-sectional area are nearly equal. The filter media of pleated and bag filters are arranged in a "zig-zagging" pattern to increase the cross-sectional surface area through which the air flows. This permits the effective filter area to be much greater than the face area of the filter cabinet. The velocity through the filter media is much lower and friction losses are smaller. The downside to this method is the higher cost of these filters and the addition of physical space in the air-handling unit. However, the filter replacement frequency can be reduced by using low-efficiency (and low-cost) disposables to pre-filter air provided the combined friction losses of the two filters in series do not exceed the static pressure capability of the fan. An additional excess filter loss precaution is to provide a high differential pressure indicator and/or alarm to signal the building owner or maintenance staff that filters should be cleaned or replaced.

Table 9.7 summarizes the performance indicators for filters with minimum efficiency reporting values (MERVs) ranging from 4 to 14 (ASHRAE 1999). In addition to MERVs, the table provides the friction losses for face velocities of 300 and 500 fpm and the recommended final friction loss for all face velocities. It is suggested that the design engineer establish a recommended final resistance based on the available static pressure capability of a fan of reasonable size rather than the filter manufacturer's recommendation. A value of two to four times the initial or clean filter loss is suggested ($\Delta h_{Final} = 2$ to $4 \times \Delta h_{Initial}$). Note the friction loss

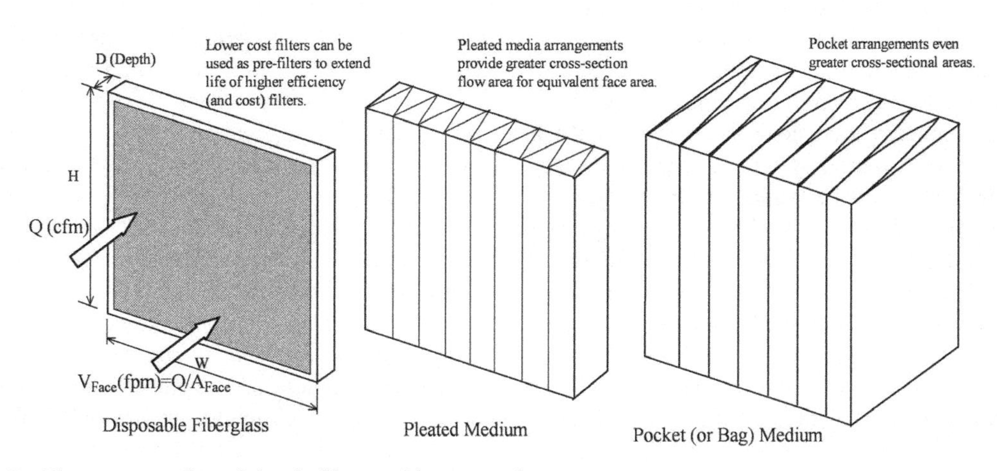

Figure 9.5 Three types of particle air filters with nomenclature.

Table 9.7 Typical Efficiencies and Pressure Drops Based on MERV and Velocity (ASHRAE 2004, chapter 24; AAF 2004; AG 2004)

Type	Depth (in.)	Standard 52.2 MERV	Δh (in.) @ 300 fpm	Δh (in.) @ 500 fpm	Rec. Final Δh (in.)	Dust Spot Eff.	Arrestance
Fiberglass disposable	1	4	0.10	0.20	0.5	–	70–75%
	2	4	0.13	0.28	0.5	–	80–85%
Pleated	1	6	0.20	0.45	1.0	25–30%	85–90%
	2	6	0.13	0.30	1.0	25–30%	85–90%
	4	6	0.12	0.25	1.0	25–30%	85–90%
Pleated	1	8	0.29	0.59	1.0	30–35%	90%
	2	8	0.20	0.40	1.0	30–35%	90%
Extended surface pleated	1	11	0.27	0.54	1.0	60–65%	95%
	2	11	0.18	0.38	1.0	60–65%	95%
	4	11	0.10	0.22	1.0	60–65%	95%
Mini–pleat	6	11	0.20	0.36	1.5	60–65%	
	6	13	0.30	0.50	1.5	80–85%	
	6	14	0.36	0.57	1.5	90–95%	
Ext. pocket $\eta = 60–65\%$	22	11	0.13	0.37	1.5	60–65%	95%
	36	11	–	0.27	1.5	60–65%	95%
Ext. pocket $\eta = 90–95\%$	22	14	0.26	0.52	1.5	90–95%	98%
	36	14	–	0.42	1.5	90–95%	98%

and arrestance across thicker disposable filters increase since the air must travel through additional filter media. However, these values decrease with thicker pleated or bag filters because the media thickness is unchanged but the cross-sectional area increases with filter thickness. Table 9.7 does not include values for high-efficiency particulate air (HEPA) filters or ultra-low-penetration air (ULPA) filters that are for clean room and toxic particulate applications and have MERVs of 17 to 20. Additional selection and maintenance recommendations are available in the *ASHRAE Handbook—HVAC Systems and Equipment* (ASHRAE 2004, chapter 24).

AIR DUCT SYSTEMS AND DESIGN CONSIDERATIONS

The routing and sizing of air distribution supply and return ducts is perhaps the most challenging activity in the design of an HVAC system. The size and cost of most ductwork systems amplifies the importance of this design task. Engineers often face balancing the conflicting constraints of lowering system installation cost by reducing the duct size with increasing air velocity, fan power requirements, and noise levels. Routing large duct runs though restricted ceiling space and fire-rated partitions and around building piping, wiring, and structural components is an additional challenge. Adequate airflow is critical to proper HVAC system performance, and situations arise that can alter the actual flow from that intended. Post-installation testing and balancing (TAB) of air distribution systems is critical, and the designer must include provisions (dampers, measurement points, avoiding parallel air distribution sections with radically different losses) that ensure the systems can be balanced. Although CAD programs have improved the designer's capability for accomplishing these tasks, the need for communication and cooperation among plumbing, electrical, and structural engineers remains very important.

The procedure described here follows the selection and location of the fan coils (air handlers), supply air diffusers, return register(s), and air filter. The *ASHRAE Handbook—Fundamentals* lists three duct design methods: equal friction, static regain, and T-method optimization (ASHRAE 2005, chapter 35). The equal friction method is the simplest and most commonly used method and will be the basis of discussion for this section. A maximum target friction loss per unit length is selected and all ducts are sized to provide this loss (or slightly less). The conventional units are inches of water per 100 feet of straight duct. Friction losses in fittings are expressed in equivalent length of straight duct. An updated set of equivalent length values for common fittings has been developed using loss coefficients (ASHRAE 2005, chapter 35; ACCA 1990) for airflow rates that result in friction losses of 0.10 and 0.30 in. of water per 100 ft of straight duct.

In addition to the items presented in this chapter, duct insulation and leakage are critical components in air distribution design. The reader is referred to the discussion of ductwork in chapter 7 of this book and to the *ASHRAE Handbook—Fundamentals* (ASHRAE 2005) for more information.

A simplified alternative method to traditional equations and graphs for friction factors and losses for fluid flow in conduit is presented for air in ducts. The friction factor (*f*) (ASHRAE 2005, chapter 35) is

$$f' = 0.11\left(\frac{12\varepsilon}{D_h} + \frac{68}{Re}\right)^{0.25}, \qquad (9.3)$$

where

f = f' if $f' \geq 0.018$; $f = 0.85f' + 0.0028$ if $f' < 0.018$,
Re = $8.56\ D_h V$ for air at standard conditions,
D_h = hydraulic diameter (in.),
ε = roughness (ft), and
V = velocity (fpm).

Traditional correlations are modified to directly compute head losses in friction units (inches of water) per unit length.

$$\Delta h\ (\text{in.})/L\ (\text{ft}) = \frac{12f}{D_h\ (\text{in.})}\rho\ (\text{lb/ft}^3)\left(\frac{V\ (\text{fpm})}{1097}\right)^2 \qquad (9.4)$$

Several tools have been developed using head loss correlations to further optimize the design of ductwork. Figure 9.6 illustrates the logarithmic chart of airflow rate (cfm) vs. friction loss (in. of water per 100 ft) that appears in the *ASHRAE Handbook—Fundamentals* (ASHRAE 2005). This chart was developed for duct with a roughness of 0.0003 ft, which is representative of metal duct. Duct with liners or flexible duct will have slightly greater losses. Round duct diameters (in.) appear as straight lines that slope up to the right. Lines of constant velocity (fpm) slope down to the left. Ductwork is sized by choosing a friction loss (in. of water per 100 ft), crossing this value with the airflow rate on the chart, and selecting the standard duct size that is closest to the intersection of the lines (usually rounded up). The velocity can be read by interpolating between constant velocity lines at the point of intersection between the duct size and flow rate.

Figure 9.7 represents a widely used air duct calculator, which is a modified circular slide rule. A rotating dial has three semicircular windows with logarithmic scales for friction and velocity and a round duct diameter indicator. A fourth log scale for rectangular duct height appears at the bottom of the dial. The values for airflow rate appear on a stationary base of the calculator in two of the windows, while round duct diameter is shown in the third window and rectangular duct width can be seen in conjunction with the duct height scale. The user rotates the scale to line up the desired friction loss with the required airflow in the upper left window. Values for required duct size and velocity are read as shown in the figure. Note that there is a slight discrep-

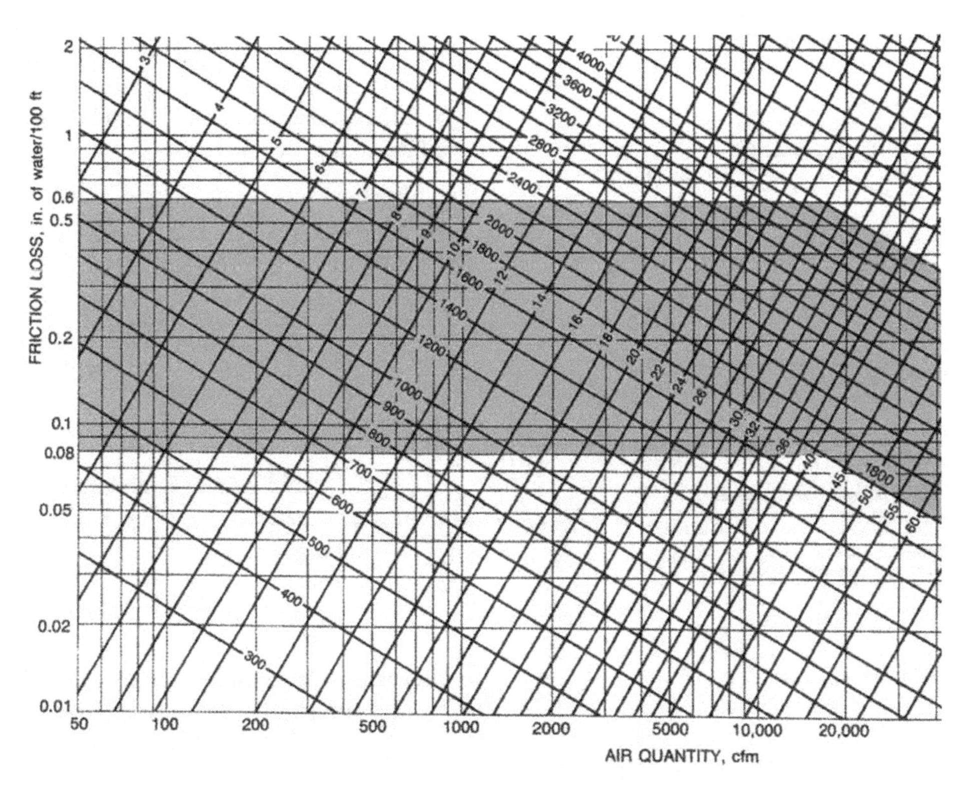

Figure 9.6 Friction chart for round metal duct (ε = 0.0003 ft), air @ 0.075 lb/ft^3 (ASHRAE 2005, chapter 35).

Example Problem 9.2

Size a duct for a flow of 1000 cfm at a friction loss of 0.1 in. of water/100 ft. Determine the actual loss and velocity for the actual duct size.

Solution:

The horizontal 0.1 in. of water/100 ft friction line intersects the vertical 1000 cfm line at a round duct diameter of 13.5 in. The next available round duct size is 14 in. When this duct is used, the loss will be 0.09 in. of water/100 ft (intersection of 14 in. and 1000 cfm), and the velocity is 950 fpm at this intersection.

ancy with the results attained using Figure 9.6 due to the difference in the higher friction value ($\varepsilon = 0.003$ ft) for a smooth-faced duct liner used to compute the head loss values of Figure 9.7.

The rectangular duct sizing feature of the air duct calculator is an important factor in the wide use of this tool. In order to size rectangular duct using the ASHRAE friction chart shown in Figure 9.6, the round duct diameter must be converted to an equivalent size. The traditional method is the hydraulic diameter concept for a rectangular duct.

$$D_H = \frac{4 \times A_{CrossSection}}{P_{Wetted}} = \frac{4 \times h \times w}{2 \times (h + w)} \tag{9.5}$$

However, an improved correlation of equivalent diameters for rectangular ducts is suggested (ASHRAE 2001, chapter 34) by

$$D_H = \frac{1.3 \times (hw)^{0.625}}{(h+w)^{0.25}}. \tag{9.6}$$

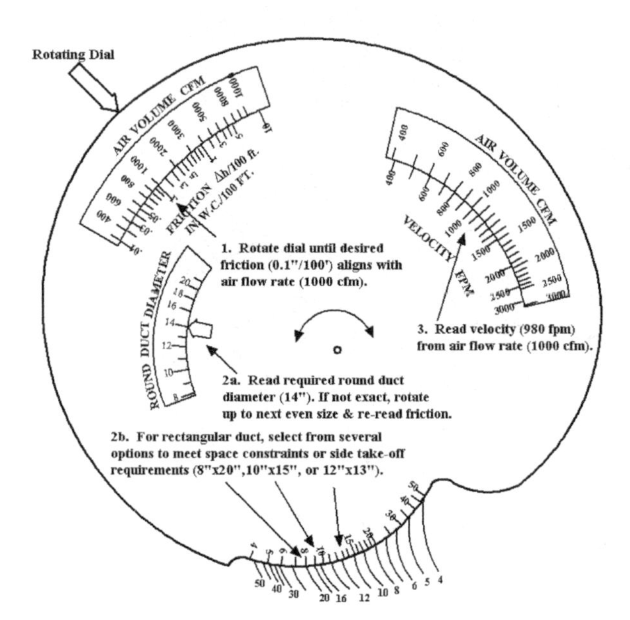

Figure 9.7 Circular slide rule type of air duct calculator (T/ASC 1996).

For flat oval ducts, as shown in Figure 9.8, the equivalent diameter is

$$D_e = \frac{1.55 \times [(\pi h^2/4) + h(w-h)]^{0.625}}{[\pi h + 2(w-h)]^{0.25}}. \tag{9.7}$$

Rectangular and flat oval ducts are widely used because ceiling heights in most buildings limit the use of round duct for large capacity main ducts. Height cannot be too small if take-off ducts are to be attached as shown in Figure 9.8. The height of rectangular and flat oval duct is often set, and the width is adjusted to provide adequate size.

Another important feature of both Figures 9.6 and 9.7 is the capability of quickly determining air velocity, which is needed for acoustic considerations. Duct noise can be minimized by lowering air velocity. Table 9.8 (ASHRAE 2003, chapter 47) provides recommended maximum velocities for main ducts, branch ducts, and the final section of duct immediately upstream of the air outlet. The table also provides values for different duct locations (behind drywall, above light ceiling, in the occupied area).

Friction losses in fittings are typically handled using loss coefficients (C) multiplied by the velocity pressure (p_v).

$$\Delta p_f = C \times p_v = C\frac{\rho V^2}{2g_c} \tag{9.8}$$

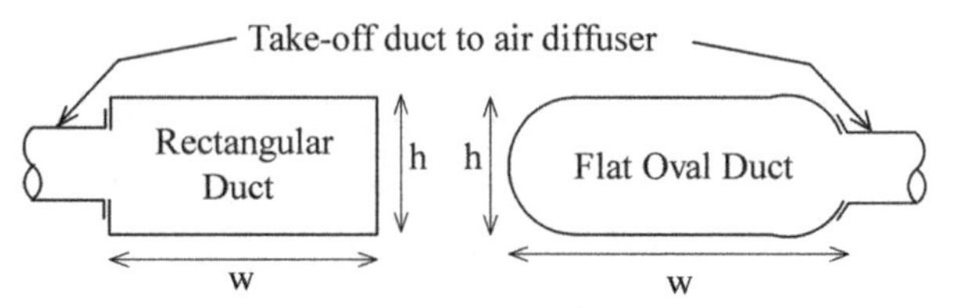

Duct height often constrained by limited ceiling heights but must be high enough to attach side take-offs.

Figure 9.8 Rectangular and flat oval duct dimensions.

Table 9.8 Maximum Recommended Air Velocities (ASHRAE 2003, chapter 47)

Duct Location	Desired RC (N)	Maximum Recommended Duct Velocity (fpm)					
		Main		Branch		Upstream of Outlet	
		Rect.	Round	Rect.	Round	Rect.	Round
In shaft or above drywall ceiling	25	1700	2500	1360	2000	850	1250
	35	2500	3500	2000	2800	1250	1750
	45	3500	5000	2800	4000	1750	2500
Above acoustical tile ceiling	25	1200	2000	960	1600	600	1000
	35	1750	3000	1400	2400	875	1500
	45	2500	4500	2000	3600	1250	2250
In occupied space	25	950	1700	760	1360	475	850
	35	1450	2600	1160	2080	725	1300
	45	2000	3900	1600	3120	1000	1950

Since the common units for friction losses are inches of water, which is head rather than pressure, and the common unit for velocity is fpm rather than fps, Equation 9.8 is adjusted for convenience.

$$\Delta h_f = C\rho\left[\frac{V\ (\text{fpm})}{1097}\right]^2$$

and when

$$\rho = 0.075\ \frac{\text{lbm}}{\text{ft}^3},\ \text{then}\ \Delta h_f = C\left[\frac{V\ (\text{fmp})}{4005}\right]^2 . \qquad (9.9)$$

A variety of loss coefficients for fittings are cataloged in other references (ASHRAE 2005, chapter 35; ACCA 1990). Unfortunately, the use of loss coefficients is cumbersome when applied to design procedures with the equal friction method. A more convenient but less accurate method is to convert the fitting losses to the losses experienced in an equivalent length (L_{eqv}) of straight duct. The loss in a section of duct becomes

$$\Delta h_{Section} = \frac{\Delta h}{100\ \text{ft}}(\text{from friction chart}) \times (L + \Sigma L_{eqv}) . \qquad (9.10)$$

Tables 9.9 and 9.10 were developed from the loss coefficient references (ASHRAE 2005, chapter 35; ACCA 1990) for round and rectangular duct fittings (Figures 9.9 and 9.10, respectively) for velocities that result in friction losses of 0.1 in. per 100 ft and 0.3 in. per 100 ft.

AIR DISTRIBUTION SYSTEM DESIGN AND ANALYSIS

This section provides a summary of the equal friction method. An example problem is given that assumes room air volumes are known and the supply air diffusers, return grilles, and filters have been selected following recommendations in the first sections of this chapter.

Duct Layout and Sizing

1. Start with a full-size plan drawing of the building and locate the air-handling unit (AHU) or fan-coil unit (FCU) for each area of the building.
2. Note the location, airflow, and friction loss of each diffuser on the drawing in each zone.
3. Plan the duct layout to minimize the length of main ducts while avoiding very long take-off lengths near the end of the main duct so that the airflow can be balanced without excessive damper losses at take-offs near the AHU.
4. Sketch the supply main duct and note size reductions after branches or take-offs. It is not necessary to reduce after every take-off, especially when several branches or take-offs are near the same location. Main ducts are typically rectangular cross sections.
5. Sketch the location of take-offs (usually round duct) from the main duct to the diffusers.
6. Size each duct section based on recommended friction per 100 ft of straight duct (i.e., 0.1 in. of water per 100 ft). Note the actual friction loss on a calculation summary sheet or sketch, as it will often be less than 0.1 in. of water since ducts are not usually available to give exact design loss. An air duct calculator (Figure 9.7) is often preferred for this task, but charts (Figure 9.6) are also available.
7. Insert duct sizes on drawings. For rectangular ducts note the size of the cross section to follow the convention of "duct width/duct height for plan view" (ASHRAE 2005, chapter 37) as shown in Figure 9.11. The height may be constrained due to limited clearance in the ceiling space. Round ducts are commonly available in diameters of 5, 6, 8, 10, 12, 14, 16, and 20 in. Avoid other sizes.
8. Note approximate length for each section and equivalent lengths for all fittings on calculation sheet or drawing. Equivalent length values are found in Tables 9.9 and 9.10.
9. Repeat steps 2 through 8 for the return duct and grilles. This may often be a single central grille and single duct, especially when the room(s), connecting corridors, or a ceiling plenum can be used to return the air to a location near the AHU.

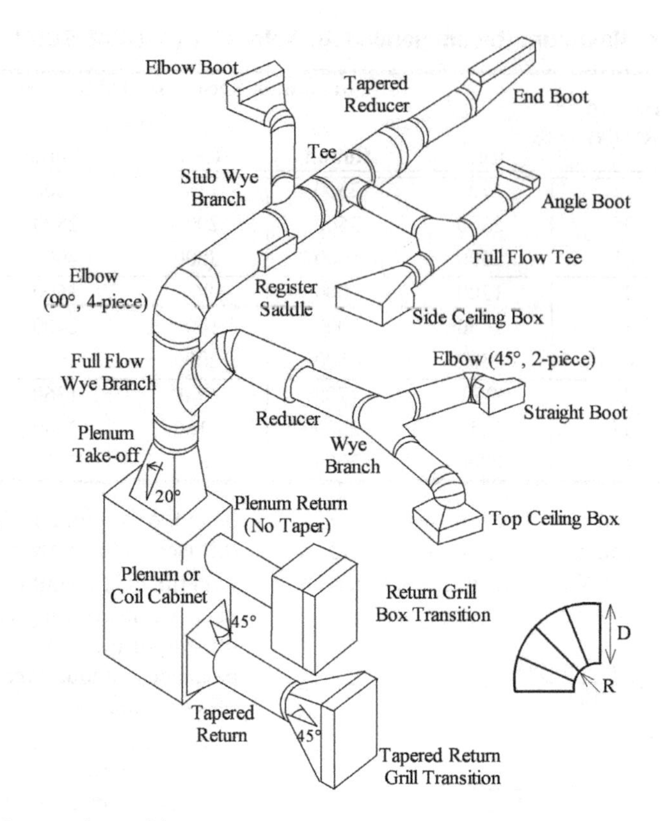

Figure 9.9 Round duct fittings and transitions.

Table 9.9 Equivalent Lengths (L_{eqv}) for Round Duct Fittings and Transitions

Fitting Type	Eqv. Length @ $\Delta h/L$ = 0.1 in./100 ft			Eqv. Length @ $\Delta h/L$ = 0.3 in./100 ft		
	8 in. Ødia.	12 in. Ødia.	16 in. Ødia.	8 in. Ødia.	12 in. Ødia.	16 in. Ødia.
Std. 90°L (R/D = 0.75)	16	30	40	18	22	44
90°L (R/D = 1.5)	8	15	21	10	16	24
Std. 45°L (R/D = 0.75)	11	18	27	12	21	30
45°L (R/D = 1.5)	4	6	9	5	7	10
Tapered (45°) reducer	4	5	10	4	6	11
Non-tapered reducer	8	14	21	9	16	23
Tee (branch)	34	60	87	39	67	97
Full flow tee (branch)	13	22	32	14	24	35
Wye	8	8	13	14	19	21
Stub wye	17	29	43	19	43	38
Full flow wye	14	24	35	16	27	39
Straight boot	5	6	11	6	8	13
End boot	30	55	80	35	60	90
Elbow boot	20	30	48	21	30	53
Register saddle	35	60	90	40	65	95
Top ceiling box	20	35	50	23	40	55
Plenum take-off (20°)	13	22	32	14	24	35
45° Tapered plen. ret.	10	18	26	12	20	30
No-taper return	28	50	70	31	55	80
Return grille box	13	23	33	15	25	37
Tapered (45°) ret. grille	6	10	15	7	11	17

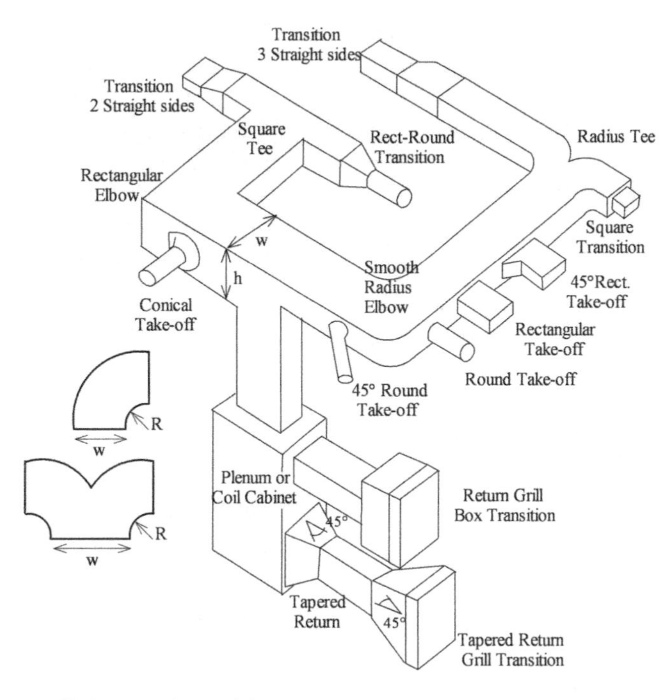

Figure 9.10 Rectangular duct fittings and transitions.

Table 9.10 Equivalent Lengths (L_{eqv}) for Rectangular Duct Fittings and Transitions

Fitting Type	Eqv. Length @ $\Delta h/L$ = 0.1 in./100 ft W/h = 1	Eqv. Length @ $\Delta h/L$ = 0.1 in./100 ft W/h = 4	Eqv. Length @ $\Delta h/L$ = 0.3 in./100 ft W/h = 1	Eqv. Length @ $\Delta h/L$ = 0.3 in./100 ft W/h = 4
Rectangular elbow	65	90	70	100
Rect. elbow with vanes	14	14	25	25
45° rectangular elbow	18	27	20	30
Radius elbow (R/w = 0.75)	24	40	27	45
Radius elbow with vanes	2	4	2	5
Square tee	50	90	55	100
Square tee with vanes	16	30	18	34
Radius tee (R/w = 1.5)	16	30	18	34
(Base L_{eqv} on outlet size)	A_{in}/A_{out} = 2	A_{in}/A_{out} = 4	A_{in}/A_{out} = 2	A_{in}/A_{out} = 4
Square trans.	12	13	19	21
3-straight side transition	5	5	5	5
2-straight side transition	5	5	5	5
Return grille box	19	20	21	22
Plenum outlet (no radius)	40	40	44	44
Rect.-round transition	13	52	14	58
Tapered (45°) return grille	8	9	9	10
	A_{in}/A_{out} = 0.5	A_{in}/A_{out} = 0.25	A_{in}/A_{out} = 0.5	A_{in}/A_{out} = 0.25
45° tapered plen. ret.	15	29	17	32
No-taper return	28	50	31	55
	8 in. ∅dia.	**12 in. ∅dia.**	**8 in. ∅dia.**	**12 in. ∅dia.**
Round take-off	35	60	40	70
45° Round take-off	17	30	20	40
Conical take-off	25	43	28	48
	Q_{TO}/Q_M = 0.1	Q_{TO}/Q_M = 0.4	Q_{TO}/Q_M = 0.1	Q_{TO}/Q_M = 0.4
Rectangular take-off	48	52	55	60
45° rect. take-off	50	55	60	65

10. Select a filter to provide required efficiency and arrestance (MERV 6 is suggested unless the client, application, or local code requires different levels). Design the system to meet the final (dirty filter) loss.*

Duct Friction Loss Calculation/Fan Requirement

1. Find the supply air path of greatest friction loss, which is normally the longest run or near-longest run with a high loss diffuser or take-off section.
2. For this path only, find the loss in each duct section (where duct size or airflow changes) by multiplying loss per unit length by the total equivalent length (straight pipe length + Σ equivalent lengths).

* *Note:* Filter losses can easily become the largest factor in the air distribution system, especially with the higher efficiency of filters required to address indoor air quality concerns. The fan should be selected based on the final recommended loss (dirty filter). However, values suggested by manufacturers are often very high, and filters with larger face areas are suggested to meet the static pressure limitations of energy-efficient fans. Larger filters will reduce face velocity to an acceptable level (300-350 fpm) compared to the velocity at which manufacturers rate losses (500 fpm).

3. Repeat steps 1 and 2 for the return air duct (which may only be a single duct).
4. Calculate the air friction loss by adding the duct losses (in the path of greatest loss): return duct loss, diffuser loss, return grille loss, and filter loss. *Do not add losses in paths that are parallel to the path of greatest loss.*
5. Select a fan that will deliver an external static pressure (ESP) that is equal to or slightly larger than the total friction loss at the required flow rate. The fan should operate near its maximum rated efficiency as noted on the fan curve or performance table.
6. Analyze results and modify the design for sections with excessively high or low losses or velocities.

System Design and Electronic Tools

Effective and thorough design is typically a result of an organized approach that follows a logical sequence and contains some method of flagging all required tasks to minimize the possibility of omissions. Table 9.11 is an example of a tool that can be used to aid the completion of the recommended steps listed in the previous section. The table is also the input and output screen of the

Table 9.11 *E-Ductulator* Spreadsheet Air Distribution System Design Tool (Values Correspond to Figure 9.11)

E-Ductulator - Equal Friction - Equivalent Length Method - See Eqv. Len. Worksheet for Leqv. of Common Fittings

Cell Color Code = Input Output

Supply Air Flow cfm	Duct Width or Dia. if round Inches	Height 0 if round Inches	Velocity fpm	Hyd. Dia. Inches	Roughness Duct Material (ft.)	Δh/100' "Wtr./100'	Length ft.	Leqv A ft.	# As	Leqv B ft.	# Bs	Ltotal	Δh (in.wtr.)
3800	16	26	1315	22.1	Galv Steel (0.00025)	0.089	35	14	2			63	0.056
							Notes-->	.- Vanes					
2800	16	20	1260	19.5	Galv Steel (0.00025)	0.096	15	5	1			20	0.019
							Notes-->						
1900	16	16	1069	17.5	Galv Steel (0.00025)	0.081	15	5	1			20	0.016
							Notes-->	Trans					
1100	14		1029	14.0	Galv Steel (0.00025)	0.099	15	10	1	8	1	33	0.033
							Notes-->	Trans		Wye			
600	12		764	12.0	Galv Steel (0.00025)	0.070	15	30	1	35	1	80	0.056
							Notes-->	90		C. Box			
500	12		637	12.0	Galv Steel (0.00025)	0.050						0	0.000
							Notes-->	Sizing	L=0				
400	10		733	10.0	Galv Steel (0.00025)	0.081						0	0.000
							Notes-->	Sizing	L=0				
300	8		859	8.0	Galv Steel (0.00025)	0.141						0	0.000
							Notes-->	Sizing	L=0				
Diffuser													
cfm	Rated cfm	Rated Δh											
600	628	0.068											0.062
Return	Duct Width or	Height											
cfm	Dia. if round	0 if round				Δh/100'	Length	Leqv A	# As	Leqv B	# Bs	L(total)	
3800	16	26	1315	22.1	Galv Steel (0.00025)	0.089	25	30	1	14	3	97	0.086
							Notes-->	Pl.Ret.					
1	1		183	1.0	Galv Steel (0.00025)	0.115						0	0.000
							Notes-->						
Grill													
cfm	Rated Δh	Rated cfm											
3800	0.067	3845											0.065
Filter	Size		Rated	Rated									
cfm	Height (in.)	Width(in.)	Δh (in.)	V(fpm)	Area (sq.ft.)	V (fpm)							
3800	48	24	0.5	500	8.0	475							0.451
												Total	**0.845**

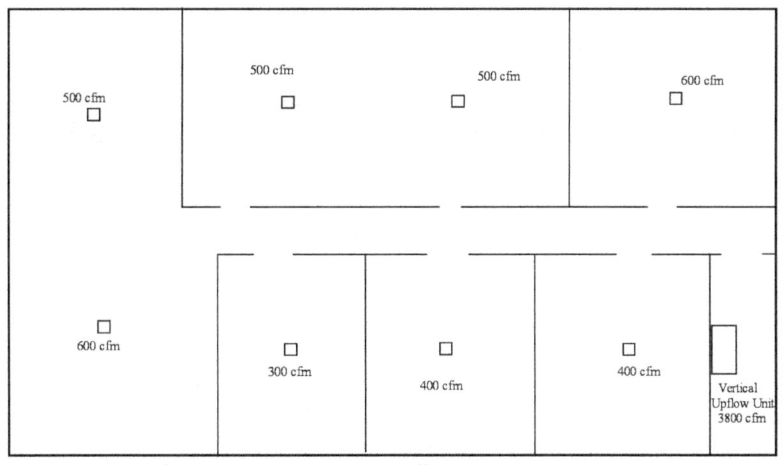

Before Air Distribution Design (Diffusers Selected and Located)

After Air Distribution Design

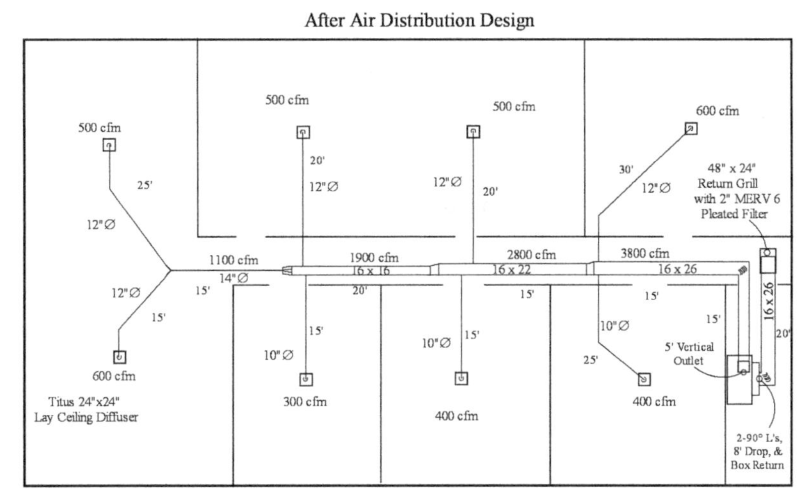

Figure 9.11 Example air distribution system.

> **Example Problem 9.3**
>
> The values shown in Table 9.11 correspond to the results of a design sequence for the building shown in Figure 9.11. This example begins after the selection of the ceiling supply diffusers, which are 24 × 24 in. lay-ins (Table 9.3) and the location of the air-handling unit (AHU). The ductwork is arranged so that the galvanized steel rectangular supply air duct exits the top of the unit, rises 5 ft, and runs over to and down the hall ceiling. Round metal take-offs connect the diffuser ceiling boxes to the main duct. The main rectangular duct terminates at a round transition and a round duct wye. This example uses galvanized steel duct sized at 0.10 in. of water /100 ft with exterior insulation.

spreadsheet program *E-Ductulator.xls* (on the CD accompanying this book), another air distribution system design tool.

Use of the *E-Ductulator* (on the CD accompanying this book) begins with entering the airflow in the first column for each section (3800 cfm in the first section) along the critical path (which is from the unit to either the 500 or 600 cfm diffuser in the large room on the left of the drawing). The height of the main duct is entered in the second column and is held at 16 in. to accommodate the take-offs, which are 10 and 12 in. diameter round ducts. A second dimension is "guessed" and entered into the third column. This column is left blank if round duct is used. Velocity and hydraulic diameter are displayed in the next two columns. The user moves to the sixth column and selects a roughness for the duct wall from a list of alternatives. Galvanized steel has a value of 0.00025 ft. A friction loss appears in the seventh column, and the user adjusts the duct dimensions in columns 2 and/or 3 (in this case, 16 and 26 in.) to provide a loss near 0.1 in. of water/100 ft (0.089). The next columns permit the user to enter the length of the duct and the equivalent length and number of all fittings. These values are summed in the next to last column and multiplied by the friction loss so that the head loss for this first section is displayed in the last column (0.056 in.).

This process is repeated for each major section along the critical path. In this case, 1000 cfm goes to the first two diffusers, so the remaining flow is 2800 cfm. The duct width is adjusted (20 in.) to provide near 0.1 in. of water/100 ft (0.096). The next sections flow 1900 cfm through a 16 × 16 in. duct, 1100 cfm through a 14 in. round duct, and 600 cfm through a 12 in. round duct. Although the path to the 500 cfm diffuser is longer (25 ft), loss is greater in the 600 cfm section because the flow is greater, which results in a greater loss per unit length of duct. Finally, airflows through ducts in the noncritical paths can be sized in the remaining rows of the supply section of the *E-Ductulator* (on the CD accompanying this book). However, the lengths are set to zero so that the results are not included in the head loss summary in the last column.

The next section of the *E-Ductulator* includes the losses of the supply diffuser on the critical path. Table 9.3 indicates the rated loss for a 24 × 24 in. lay-in diffuser is 0.068 in. of water at the nearest rated flow of 628 cfm. Equation 9.1 is applied for the design flow of 600 cfm and a result of 0.062 in. of water appears in the right column.

The third section of the *E-Ductulator* is for the return air duct, which is similar to the arrangement for the supply section. In this case, a single return filter-grille is located in the hall ceiling and the duct travels over the equipment room and down to the unit return air inlet. The duct is the same size as the first section of supply duct and the loss is 0.086 when the friction is multiplied by the equivalent length of return duct.

A 48 × 24 in. filter-grille is selected since even multiples are convenient for ceiling grid installations. Equation 9.1 is applied using the design airflow rate (3800 cfm) and the nearest rated head loss and flow (0.067 in., 3845 cfm) from Table 9.6 to determine the head loss for the filter-grille (0.065 in.).

A MERV 6, 2 in. deep pleated filter is located in the filter-grille. The recommended final loss is 1.0 in. of water. However, this is excessive, so a lower limit of 0.5 in. is set at the rated flow, which results in an actual maximum allowable loss of 0.45 in. when Equation 9.2 is applied.

The final step before fan selection is analysis. The total loss is 0.845 in. of water. Over 50% of this loss is across the filter. Therefore, the first step in reducing head loss should be to provide a filter with a larger face area or to set a lower limit on final head loss. The larger filter could be located in the return near the unit, which can accommodate a larger face area than the selected ceiling filter-grille. Velocity values in the fourth column of the *E-Ductulator* should be compared with the limits of Table 9.8 for ductwork located above an acoustical tile ceiling. All velocities are below the limits for achieving an RC(N) < 35, and only the first two main rectangular ducts do not comply with the limit RC(N) < 25. All round ducts comply with the RC(N) < 25 limit for round ducts connected to the air outlets.

FANS AND FLOW CONTROL METHODS

Fans used in air distribution systems are categorized as either centrifugal or axial types, as shown in Figure 9.12. Centrifugal fans are more typically used in ducted systems that require higher static pressure to overcome friction losses. Axial fans can also be used in ducted systems since tube axial and vane axial designs can provide higher pressures than propeller-type fans. Propeller fans are widely used to provide high flow rates in applications where little pressure is needed, such as condenser, cooling tower, and room ventilation fans.

Figure 9.13 provides a more detailed listing of centrifugal and axial fan types and their performance characteristics, including general pressure vs. flow and efficiency vs. flow curves. The performance curves of total static fan pressure (*P*) vs. airflow rate (*Q*) and fan efficiency vs. airflow rate exhibit a general trend. For

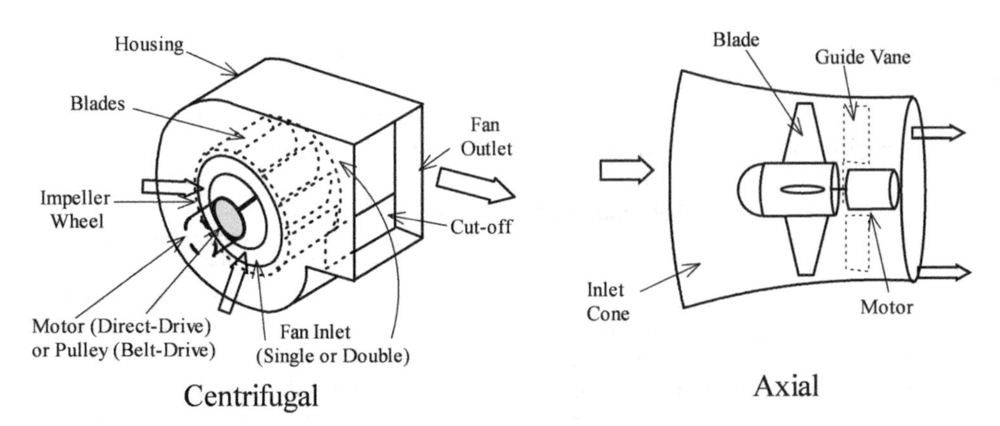

Figure 9.12 Centrifugal and axial fan components.

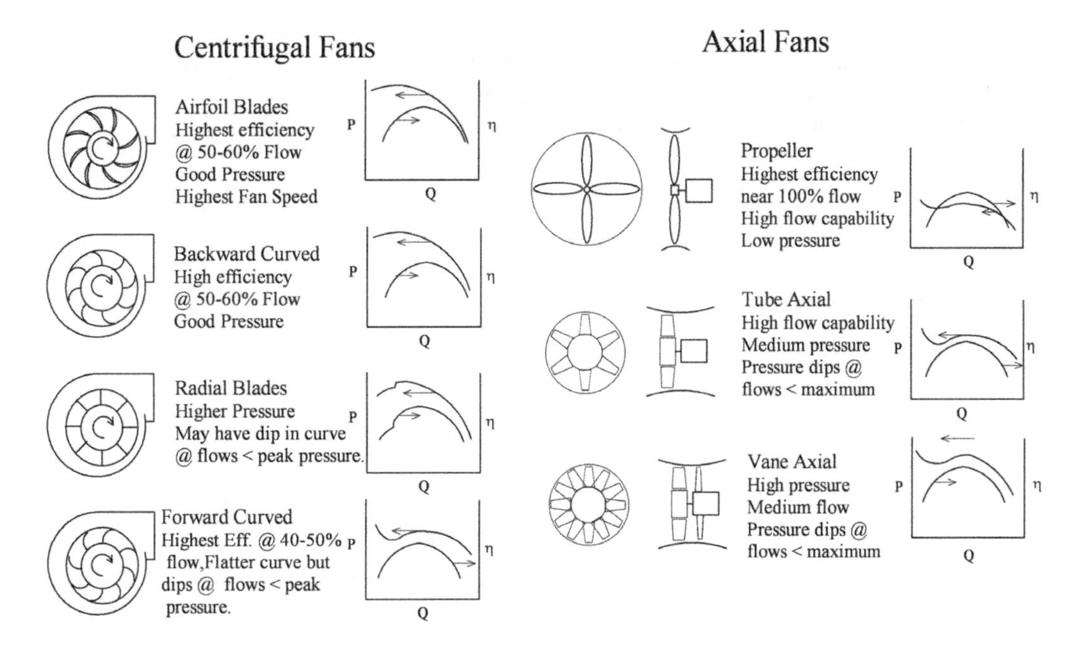

Figure 9.13 Centrifugal and axial fan types and characteristics.

centrifugal fans with backward-curved blades, the pressure rises slightly as flow increases from zero until about 10% to 25% of maximum flow. Efficiency increases over this range more dramatically. As flow continues to increase past the point of maximum pressure, the pressure decreases but the efficiency continues to increase until about 50% to 60% of maximum flow is reached. *This is the range in which fans should be selected to operate.* After the point of maximum efficiency, the efficiency and the pressure both decline. Centrifugal fans with radial blade, forward-curved vanes and axial fans have similar curves except for a noticeable dip in fan pressures at flows slightly less than the flow at which maximum pressures occur. These fans should not be selected to operate below the flow rate where the dip occurs.

Fans are typically selected to overcome the static pressure of the air distribution system, and performance data often are presented in the form of total static pressure (TSP = p_s). The common terminology is somewhat misleading since this value is actually a change or rise in pressure. However, some manufacturers may express performance in terms of total pressure (TP or p), which is the sum of the change (Δ) in static pressure (TSP or p_s), velocity pressure (VP or p_v), and elevation.

$$TP(\text{in. of water}) = \Delta p = TSP + VP = (\Delta p_s + \Delta p_v)$$
$$= (p_{s,out} - p_{s,in}) + \left(\frac{V_{out}\,(\text{fpm})}{4005}\right)^2 - \left(\frac{V_{in}\,(\text{fpm})}{4005}\right)^2 \quad (9.11)$$

In many cases the difference in the outlet and inlet velocity is small, so the velocity pressure is negligible.

Another common practice is to express performance data for air handlers as external static pressure (ESP), which is the TSP of the fan less the losses in the coil (Δh_{coil}), cabinet ($\Delta h_{cabinet}$), internal dampers ($\Delta h_{dampers}$), and *clean* filter (Δh_{filter}).

$$ESP\ (\text{in. of water}) = TSP - \Delta h_{coil} - \Delta h_{cabinet} - \Delta h_{dampers} - \Delta h_{clean\ filter} \quad (9.12)$$

The total pressure is needed to determine the required input power to a fan, which is the product of the pressure difference and airflow. When the traditional units for pressure difference (in. of water) and airflow rate (cfm) are used, the required fan power is

$$w_{Req'd}\ (\text{hp}) = \frac{Q\ (\text{cfm}) \times \Delta p\ (\text{in. of water})}{6350 \times \eta_{fan}}$$
(a.k.a. bhp in fan data). (9.13)

The electrical power input to the fan ($w_{M,In}$) must also consider the efficiency of the motor (η_M) and the variable-speed drive (η_{VSD}) if one is used. The traditional units for fan power (hp) are converted to electrical demand (kW).

$$w_{M,In}\ (\text{kW}) = \frac{0.746\ (\text{kW/hp}) \times w_{Reqd}\ (\text{hp})}{\eta_M \times \eta_{VSD}}$$
(if no VSD is used, $\eta_{VSD} = 100\%$). (9.14)

An often neglected consideration is the fan power that is converted to heat. In the cooling mode, this item can be a significant penalty, especially in systems with large, central, high static pressure air-handlers and parasitic fan power, variable air volume (FPVAV) terminals. Traditional practice is to locate the motor in the airstream so that all motor and fan inefficiency is converted immediately to heat and the remaining useful fan power is converted into heat via the friction of duct, dampers, coils, and other restrictions. The fan heat is a penalty in the cooling mode (although some reheat may be beneficial in high humidity applications) and a benefit in heating. Fan heat is expressed in traditional units (MBtu/h) using the conversion of 2.545 MBtu/hp·h.

$$q_{Fan}\ (\text{MBtu/h}) = \frac{Q\ (\text{cfm}) \times \Delta p\ (\text{in. of water})}{2495 \times \eta_{fan} \times \eta_M} \quad (9.15)$$

Note that the impact of the VSD drive efficiency (η_{VSD}) is not considered in Equation 9.15 since this component is typically located outside the air distribution system.

The increasing use of VSDs is a result of the significant energy-saving potential of these devices. More details will be discussed in chapter 11, but the fan laws are presented here to demonstrate the connection of these devices to fans (and pumps). Fan airflow rate, pressure loss (or increase), and power are functions of fan speed (rpm). Equations 9.16, 9.17, and 9.18 are referred to as the basic fan laws. Additional laws are listed in the *ASHRAE Handbook—HVAC Systems and Equipment* (ASHRAE 2004, chapter 18).

$$Q_2\ (\text{cfm}) \approx Q_1\ (\text{cfm}) \times \left(\frac{\text{rpm}_2}{\text{rpm}_1}\right) \quad (9.16)$$

$$\Delta p_2\ (\text{in. of water}) \approx \Delta p_1\ (\text{in. of water}) \times \left(\frac{\text{rpm}_2}{\text{rpm}_1}\right)^2 \quad (9.17)$$

$$w_2\ (\text{hp}) \approx w_1\ (\text{hp}) \times \left(\frac{\text{rpm}_2}{\text{rpm}_1}\right)^3 \quad (9.18)$$

The equations are theoretical approximations since the implied assumption is that the efficiencies of the fan, motor, and drive are constant. In most applications of modern equipment, this assumption does not result in significant error for fan speeds between 50% and 100% of rated speed. However, below this speed, error will increase and the equations are likely to be invalid, especially below 30% of full speed. Figure 9.14 is a plot of system efficiencies ($\eta_S = \eta_{Fan} \times \eta_M \times \eta_{VSD}$) for four different variable-speed drive systems. An equation that can be used to express system efficiency (η_s) as a function of system full efficiency has been derived to express this relationship (Kavanaugh and Lambert 2004).

$$\eta_s = \left(\frac{\eta_{s\ \text{at}\ 100\%}}{0.58}\right) \times \left(-0.48\left(\frac{\text{rpm}}{\text{rpm}_{100\%}}\right)^2 + 0.924\left(\frac{\text{rpm}}{\text{rpm}_{100\%}}\right) + 0.125\right) \quad (9.19)$$

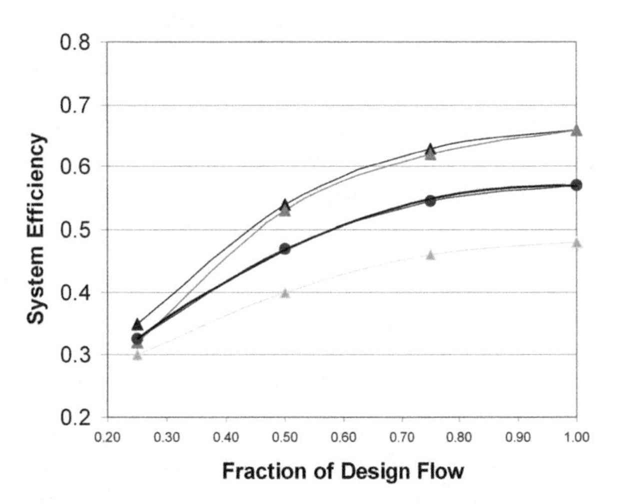

Figure 9.14 Variable-speed system efficiency.

Tables 9.12 and 9.13 present fan performance in a traditional format for engineering design and selection

Table 9.12 Performance of Small Direct-Drive and Belt-Drive Centrifugal Fans

	Direct-Drive Fan Airflow Rate (cfm)					
rpm	1/3 hp Fan—TSP (in. of water)					bhp
	0.3	0.5	0.7	0.9		
		1160	1030	510		0.14
	1325	1320	1210	720		0.17
	1500	1460	1340	800		0.25
1075	1940	1740	1540	1080		0.33
rpm	1/2 hp Fan—TSP (in. of water)					bhp
	0.5	0.7	0.9	1.1		
		1120	1030	660		0.17
		1240	1170	760		0.25
		1360	1290	990		0.33
1075	2000	1900	1765	1540		0.5
	Belt-Drive Fan Airflow Rate (cfm)					
rpm	1-1/2 hp Fan—TSP (in. of water)					bhp
	0.5	0.75	1	1.25	1.5	
990	1640	1140				0.5
1110	1950	1660	960			0.75
1210	2205	2010	1660			1
1360	2540	2400	2190	1870	1260	1.5
rpm	2 hp Fan—TSP (in. of water)					bhp
	0.5	0.75	1	1.25	1.5	
1070	2290	1870				0.75
1150	2550	2230	1642			1
1300	3000	2780	2460	1950		1.5
1410	3310	3120	2875	2540	1990	2
rpm	3 hp Fan—TSP (in. of water)					bhp
	0.5	0.75	1	1.25	1.5	
735	3130	2490				1
830	3750	3300	2630			1.5
930	4400	4050	3620	3020		2
1010	4904	4750	4240	3820	3220	3

Table 9.13 Performance of 4,000 through 20,000 cfm Drive Centrifugal Fans

Unit Size 10 FG-12 in. AF														
	1.0 in. TSP		**1.5 in. TSP**		**2.0 in. TSP**		**2.5 in. TSP**		**3.0 in. TSP**		**3.5 in. TSP**		**4.0 in. TSP**	
cfm	rpm	bhp	rpm	bhp	rpm	bhp	rpm	bhp	rpm	bhp	rpm	bhp	rpm	bhp
4000	2530	2.59	2640	2.95	2760	3.29	2890	3.69	3000	4.08	3100	4.47	3190	4.85
5000	3060	4.67	3150	5.04	3240	5.47	3330	5.91	3430	6.35	3530	6.81	3630	7.3
6000	3600	7.4	3680	7.95	3760	8.49	3830	9.03	3900	9.55				
Unit Size 17 R-24.5 in. AF (Note TSP Scale Change [2.0 to 5.0 in. of water])														
	2.0 in. TSP		**2.5 in. TSP**		**3.0 in. TSP**		**3.5 in. TSP**		**4.0 in. TSP**		**4.5 in. TSP**		**5.0 in. TSP**	
cfm	rpm	bhp	rpm	bhp	rpm	bhp	rpm	bhp	rpm	bhp	rpm	bhp	rpm	bhp
6000	1550	2.99	1640	3.64	1730	4.33	1820	5.06	1900	5.85				
7500	1770	4.18	1850	4.91	1920	5.68	1990	6.41	2050	7.25	2130	8.12	2310	9.05
9000	2000	5.63	2080	6.52	2140	7.41	2210	8.29	2270	9.19	2330	10.1	2400	11
Unit Size 30 DE-25 in. AF (Note TSP Scale Change [2.0 to 5.0 in. of water])														
	2.0 in. TSP		**2.5 in. TSP**		**3.0 in. TSP**		**3.5 in. TSP**		**4.0 in. TSP**		**4.5 in. TSP**		**5.0 in. TSP**	
cfm	rpm	bhp	rpm	bhp	rpm	bhp	rpm	bhp	rpm	bhp	rpm	bhp	rpm	bhp
10000	984	4.67	1055	5.66	1126	6.72	1195	7.84	1266	9.09	1335	10.4		
15000	1207	8.95	1270	10.2	1326	11.6	1379	13	1428	14.3	1476	15.8	1524	17.2
20000	1460	16.2	1507	17.8	1554	19.6	1603	21.1	1653	22.7	1699	24.5	1741	26.3

purposes. Table 9.12 gives the performance of small direct-drive fans coupled to multi-speed motors. The high-speed rpm is provided for the direct-drive fans with the required fan power at each speed. Note the traditional unit for required fan input power is brake horsepower (bhp). Airflow rate (cfm) is given as a function of static pressure. Information for belt-drive fans is given in a similar format, and the actual fan speeds at lower values are given. This permits the designer to match the fan speed and motor power to system requirements by using various fan and motor pulley diameter ratios.

$$rpm_{Fan} = rpm_M \times \left(\frac{D_{MotorPulley}}{D_{FanPulley}}\right) \quad (9.20)$$

Table 9.13 provides similar information for larger belt-drive fans. Note also that Table 5.2 of this book provides flow rate, bhp, and electrical demand as a function of external static pressure (ESP) for a series of rooftop units (RTUs).

Performance of fans can also be provided in the form of curves of airflow and total static pressure, as shown in Figure 9.15. These curves are plotted for several speeds as shown (1800, 1500, and 1200 rpm). Lines of constant fan power are also shown. The data in the figure are for the largest fan (25 in. wheel diameter) listed in Table 9.13 but with higher TSP values. While there is some loss of accuracy with this format, system curves can also be plotted on the same graph. This will allow easy interpolation of required fan speed and required fan motor size. Fan manufacturers also provide fan performance in electronic format. This includes fan selection programs.

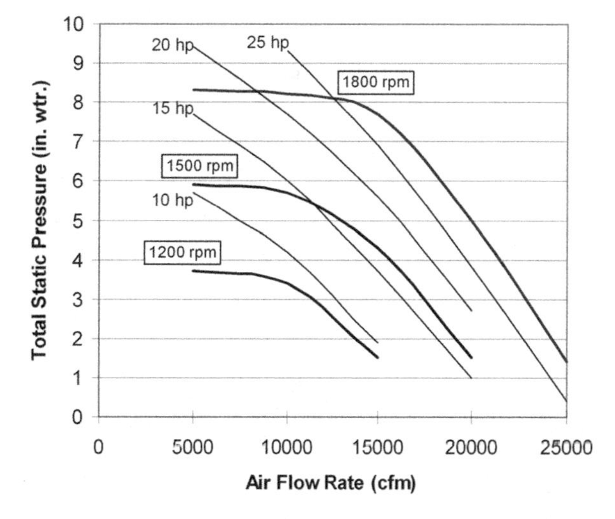

Figure 9.15 Total static pressure and power curves for three speeds of 25 in. fan.

System curves can be generated from a single design operating point (Δh, cfm) by applying Equation 9.2 to find the system loss (Δh) for airflow rates above and below the design flow and drawing a curve intersecting the three points. Curves of constant fan speed will intersect the system curve to generate operating points for off-design fan speeds. The required motor power can be determined by interpolating the resulting operating points between lines of constant power.

Example Problem 9.4

A ducted 16,000 cfm air distribution system with the fan whose performance is given in Figure 9.15 has an external static pressure loss of 3.0 in. of water and an air handler loss of 2.1 in. of water.

a. Find the required fan power, fan speed, and fan pulley diameter if the motor speed is 1750 rpm and the drive pulley diameter is 6 in.
b. Generate a system curve to find the pressure loss, required fan power, and fan speed at 13,000 and 17,000 cfm.
c. If dampers are used to produce a loss of 4.0 TSP at 10,000, find the required fan power and speed and generate a system curve for the dampened system.

Solution:

a. The total static pressure is the sum of the ESP and the air handler losses.

$$TSP = 3.0 \text{ in.} + 2.1 \text{ in.} = 5.1 \text{ in. of water at } 16{,}000 \text{ cfm}$$

The operating point is the intersection of 16,000 cfm and 5.1 in. of water, and a triangle marker is located on Figure 9.16. The fan power required is 20 hp since the 20 hp line exactly intersects the operating point. By interpolation between 1500 and 1800 rpm, the required fan speed is approximately 1600 rpm. Using this speed and Equation 9.20, the fan pulley size is

$$D_{FanPulley} = D_{MotorPulley} \times (\text{rpm}_M \div \text{rpm}_{Fan}) = 6.0 \text{ in.} \times (1750 \div 1600) = 6.5625 \text{ in.} = 6\text{-}9/16 \text{ in.}$$

b. To generate a system curve, the design operating point is used with Equation 9.2 for 12,000 and 20,000 cfm.

$$\Delta h_2 \approx \Delta h_1 \times \left(\frac{Q_2}{Q_1}\right)^2 : \ \Delta h_{12000} \approx 5.1 \times \left(\frac{12000}{16000}\right)^2 = 2.9 \text{ in.}, \ \Delta h_{20000} \approx 5.1 \times \left(\frac{20000}{16000}\right)^2 = 8.0 \text{ in.}$$

A curve is drawn through the points (12,000 cfm/2.9 in.), (16,000 cfm/5.1 in.), and (20,000 cfm/8.0 in.) as shown in Figure 9.16.

The curve passes through the point of 13,000 cfm at TSP ≈ 3.6 in., w ≈ 12.5 hp, and rpm ≈ 1350. At 17,000 cfm the values are TSP ≈ 5.8 in., w ≈ 25 hp, and rpm ≈ 1700.

c. At the operating point of 10,000 cfm and 4.0 in., the required fan power and speed are approximately 10 hp and 1300 rpm. A dampened curve is generated using Equation 9.2.

$$\Delta h_2 \approx \Delta h_1 \times \left(\frac{Q_2}{Q_1}\right)^2 : \ \Delta h_{6000} \approx 4.0 \times \left(\frac{6000}{10000}\right)^2 = 1.44 \text{ in.}, \ \Delta h_{14000} \approx 4.0 \times \left(\frac{14000}{10000}\right)^2 = 7.8 \text{ in.}$$

A second system curve is drawn on Figure 9.16 through the points (6,000 cfm/1.44 in.), (10,000 cfm/4.0 in.), and (14,000 cfm/7.8 in.) as shown.

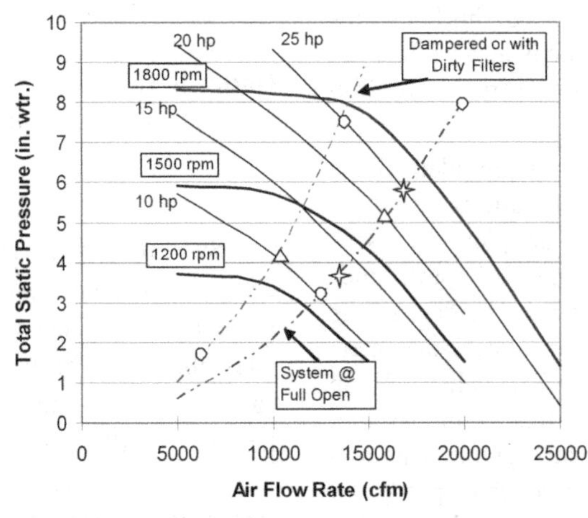

Figure 9.16 Full-open and restricted system curves plotted on fan curve.

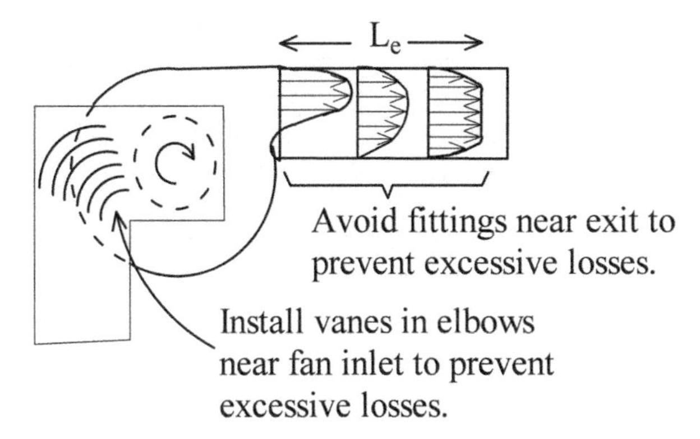

Figure 9.17 Fan inlet swirl and outlet system effect.

Design Fan—Duct Interface

The inlet and outlet duct connections are extremely critical to achieving rated fan performance. Inlet flow should be near-axial to the shaft of a centrifugal fan. Elbows placed near the inlet will create swirl unless turning vanes are installed to induce axial flow, as shown in Figure 9.17. Velocity profiles of exit duct airflow that contribute to the fan system effect are also shown in the figure. Note the exit velocity profile is non-uniform and high losses will result if fittings are located immediately downstream before the profile becomes uniform. The recommended length (L_e) of straight duct at the fan exit (ASHRAE 2005, chapter 35) is

$$L_e\ (\text{ft}) = \frac{V_o\sqrt{A_o\ (\text{in.}^2)}}{10,600}(V_o > 2500\ \text{fpm})$$

$$\text{or} \quad L_e\ (\text{ft}) = \frac{\sqrt{A_o\ (\text{in.}^2)}}{4.3}(V_o \le 2500\ \text{fpm}), \tag{9.21}$$

where

V_o = average outlet velocity (fpm) and
A_o = fan outlet area (in.^{2}).

Installations with smaller values of L_e should consult the references (AMCA 1990; ASHRAE 2005, chapter 35) for flow coefficients that permit local losses for various outlet duct arrangements to be determined. Corrections are also provided for other abrupt duct outlet arrangements.

CHAPTER 9 PROBLEMS

9.1 Room 255 (Appendix C, Figure C.7) has a 9 ft ceiling and an 18 × 40 ft floor. Select square ceiling diffusers to distribute 1400 cfm of air and provide an ADPI > 90%, an NC < 30, and a TP < 0.1 in. of water. The solution should include a sketch.

9.2 Repeat Problem 9.1 for room 256 (Appendix C, Figure C.7), which has a 40 × 40 ft floor with 2200 cfm.

9.3 Size a MERV 6 filter and a matching filter-grille for room 255 (Appendix C, Figure C.7) that will limit the final resistance to 0.5 in. of water. One dimension should be a multiple of 24 in. if possible.

9.4 Repeat Problem 9.3 for room 256 (Appendix C, Figure C.7).

9.5 Select a unit to either fit in the hallway outside room 255 (Appendix C, Figure C.7) or above the ceiling (42 in. vertical space) and route metal supply ductwork with round take-offs and metal return ductwork.

9.6 Repeat Problem 9.5 for room 256 (Appendix C, Figure C.7).

9.7 Use the *E-Ductulator* program on the CD accompanying this book to design an air distribution system to deliver 6000 cfm evenly in the building below. Provide a MERV 6 filter and limit total losses to less than 1.2 in. of water.

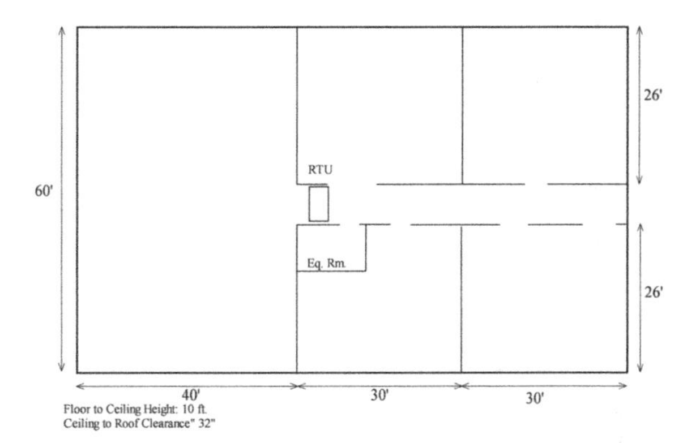

9.8 Select a fan with a direct-drive motor to provide 1200 cfm at a TSP of 0.8 in. of water. Specify the required speed tap setting, resulting motor power output, and estimated required power demand.

9.9 Select a belt-drive fan and motor to provide 2000 cfm at a TSP of 1.2 in. of water. Specify the required fan pulley diameter for a 1750 rpm motor with a 6 in. diameter drive pulley, resulting motor power output, and estimated required power demand.

9.10 Select a belt-drive fan and motor to provide 7500 cfm at a TSP of 3.25 in. of water. Specify the required fan pulley diameter for a 1750 rpm motor with an 8 in. diameter drive pulley, resulting motor power output, and estimated required power demand.

9.11 Select a motor and specify the resulting bhp and fan speed to provide 1.2 in. of water external static pressure (ESP) and 6000 cfm for a Model 180 (Table 5.2 of this book) rooftop unit. Provide the fan pulley diameter for a 1750 rpm motor with a 4 in. diameter drive pulley.

9.12 A centrifugal fan has the following characteristics at 1000 rpm:

Δh = 2.96 in. of water at 2400 cfm
Δh = 2.94 in. of water at 4800 cfm
Δh = 2.57 in. of water at 7200 cfm
Δh = 2.09 in. of water at 9600 cfm
Δh = 1.35 in. of water at 12000 cfm

Develop a fan curve for 1000 rpm and calculate the required input power (hp) assuming a 75% efficiency at 7200 cfm, 70% at 4800 and 9600 cfm, and 65% at 2400 and 12,000 cfm. Calculate the required motor input (kW) assuming a 93% efficient motor.

9.13 Using the fan laws, develop fan curves for 800 and 600 rpm.

9.14 The fan described in Problem 9.12 is connected to an air distribution system that has a loss of 1.8 in. of water at 9500 cfm when the dampers are open and 3.0 in. of water at 5000 cfm when the dampers are set at a minimum opening. Develop a system curve for both situations (dampers full open and minimum) and find the resulting flow when the fan is turning 1000, 800, and 600 rpm for both situations (dampers full open and minimum). Estimate the required power at all six points. ***Note:*** efficiency will remain nearly constant with varying fan speed when the fan law $[\Delta h_2 = \Delta h_1 \times (rpm_2/rpm_1)^2]$ is applied.

REFERENCES

AAF. 2004. Product Data. Louisville, KY: American Air Filter.

ACCA. 1990. *Low Pressure, Low Velocity Duct System Design, Manual Q*. Washington, DC: Air-Conditioning Contractors of America.

AG. 2004. *Extended Surface Pocket Filters—Technical Data*, TD-109-B. Louisville, KY: AirGuard.

AMCA. 1990. *Fans and Systems*, Publication 201-90. Arlington Heights, IL: Air Movement and Control Association, Inc.

ASHRAE. 1999. *ANSI/ASHRAE Standard 52.2-1999, Method of Testing General Air-Cleaning Devices for Removal Efficiency by Particle Size*. Atlanta: American Society of Heating, Refrigerating and Air-Conditioning Engineers, Inc.

ASHRAE. 2003. *2003 ASHRAE Handbook—HVAC Applications,* chapter 47, Sound and vibration control. Atlanta: American Society of Heating, Refrigerating and Air-Conditioning Engineers, Inc.

ASHRAE. 2004. *ANSI/ASHRAE Standard 62.1-2004, Ventilation for Acceptable Indoor Air Quality*. Atlanta: American Society of Heating, Refrigerating and Air-Conditioning Engineers, Inc.

ASHRAE. 2004. *ASHRAE Handbook—HVAC Systems and Equipment,* chapter 18, Fans; chapter 24, Air cleaners for particulate contaminants. Atlanta: American Society of Heating, Refrigerating and Air-Conditioning Engineers, Inc.

ASHRAE. 2005. *2005 ASHRAE Handbook—Fundamentals,* chapter 33, Space air diffusion; chapter 35, Duct design; chapter 37, Abbreviations and symbols. Atlanta: American Society of Heating, Refrigerating and Air-Conditioning Engineers, Inc.

Kavanaugh, S.P., and S.E. Lambert. 2004. A bin method energy analysis for ground-coupled heat pumps. *ASHRAE Transactions* 110(1):535–542.

Nailor. 2003. *Grilles and Registers*. Toronto: Nailor Industries, Inc.

Rock, B.A., and D. Zhu. 2002. *Designer's Guide to Ceiling-Based Air Diffusion*. Atlanta: American Society of Heating, Refrigerating and Air-Conditioning Engineers, Inc.

Schaffer, M.E. 1991. *A Practical Guide to Noise and Vibration Control for HVAC Systems*. Atlanta: American Society of Heating, Refrigerating and Air-Conditioning Engineers, Inc.

T/ASC. 1996. "Ductulator." The Trane Company, American Standard Company, LaCrosse, WI.

Titus. 1996. Model TMS High Performance Square Ceiling Diffusers. Titus Co., Richardson, TX.

10 Water Distribution System Design

Water distribution systems have many similarities to air distribution systems. Centrifugal pump performance is similar to fan performance; piping design follows procedures similar to duct design; and electric motors are the drivers of choice. Design tasks include the following.

- Determine the location, liquid flow rate requirements, and corresponding friction losses of the terminal units (fan coils, air-handling units, water-source heat pumps).
- Determine the location of the chilling and water heating equipment, liquid flow rate requirements, and corresponding friction losses.
- In the case of the water-cooled equipment, locate the cooling tower, fluid cooler, or ground heat exchanger. Determine the flow rate requirements, corresponding friction losses in the devices and the chiller condenser, and the elevation head for open cooling towers. Follow guidelines (ASHRAE 2000) when locating cooling towers and fluid coolers.
- Select the most appropriate piping material (considering cost, ease of installation, resistance to corrosion, pressure requirements, and wall thickness).
- Determine the most appropriate piping routes, and determine the required flow rate in each section in all paths.
- Size the piping in each section to minimize required diameters without exceeding recommended ranges for friction losses and velocities (ASHRAE 2005, chapter 36).
- Select and locate flow control and/or flow balancing devices (balance valves, automatic flow control valves, two- and three-way valves, etc.).
- Determine the total friction losses (and open system elevation heads) for piping critical paths (loops in which friction losses are maximum or near maximum).
- Sum component friction losses (piping, fittings, valves, water coils, etc.) in critical paths.
- Locate and select pumps to provide required flow rate and adequate head to overcome critical path friction losses under specified operating conditions (temperature, fluid type, available inlet pressure, etc.).
- Provide a means of controlling flow rate and/or pumps to minimize energy consumption during the many hours when full pump capacity is not required.
- Specify pipe insulation thicknesses to minimize heat losses and gains, prevent pipe freezing, and eliminate condensation on chilled water lines.
- Select and specify pipe hanging systems and thermal expansion precautions.
- Select and specify environmentally acceptable water treatment products to minimize corrosion and fouling and to avoid freezing.
- Locate and size necessary accessories (expansion tanks to avoid excessive system pressure deviations, air venting devices, flow and pressure measurement ports and/or instruments, pump inlet diffusers when pipe elbows near inlets are necessary, etc.).

A complete description of all these tasks is not possible in this chapter. However, several of the primary tasks are discussed, and references are provided for the remaining items.

PIPING LOOP SYSTEMS

Several types of water loops have already been discussed in chapter 5. Figure 5.10, a two-pipe chilled water system (CWS) with a water heater (boiler), is repeated as Figure 10.1 in this chapter. A supply and return piping network (two-pipe) circulates water from either the chiller or water heater to the fan-coil units (FCUs) in the zones. Changeover valves are used to switch the entire system from heating to cooling. However, *simultaneous heating and cooling in this configuration is not possible.* Note that some heating can be provided with the addition of an auxiliary electric heater

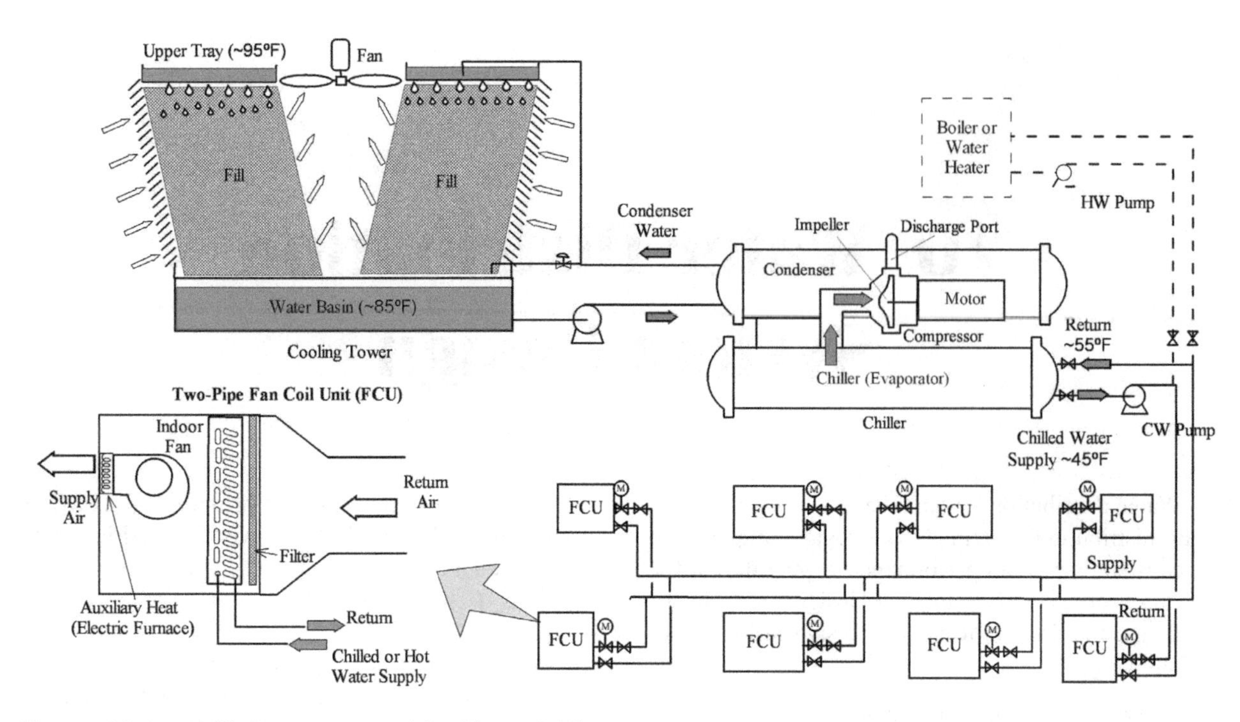

Figure 10.1 Chilled water system (also Figure 5.10).

in the fan-coil unit as shown. The condenser water loop shown in Figure 10.1 consists of an open cooling tower, which cools water that has been used to absorb heat rejected by the chiller system, a pump, and a piping loop to circulate cooling water from the tower basin to the compressor condenser.

Additional water-loop systems were presented in chapter 5, including several exterior ground-source heat pump (GSHP) water distribution systems in Figure 5.8. Figure 5.9 is a diagram of a piping loop for a water-source heat pump (WSHP) system connected to a boiler and cooling tower. This type of loop is similar to the interior piping of a closed-loop GSHP.

The direct-return piping diagram in Figure 10.1 is a schematic of a chilled water loop connecting several fan-coil units (FCUs) in a direct-return loop. Figure 10.2a provides more details of a direct-return piping loop for several fan-coil units. Each fan coil is connected to the main piping loop using two isolation valves, which permit the unit to be closed off and serviced. A motorized two-way valve may be included that can vary flow to the unit to control cooling or heating. This type of valve is essential when variable-volume pumping is incorporated. Note that one valve serves a dual function of being an isolation valve and a manual flow-balancing valve. The flow balancing, or regulating, function can be accomplished by a variety of devices including automatic flow control valves (ASHRAE 2004, chapter 42). Connections are typically made through flexible hoses with swivel connectors with ports to measure water temperature and pressure.

In longer piping runs, the friction losses may be large compared to the losses through the FCUs or heat pump water coils. Units at greater distances from the pump(s) may receive insufficient flow, and coils closer to the pump will receive the excess. In some cases it is advisable to arrange the piping in a reverse-return loop as shown in Figure 10.2b. This will minimize the flow imbalance because all piping runs are nearly equal. However, the added return piping run will increase cost, and reverse-return layouts are no longer common except when the terminal units can be located in a near circular layout as shown in Figure 10.2c. In these cases, the added required length of the return piping is minimal.

Historically, chilled water systems used three-way control valves to achieve constant water flow through the chiller (ASHRAE 2004, chapter 12). Figure 10.3 is a diagram of a configuration of a system of two chillers and four air-handling units (AHUs) with three-way valves. At full-load operating conditions, the system would typically be designed to operate with 44°F or 45°F water leaving the chiller through the supply piping. The bulk of the water exits the AHU water coil near 54°F or 55°F and returns through the left port of the three-way valve. Little or no liquid is bypassed around the coil through the center port. As the load on the AHU coil decreases, more liquid is bypassed around the water coil. This liquid mixes with the warmer water that is cir-

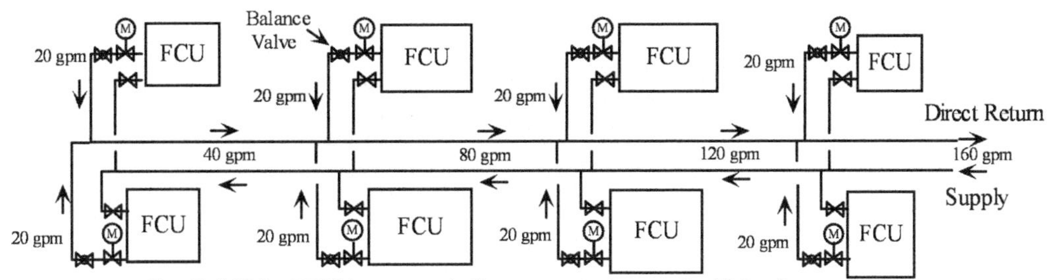

Fan Coil Units (FCUs) near supply from pump tend to receive higher flow rate since overall flow path is shorter than coils at the end. Balancing required.

(a) Direct-Return Piping

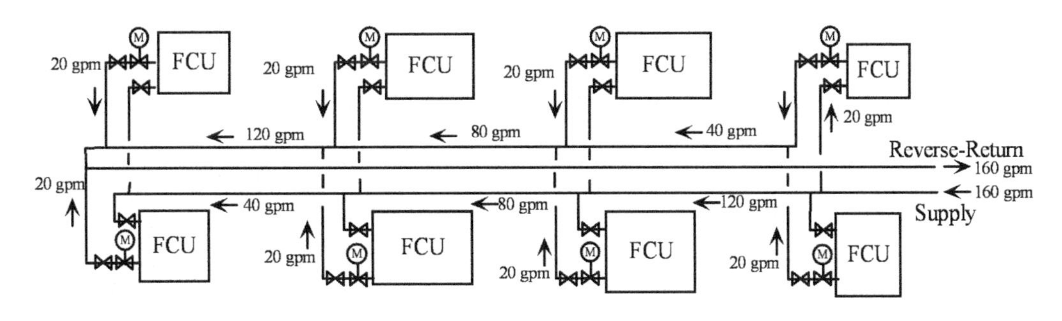

Fan Coil Units (FCUs) tend to get equivalent flow rates since overall flow paths through each coil is nearly equal. However, piping cost will increase because of added length of return header.

(b) Reverse-Return Piping

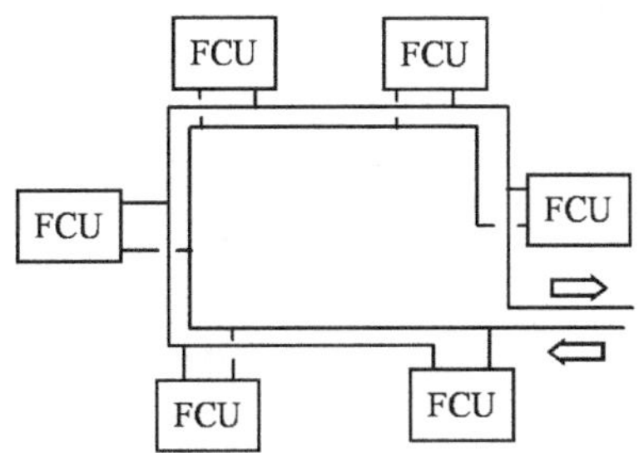

(c) Modified "Reverse-Return" with Minimal Added Piping

Figure 10.2 Direct-return and reverse-return water loop piping.

culated through the coil. The temperature of the mixture returning to the chiller is below the full-load return temperature of 54°F or 55°F. However, the flow through the chillers can be held to a near constant value. Chiller control is achieved by sensing return water temperature and turning off one of the chillers when this temperature indicates the loads on the AHUs can be handled by a single chiller. This condition is typically indicated when the supply-return temperature difference is equal to or less than half of the design temperature difference (~5°F for the conditions used in this discussion).

Figure 10.3 includes a second pump. One option is to install both pumps so that either one is capable of handling the full capacity of the system. This would permit one pump to operate while the second serves as the backup pump. Another option is to specify pumps so that both are required to meet the system's full load. If one fails or is taken out of service for maintenance, approximately 70% of the full-load flow rate capacity is available because the remaining pump will be able to operate at reduced heads due to the lower flow rate.

Figure 10.3 also includes an expansion tank that will moderate the pressure variations in the piping loop that result from the expansion of water with increasing temperatures relative to the expansion capacity of the metal piping and other components in the loop. Recommendations for sizing and locating these tanks are available in the *2004 ASHRAE Handbook—HVAC Systems and Equipment*, chapter 12.

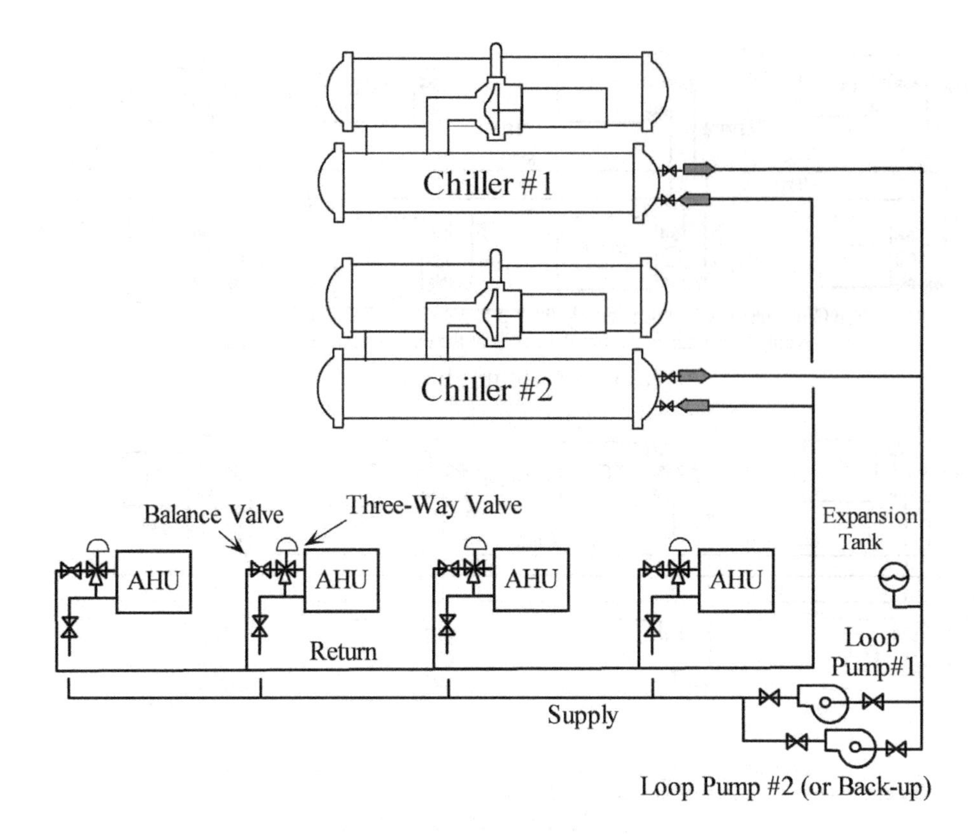

Figure 10.3 Constant-flow chilled water system.

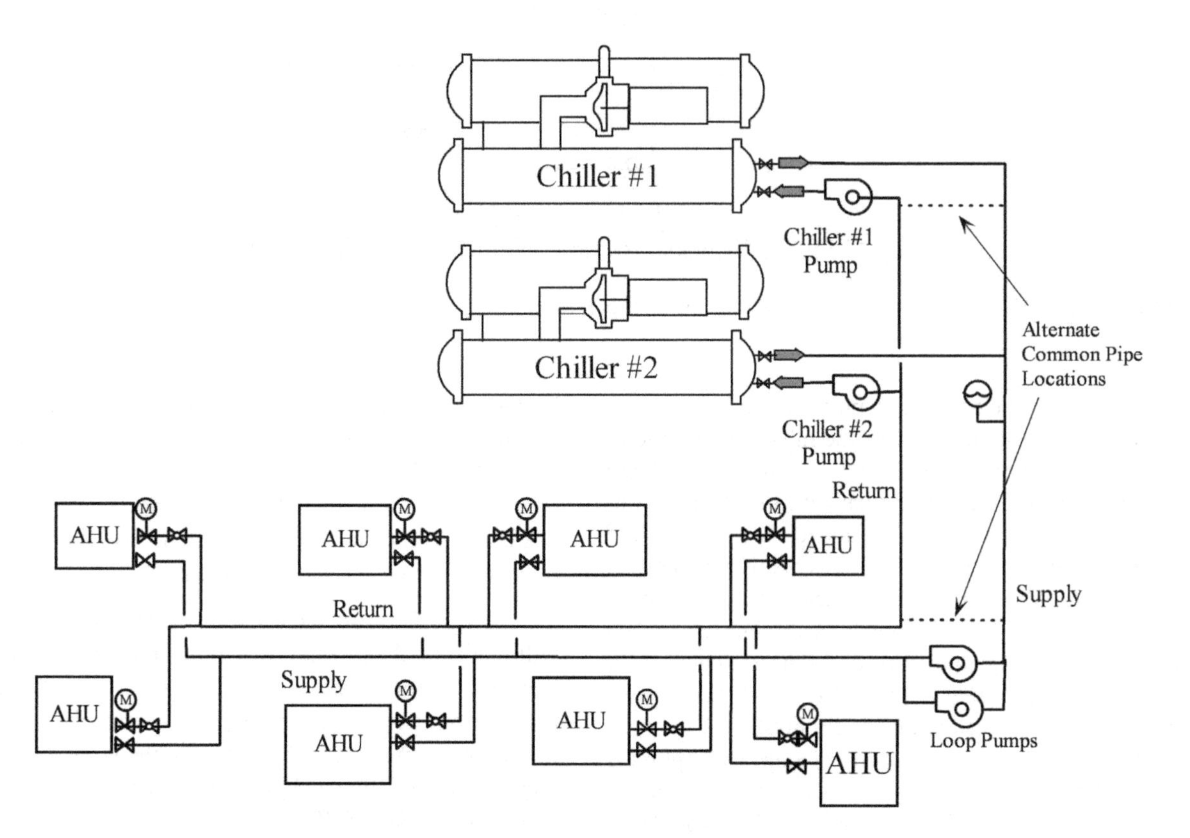

Figure 10.4 Variable-flow chilled water system with primary-secondary pumps.

Figure 10.4 represents a variable-flow chilled water system configuration that incorporates two-way valves on the air handlers. This permits improved pump energy savings resulting from lower flow rates and the accompanying reduced friction losses (which theoretically vary according to the flow rate reduction squared). Typical (but not universally adopted) liquid temperatures are 44°F or 45°F supply and 54°F or 55°F return at full load and full flow. As the loads on the air-handling units decrease, the two-way valves are used to modulate flow rates to provide constant temperature rise across the water coil. If variable-speed drive capability is provided for the loop (secondary) pump, motor speed can be reduced to maintain pressure difference across the supply and return header. This will enable the flow rates through the AHU water coils to be maintained, and substantial pump energy savings are possible. It should be noted that this arrangement is also effective for systems that employ on-off control in the terminal units (fan coil units, water-source heat pumps) if on-off two-way valves are installed.

Figure 10.4 also demonstrates a method (although not a universally employed option) for primary pump control using a common pipe. Two options for the common pipe location are shown. At full load the capacity of the primary (chiller) pumps and the secondary loop pump(s) is equal. No flow enters the common pipe. If the flow rates of the primary pumps exceed the secondary pump(s), liquid will flow from right to left in the common pipe and will mix with the liquid returning to the chillers from the air-handling units. If the flow rate of the secondary pump exceeds the flow rates of the primary pumps, water will flow from right to left in the common pipe and be mixed with the liquid circulating to the AHUs. Chiller operation can be controlled by sensing the supply-return temperature difference.

Compelling reasons have also been provided for primary-only variable-volume water distribution systems with variable-speed drive pumps, as shown in Figure 10.5 (Stein and Taylor 2005; Rishel 2005). Pumps with sufficient head capacity to overcome friction losses in the entire system are equipped with variable-speed motor drives. Two-way valves on the AHUs control liquid flow through the coils according to loads, and the pump motor speeds are modulated to provide adequate differential head across the supply and return headers at the AHUs. Some method of bypassing flow at minimum loads is required. Simplicity is one advantage that has been suggested for this configuration.

Figures 10.1 through 10.5 only provide a description of a single-mode cooling or heating piping loop. This two-pipe arrangement cannot provide simultaneous cooling and heating with the water loop. This arrangement is typically unacceptable in modern applications in which heating and cooling are frequently required. In some applications, heating requirements are modest even when outdoor temperatures are well below indoor temperatures. In these cases adequate heating can be

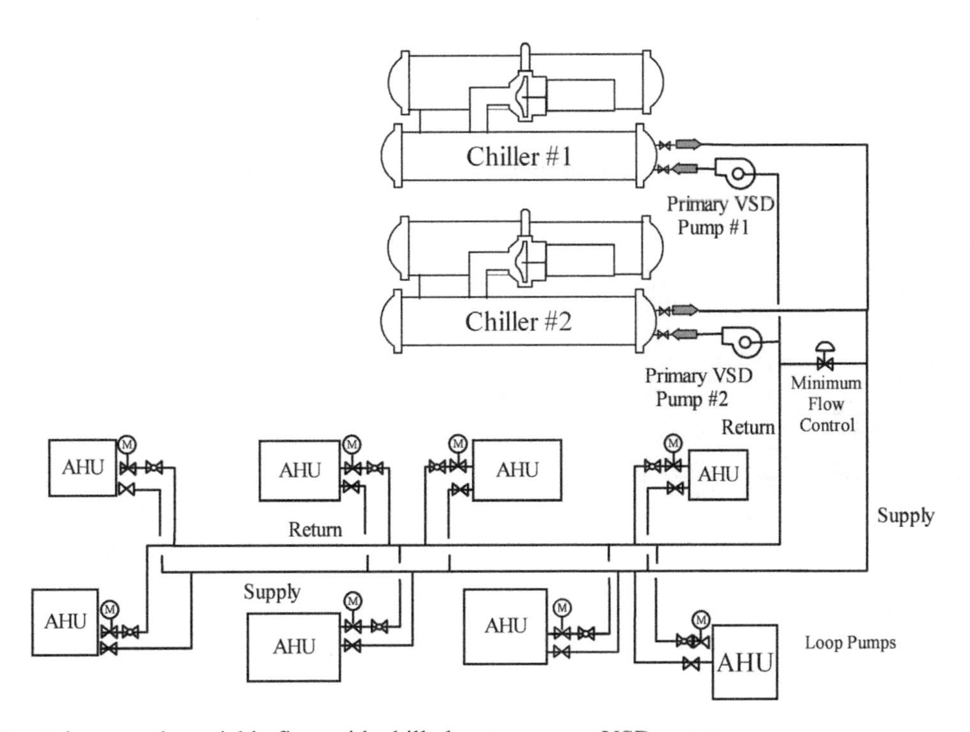

Figure 10.5 Primary-only variable flow with chilled water system VSD pumps.

provided with heating coils (electric resistance) in the zones without significant added electrical distribution capacity. In many other applications, a second water loop is necessary to meet the heating requirements. These systems are referred to as *four-pipe systems*.

Figure 10.6 shows fan-coil options that include a single coil connected to a two-pipe system, a single coil connected to four-pipe system, and a dual coil connected to a four-pipe system. The dual-coil four-pipe system can provide reheat with the heating coil located downstream for applications with high humidity loads or airflow requirements. To accommodate the four-pipe coils, the configurations shown in Figures 10.1 through 10.5 would require a second set of supply and return piping loops connected to a water heater (boiler).

A cooling tower condensing water loop is shown in Figure 10.7. Water is drawn from the basin of an external cooling tower and pumped through the condenser of a chiller to extract the heat of rejection from the refrigeration circuit. The heated water is pumped to a shallow tray or a network of distribution headers and spray nozzles at the top of the tower. The water leaves the top of the tower and is further atomized into smaller droplets by the tower "fill." This increases the evaporation rate of water to the air that is being drawn through the tower by the fan. A small amount of water evaporates, which low-

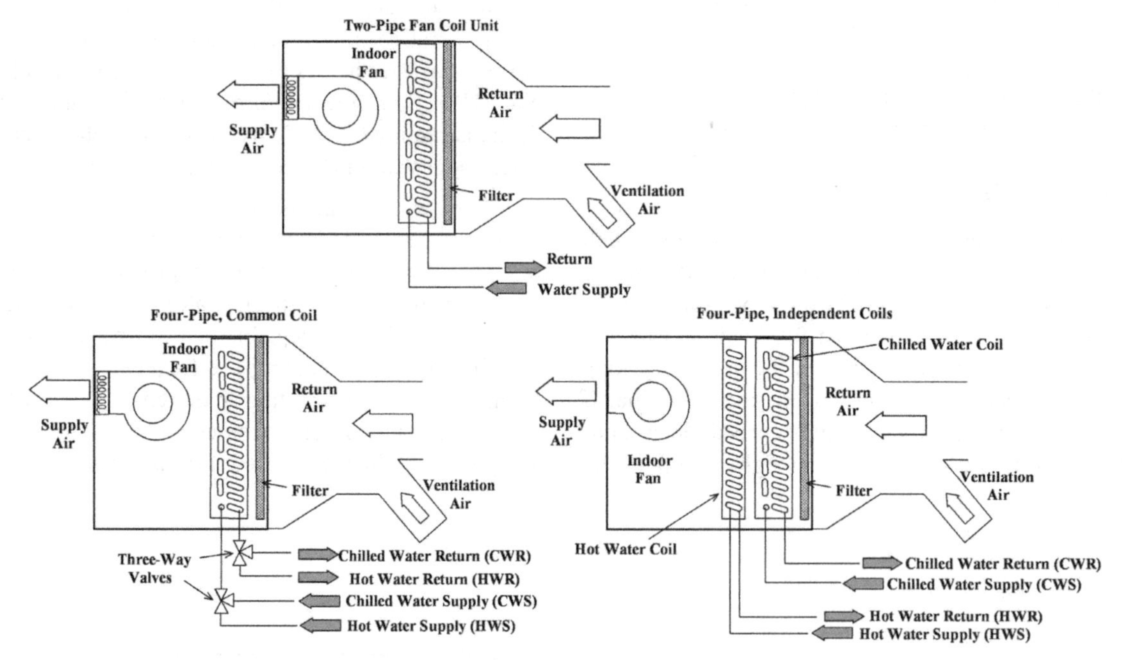

Figure 10.6 Two-pipe cooling only, four-pipe heating or cooling, and four-pipe coils.

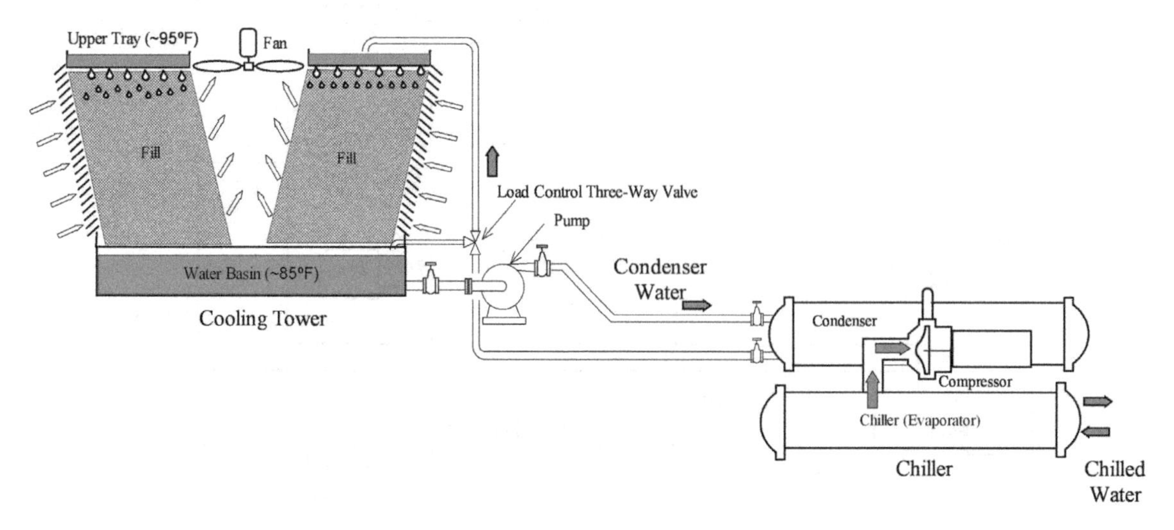

Figure 10.7 Open cooling tower, condenser water loop.

ers the temperature of the remaining water 10°F to 15°F on a design day. A three-way bypass valve provides an added measure of control during periods of low loads and/or low ambient temperatures. Note that this is an open loop, so heightened water treatment precautions are required and the pump must overcome the elevation head to the top of the tower in addition to the pipe friction losses and the condenser pressure drop. Closed-loop cooling towers (often called *fluid coolers*) are also available. In these coolers, water returns from the condenser and is circulated through a closed coil in the tower. A water spray is directed over the outside of the tubes to generate the evaporative cooling effect. This results in slightly higher condenser water temperatures compared to an open tower of equivalent size (+5°F to 10°F), but water treatment needs will be reduced and the condenser pump does not have to overcome the elevation head since it is a closed-loop circuit.

The examples presented in this section describe central chilled water systems that are often the system of choice in medium-sized and large buildings. Design engineers should consider the overall net system efficiency during the evaluation phase. Extensive data gathered by the Energy Information Administration of the US Department of Energy indicate auxiliary energy use of fans and pumps exceeds the energy consumption of the primary chilling equipment (Broderick and Westphalen 2001). Therefore, efficiency evaluations should include the impact of many pumps and fans with sufficient capacity to overcome control device restrictions and friction losses in extensive piping and duct networks. It is often possible to minimize these auxiliary requirements with adjustments to the design of central systems. In other cases, multiple smaller systems with low auxiliary requirements may be a lower energy consumption option.

HEAD LOSS CHARACTERISTICS

As noted in the previous chapter, the terms *head loss* and *pressure drop* are often used interchangeably. This chapter will use *head loss* since most of the tables and charts used in the HVAC industry are done in terms of head loss. It is important to note that the units should be "feet of liquid" for liquids other than water. Head loss is related to pressure drop ($\Delta p \equiv$ psi) by

$$\Delta h \text{ (ft of liquid)} = \frac{\Delta p g}{\rho g_c} \equiv \frac{\frac{\text{lb}_f}{\text{in.}^2} \times 144 \ (\text{in.}^2/\text{ft}^2)}{\text{lb}_m/\text{ft}^3} \times \frac{32.2 \left(\frac{\text{ft}}{\text{s}^2}\right)}{32.2 \left(\frac{\text{ft}-\text{lb}_f}{\text{lb}_m-\text{s}^2}\right)} . \quad (10.1)$$

For water at 60°F (ρ = 62.3 lb/ft^3), Equation 10.1 reduces to

$$\Delta h \text{ (ft of water)} = 2.31 \Delta p \text{ (psi)} . \quad (10.2)$$

Head loss (or pressure drop) calculations are typically performed using either tables or graphs of head loss per 100 linear feet. These losses are functions of velocity, pipe diameter, fluid properties, and pipe wall friction factor. Friction factors are difficult to determine and predict, especially in aged piping systems where interior roughness varies with time and quality of water treatment programs. To reduce the amount of computation, these figures and tables have been developed to eliminate many of the detailed computations; however, the tables and graphs are for specific conditions (water at 60°F, new clean pipe, standard pipe diameter, etc.) and must be adjusted for variations.

The "head loss per 100 feet of linear pipe" values from the graphs and tables are multiplied by the sum of straight pipe plus the equivalent length of fittings in the pipe section. A more exact computation for fitting losses with K-factors is preferred in some cases.

$$\Delta h = K\frac{V^2}{2} \quad (10.3)$$

This method is more cumbersome in design practice and is often discarded in favor of equivalent lengths. Published data regarding the high degree of uncertainty of values for K for fittings (ASHRAE 2005, chapter 36) indicate this is acceptable. This text will emphasize the equivalent length method that was used in the previous chapter to compute duct losses. For example, a 150 ft section of 2 in. diameter pipe with two 90° elbows (L_{eqv} = 2.8 ft) flowing at 40 gpm will have a head loss of 3.2 ft of water per 100 ft. The total loss for the section is found using Equation 10.4.

$$\begin{aligned} \Delta h \text{ (ft)} &= \Delta h/100 \text{ ft (from head loss tables or charts)} \times \left(L + \sum L_{eqv}\right) \\ &= 3.2 \text{ ft}/100 \text{ ft} \times (150 \text{ ft} + 2.8 \text{ Eqv. ft} \times 2 \text{ Elbows}) = 5.0 \text{ ft of water} \end{aligned} \quad (10.4)$$

Figure 10.8 (ASHRAE 2005, chapter 36) and Figure 10.9 (Kavanaugh and Rafferty 1997) are log-log plots of liquid flow in gpm vs. head loss in *feet of head* per 100 *linear feet of pipe*. Note that the pipe velocity can also be determined from the figures as values that appear as straight lines sloping downward with increasing flow rate. Figure 10.8 represents the head loss of water at 60°F in new, clean, schedule 40 steel and copper and schedule 80 PVC pipe. Although these three types of pipe are common options, a variety of materials and sizes are available. The recommended maximum head loss is 4 ft of water per 100 ft of pipe. Velocity in small-diameter pipe (≤ 2.0 in.) should not exceed 4 fps. Noise and erosion dictate that velocity should never exceed 15 fps in any pipe (ASHRAE 2005, chapter 36).

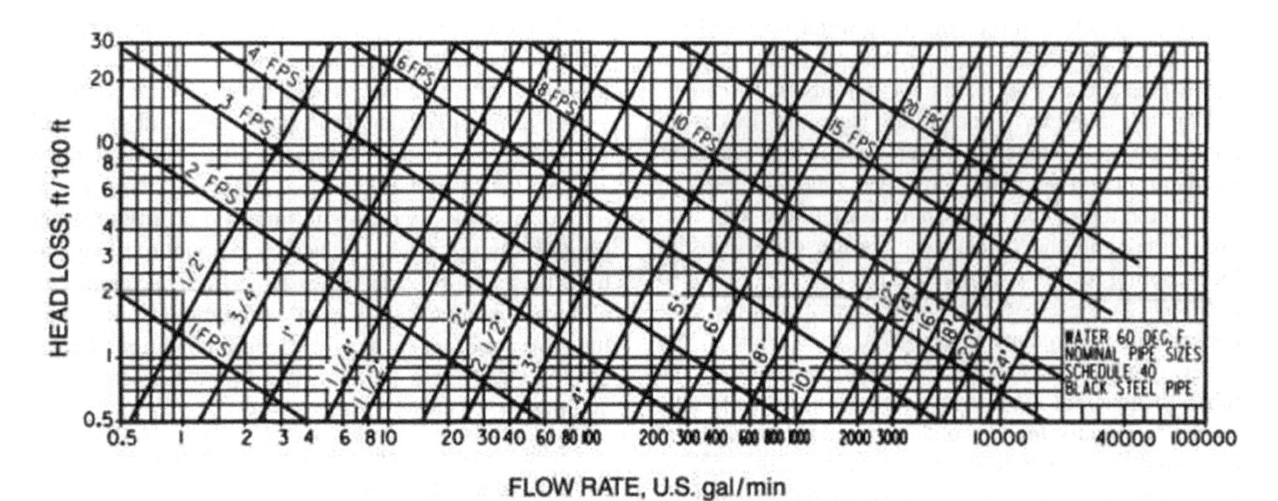

Fig. 1 Friction Loss for Water in Commercial Steel Pipe (Schedule 40)

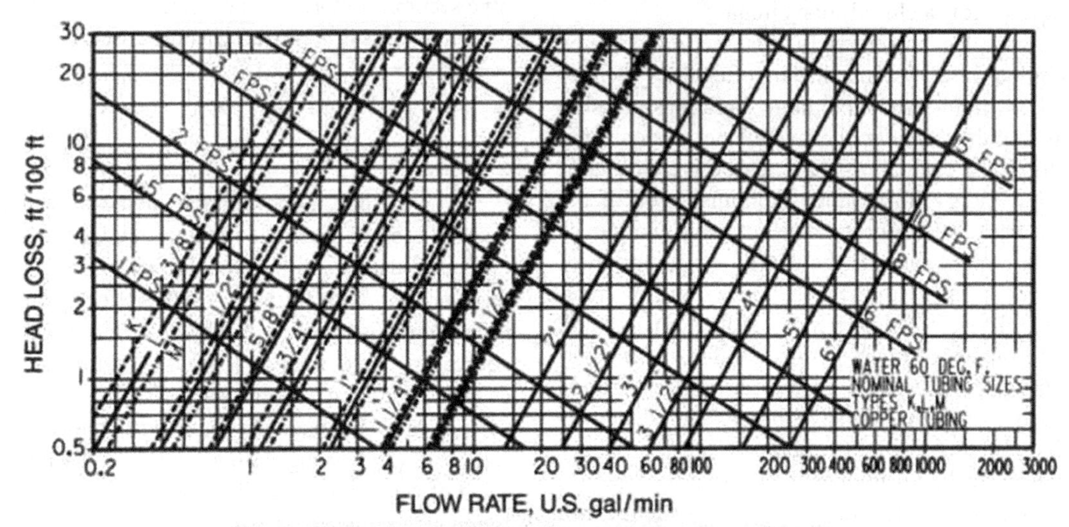

Fig. 2 Friction Loss for Water in Copper Tubing (Types K, L, M)

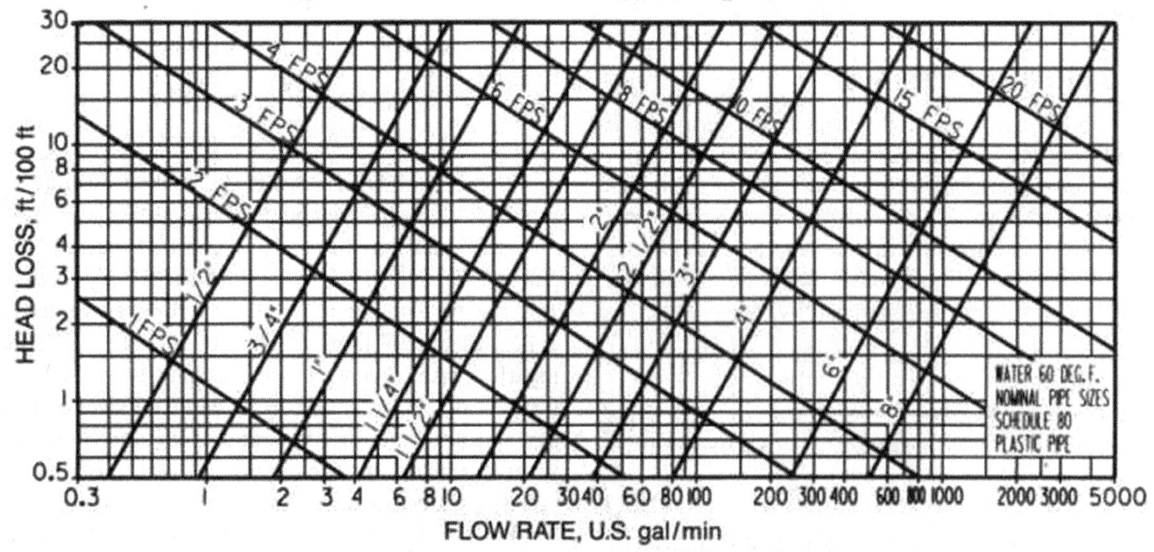

Fig. 3 Friction Loss for Water in Plastic Pipe (Schedule 80)

Figure 10.8 Head loss charts for steel, copper, and PVC piping (ASHRAE 2005, chapter 36).

One newer option is high-density polyethylene (HDPE), in which fittings and connections are made by thermal fusion (melted). This type of pipe is widely used for natural gas distribution and is becoming more widely used in the HVAC industry, especially with ground-source heat pumps. Figure 10.9 is a log-log plot for HDPE pipe for water for two common dimension ratios ($DR = d_{outside}/t_{wall}$). Equation 10.5 can be used to estimate losses for flow rates between the values in the table.

Aging of metal pipe significantly increases the head losses compared to the values in tables for new, clean pipe. This is especially true for open systems that continuously supply fresh oxygen for corrosion and fresh minerals for deposition. To a lesser extent, this is also true for closed systems with poor or nonexistent water treatment procedures. One conservative recommendation is to increase the expected head loss by 15% to 20% for closed piping and by 75% to 90% for open loops (*HDR Design Guide* 1981). Another recommendation is to increase the design flow rate by 1.55 for cast iron pipe, by 1.08 for iron and steel, and by 1.06 for cement-lined pipe (Smith 1983). However, a study cited by the Plastic Pipe Institute suggests these recommendations are unnecessary for plastic pipe (PVC, HDPE, etc.) since no corrosion-related degradation was observed (PPI 1971).

Head loss data are also often presented in tabular format at fixed increments of flow rate (gpm). Since head loss varies with the velocity (and flow rate) squared, the head loss for other flow rates can be approximated from Equation 10.5. However, the equation ignores the impact of friction factors and will provide unacceptable error if flow rate (gpm) varies from the tabular value (gpm_{Table}) more than 20%.

$$\Delta h \approx \Delta h_{Table} \times [gpm/gpm_{Table}]^2 \qquad (10.5)$$

In addition to the variation in material types, there are several different designations for pipe diameters and corresponding pressure ratings. For this reason, Table 10.1 is provided to show the relationship between nominal diameter and actual diameter for three of the pipe size classifications. Iron pipe sizes (IPS) are often given in schedules. Schedule 40 is common, and if higher pressure ratings are required, the outside diameter is held constant and the wall thickness is increased. Schedule 80 is the next higher pressure rating available. There is no standard correlation between nominal, inside, and outside diameters, and engineers must consult manufacturer's data or tables such as Table 10.1 for exact values.

Copper tube sizes (CTS) have more consistent increments, but nominal diameter denotes (but is not equal to) an outer diameter. Pressure rating and wall thickness are indicated by the letter designations of K, L, and M. ***Caution:*** Copper tubing for refrigeration applications follows another designation in which nominal diameter is actually the outside diameter.

The dimension ratio (DR) or standard dimension ratio (SDR) designation for HDPE pipe has the same outside dimensions as schedule-designated pipe, but the inside diameter (d_i) is determined from the DR value, the outside diameter (d_o), and the pipe wall thickness (t_{wall}). Table 10.1 lists inside and outside diameters for common water pipe designations.

$$d_i = d_o - 2 \div t_{wall} = d_o(1 - 2/DR) \qquad (10.6)$$

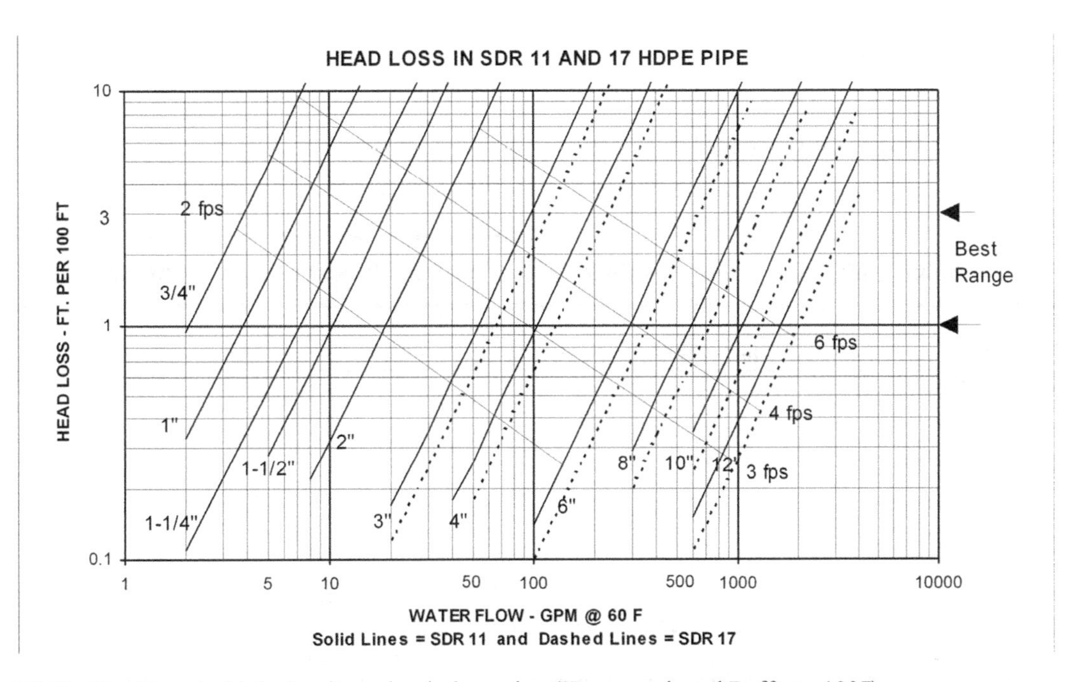

Figure 10.9 Head loss in high-density polyethylene pipe (Kavanaugh and Rafferty 1997).

FITTINGS, VALVES, AND COMPONENTS

Friction losses in fittings are also expressed in a variety of methods. In addition to the K-factor values and equivalent length methods previously mentioned, a flow coefficient (C_v) is also used. Flow coefficient is defined as the flow rate in gallons per minute (gpm) that produces a 1.0 psi (2.31 ft of water) loss through the fitting. Table 10.2 includes examples of flow coefficients provided by the manufacturers of various valves and accessories. The head loss or pressure drop for other flow rates is estimated from Equation 10.7

$$\Delta p \text{ (psi)} \approx [\text{gpm}/C_v]^2 \Rightarrow \qquad (10.7)$$
$$\Delta h \text{ (ft of water)} \approx 2.31 \times [\text{gpm}/C_v]^2$$

Table 10.3 is a set of equivalent length values developed from the more detailed table in the *2005 ASHRAE Handbook—Fundamentals* (ASHRAE 2005, chapter 36), which requires the velocity of the liquid to be known. The values in Table 10.3 are the equivalent lengths for the velocity that results in a head loss near 3 ft of water per 100 ft of pipe. Since the recommended design values range between 1 ft and 4 ft per 100 ft, the values in Table 10.3 represent an average equivalent length in the typical design range. Table 10.4 is a similar table for HDPE fittings for the two most common thermal fusion joints of socket fusion (typically used for 1.25 in. and smaller) and butt fusion (used for 1.25 in. and larger).

Head losses through fan coils, chillers, and heat pumps are a significant component, and this information is available in chapter 5. Figures 5.13, 5.15, and 5.16 provide the head loss vs. flow rate curves for the condenser and evaporators of the chillers. Table 5.12 includes this information for a product line of chilled water coils, and Table 5.6 includes head loss data for the water coils of a product line of water-to-air heat pumps.

DESIGN METHODS AND HEAD LOSS COMPUTATIONS

Water system distribution design includes the selection and specification of the primary heat exchangers (chiller, cooling tower, heat exchanger, etc.); the selection and specification of the terminal units (fan-coil units, water-source heat pumps, etc.); the routing and sizing of piping; the selection of valves, strainers, and water flow control devices; the determination of system head loss through the critical (high head loss) path(s); and the selection and specification of a pump to meet the required flow rate and critical path head loss. This section emphasizes pipe routing/sizing and critical head loss computation after the terminal units and primary heat exchangers have been selected as described in chapter 5.

Table 10.1 Inside and Outside Diameter for Common Water Pipe Designations

Nominal Diameter (in.)	SDR 11 OD/ID (in.)	SDR 17 OD/ID (in.)	Schedule 40 OD/ID (in.)	Schedule 80 OD/ID (in.)	Copper (L) OD/ID (in)
3/4	1.05/ 0.86	1.05/ 0.93	1.05/ 0.824	1.05/ 0.742	0.875/ 0.785
1	1.315/ 1.08	1.315/ 1.16	1.315/ 1.049	1.315/ 0.957	1.125/ 1.025
1.25	1.66/ 1.36	1.66/ 1.46	1.66/ 1.38	1.66/ 1.278	1.375/ 1.265
1.5	1.90/ 1.55	1.90/ 1.68	1.90/ 1.61	1.90/ 1.50	1.625/ 1.505
2	2.375/ 1.94	2.375/ 2.10	2.375/ 2.067	2.375/ 1.939	2.125/ 1.985
3	3.5/ 2.86	3.5/ 3.09	3.5/ 3.068	3.5/ 2.90	3.125/ 2.945
4	4.5/ 3.68	4.5/ 3.97	4.5/ 4.026	4.5/ 3.826	4.125/ 3.905
6	6.625/ 5.42	6.625/ 5.84	6.625/ 6.065	6.625/ 5.761	
8	8.625/ 7.06	8.625/ 7.61	8.625/ 7.98	8.625/ 7.625	
10	10.75/ 8.80	10.75/ 9.49	10.75/ 10.02	10.75/ 9.562	
12	12.75/ 10.43	12.75/ 11.25	12.75/ 11.94	12.75/ 11.374	

Table 10.2 Typical Flow Coefficients (C_v) for Fully Open Valves and Fittings

	Nominal Pipe Diameter										
Valve/Fitting	3/4 in.	1 in.	1¼ in.	1½ in.	2 in.	3 in.	4 in.	6 in.	8 in.	10 in.	12 in.
Zone valve (WR)	23.5	37									
Zone valve (J)	8.6	13.9		27.5	41						
Zone valve (HW)	3.5	3.5									
Zone valve-Ball	25	35	47	81							
Ball valve (App)	25	35	47	81	105	390	830	1250	2010	3195	
Butterfly valve					144	461	841	1850	3316	5430	
Swing check	13	21	35	45	75	195	350	990	1700	2400	
Y-strainer—FPT	18	28	43	60	95	155	250				
Y-strainer—Flange				30	70	160	260	550	920	1600	2200

C_v = flow rate in gpm for 1.0 psi pressure loss.

Table 10.3 Equivalent Lengths* for Iron and Copper Fittings
(Based on Velocity Values That Result in Δh ≈ 3 ft of water/100 ft from ASHRAE 2005, chapter 36)

	90°Elbow Screwed	90°Elbow Welded	45°Elbow Screwed and Welded	Reducer	Straight Run Tee	Branch Run Tee	Open Gate Valve
3/4 in.	2	1	1.4	0.8	1.2	8	1
1 in.	2.5	1.3	1.75	1	1.5	10	1.3
1.25 in.	3.6	1.8	2.5	1.4	2.2	14	1.8
1.5 in.	4.2	2.1	2.9	1.7	2.5	17	2.1
2 in.	5.6	2.8	3.9	2.2	3.4	22	2.8
3 in.	9	4.4	6.2	3.5	5.2	35	4.4
4 in.	11	5.7	8	4.6	6.8	46	5.7
5 in.	14	7.2	10	5.7	8.6	57	7.2
6 in.	17	8.6	12	7	10	68	8.6
8 in.		11	15	9	13	88	11
10 in.		14	19	11	16	108	14

* Equivalent lengths are in feet.

Table 10.4 Equivalent Lengths* For HDPE Pipe Fittings

Fitting	0.75 in.	1 in.	1.25 in.	1.5 in.	2 in.	3 in.	4 in.	6 in.	8 in.	10 in.	12 in.
Socket U-bend	12	6.4	11	–	–	–	–				
Socket U-Do	8.5										
Socket 90° L	3.4	2.5	6.3	6.5	6.8	–	–				
Socket tee—Branch	4.1	5.2	6.4	10.0	13	–	–				
Socket tee—Straight	1.2	1.2	0.9	2.0	2.8	–	–				
Soc. reducer (1 step)	–	6.1	4.0	3.9	4.2						
Soc. reducer (2 step)	–	–	4.2		5.1	–	–				
UniCoil^{TM}	8.7	10.2		–	–	–	–				
Butt U-bend	12.4	22.4	35	43		–	–				
Butt 90° L	7.2	10.0	18.5	10.7	12.3	32	38	51	63	75	87
Butt tee—Branch	7.5	7.1	17.2	10.7	15.2	31	37	50	62	74	86
Butt tee—Straight	4.5	2.7	5.5	2.9	4.1	6.8	7.1	7.7	8.3	9	10
Butt reducer	–	4.8	5.5	6.0	6.8	10.3	13.4	20	26	33	39
Butt joint	2.0	1.2	1.3	1.3	1.2	0.8	1.0				
5 ton close header (First/last take-off)		17/30									
10 ton close header		20/34									

* Equivalent lengths are in feet.

Piping Layout and Sizing

The following steps are taken when designing piping layout and indicating size. This description is an abbreviated version of the procedure and does not include details that are often required, such as addition of piping to route around structural and other utility piping, transitions from piping to equipment, measurement points, thermal expansion joints, strainers, water treatment fittings, etc.

Note: Steps 2–5 apply to loops with multiple terminal units. System design for loops with only two primary heat exchangers (chiller condenser-cooling tower, etc.) can omit steps 2–5.

1. Start with a full-size plan drawing of the building or building section.
2. Select the primary heat exchanger(s), fan coils, and/or water-source heat pumps. Locate each component on the sketch and note flow requirements and head losses.
3. Note design flow rate and head loss for each coil/heat pump.
4. Sketch piping diagram to connect each coil to the primary heat exchanger.
5. Use a direct return piping arrangement if piping losses are less than 30% to 40% of the total losses or if automatic flow control valves will be used. Use of modified direct-return headers are commonly

Example Problem 10.1

Design a direct-return, closed-loop water distribution system connecting the plate heat exchanger (PHE) to the classroom water-to-air heat pumps shown in Figure 10.10. The design-day water-loop temperature is 75°F. Use DR 11, HDPE piping with thermally fused butt joints. The PHE has a head loss of 12.9 ft of water at 90 gpm. Head loss values for the water-to-air heat pumps are given in Table 5.6. Provide isolation ball valves and a zone valve for each pump. Use gate valves for piping 3 in. and larger and ball valves for smaller pipes.

Solution:
See Figure 10.10 (before [left diagram] and after [right diagram]) and follow steps previously listed under "Pipe Layout and Sizing."

Steps 1-3: The left diagram contains a floor plan with the location of six water-to-air heat pumps, with required flow rates. The plate heat exchanger rated head loss is provided in the problem statement (12.9 ft of water at 90 gpm). The six heat pump head losses are found from Table 5.6 as noted in the problem statement.

1 – SX042 @ 11 gpm, h = 13.0 ft of water
3 – SX048 @ 12 gpm, h = 22.2 ft of water
2 – SX060 @ 14 gpm, h = 18.0 ft of water

Step 4: The direct-return piping diagram is sketched on the right diagram to connect the plate heat exchanger (PHE) to each heat pump.

Step 5: Unnecessary, since problem statement calls for direct-return piping.

Step 6: The flow rates in the most remote sections (14 gpm) for both the supply and return piping are noted. In the next section this flow is added to the flow that branches to the second most remote unit. The process is repeated until all section flows are noted as shown in the right diagram of Figure 10.10.

Step 7: Isolation valves are inserted on the inlet and outlet of each heat pump, the PHE, and the pump. Flow control valves (motorized zone valves) are inserted on each heat pump.

Step 8: Size each piping section to provide $h/L < 4.0$ ft of water/100 ft of pipe. Use Figure 10.9 and neglect the difference between the losses at actual temperature of 75°F compared to the table value of 60°F.

75 gpm section: use 3 in. @ 1.8 ft/100 ft; 63 gpm section: use 3 in. @ 1.4 ft/100 ft;
51 gpm section: use 3 in. @ 0.9 ft/100 ft; 39 gpm section: use 2 in. @ 3.8 ft/100 ft;
25 gpm section: use 2 in. @ 1.7 ft/100 ft; 14 gpm section: use 1.25 in.@ 3.0 ft/100 ft;
12 gpm section: use 1.25 in. @ 2.2 ft/100 ft; 11 gpm section: use 1.25 in. @ 2.0 ft/100 ft

System head loss and pump selection:
Step 1: The critical path is from the PHE to the last heat pump, which has a relatively high loss (18 ft). Although the SX048 units have a higher loss, they are closer to the PHE.

Step 2: Compute head losses for each section from: $h = h/100 \times L + L_{eqv}$:

3 in. @ 75 gpm: h = 1.8 ft/100 × (120 ft + 2 × 32 ft [butt L] + 4 × 4.4 [gate valves])	= 3.6 ft
3 in. @ 63 gpm: h = 1.4 ft/100 × (60 ft + 2 × 6.8 ft [T-straight])	= 1.0 ft
3 in. @ 51 gpm: h = 0.9 ft/100 × (60 ft + 2 × 6.8 ft [T-straight])	= 0.7 ft
2 in. @ 39 gpm: h = 3.8 ft/100 × (60 ft + 2 × 6.8 ft [T-straight] + 2 × 10.3 [red])	= 3.1 ft
2 in. @ 25 gpm: h = 1.7 ft/100 × (50 ft + 2 × 4.1 ft [T-straight])	= 1.0 ft
1.25 in. @ 14 gpm: h = 3.3 ft/100 × (60 ft + 2 × 10.7 ft [butt-L] + 2 × 6.8 [red])	= 3.1 ft
	12.5 ft

Step 3: Sum total losses:

Piping loss (from Step 2):		12.5 ft
Plate heat exchanger:	$\Delta h \approx \Delta h_{Rated} \times [\text{gpm}/\text{gpm}_{Rated}]^2 \approx 12.9 \times [75/90]^2 \approx$	9.0 ft
Heat Pump:		18.0 ft
2-ball valves (C_v = 47):	$\Delta h \approx 2 \times 2.31 \times [14/47]^2 \approx$	0.4 ft
Zone valve (C_v = 13.9):	$\Delta h \approx 2.31 \times [14/13.9]^2 \approx$	2.3 ft
Total (note Δh for gate valves were included in piping losses)		**42.2 ft**

Step 4: See Problems 10.2 and 10.3 in "Pumps and Pump Curves" section.

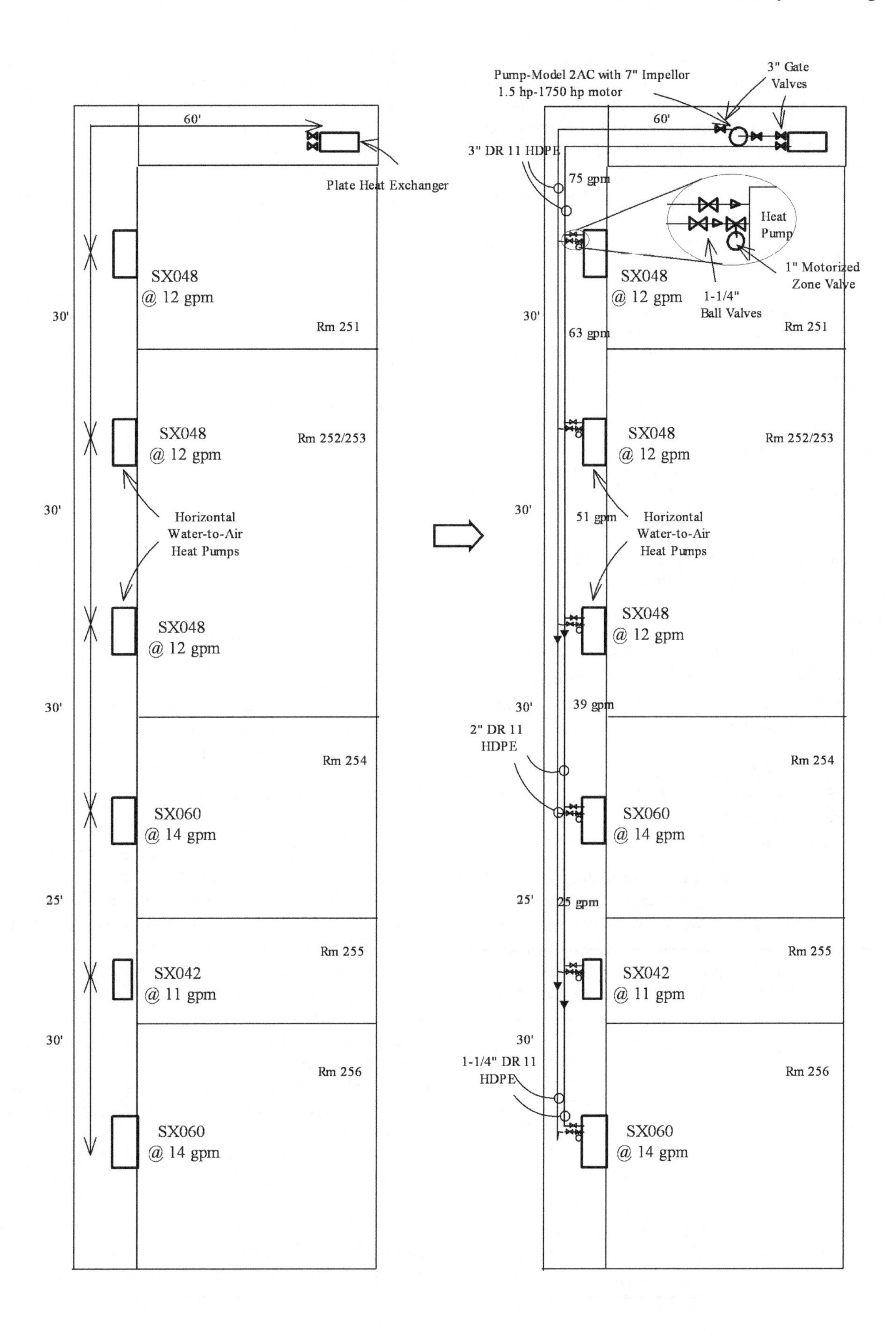

60'
Plate Heat Exchanger
SX048
@ 12 gpm
Rm 251
30'
SX048
@ 12 gpm
Rm 252/253
Horizontal
Water-to-Air
Heat Pumps
30'
SX048
@ 12 gpm
30'
Rm 254
SX060
@ 14 gpm
25'
Rm 255
SX042
@ 11 gpm
30'
Rm 256
SX060
@ 14 gpm
Pump-Model 2AC with 7" Impellor
1.5 hp-1750 hp motor
3" Gate
Valves
60'
3" DR 11 HDPE
75 gpm
Heat
Pump
1" Motorized
Zone Valve
1-1/4"
Ball Valves
63 gpm
51 gpm
39 gpm
2" DR 11
HDPE
25 gpm
1-1/4" DR 11
HDPE

employed. Use of a modified reverse-return header (Figure 10.2) should be considered if the losses in the piping are expected to be 30% to 40% or more of the total loss. Piping losses can be initially estimated by multiplying header lengths by the design loss per unit length.

6. Insert the flow rate of each piping section on the piping diagram.
7. Sketch valves, control devices, strainers, and other accessories on the piping diagram and note head loss or C_V of components.
8. Size each piping section based on recommended head loss per 100 ft of piping. ASHRAE (2005, chapter 36) recommends 1 to 4 ft of water per 100 ft of pipe, but note actual loss (actual loss will often be less since piping to give exact loss is not usually available). In addition to the 4 ft per 100 ft limit, liquid velocity should not exceed 4 fps in piping with diameters of 2 in. and less (ASHRAE 2005, chapter 36).

Table 10.5 is provided as an aid in selecting the pipe size. It lists the maximum flow rate for piping of various types and sizes that will comply with the recommended 4 ft per 100 ft head loss value. It should not be used to compute head loss for other flow rates.

System Head Loss Calculation and Pump Selection

The following steps are necessary to calculate system head loss and to select the appropriate pumps.

1. Find critical path of greatest head loss, which is normally the longest run or near-longest run with a high loss terminal unit.
2. Find loss in each pipe section of the critical path (where pipe size or water flow changes) by multiplying loss per unit length by the total equivalent length (straight pipe length + Σ equivalent lengths).

Table 10.5 Maximum Recommended Water Flow Rates (gpm) Based on ASHRAE Recommended Head Loss of 4 ft Water* per 100 ft of Pipe

Nominal Dia. (in.)	DR 11 HDPE	DR 17 HDPE	Sch. 40 Steel	Sch. 80 Plastic	Copper Type L
3/4	4.5	–	4	3	3.5
1	8	–	7	6	7
1¼	15	–	15	13	13
1½	22	–	23	21	20
2	40	–	45	40	44
3	110	140	150	125	130
4	220	300	260	250	260
6	600	750	800	750	800
8	1200	1500	1600	1500	–
10	2200	2600	3000	–	–
12	3500	4200	4600	–	–

* Multipliers for antifreeze mixtures: 20% propylene glycol = 0.85, 20% methanol = 0.90.

3. Calculate the total head loss by adding the piping losses in the critical path of greatest loss to losses through primary heat exchanger, terminal unit, control valve, and other fittings not accounted for with equivalent lengths. Although the flow rates through all parallel paths should be summed to arrive at the required flow rate, *do not add losses in paths that are parallel to the path of greatest loss. Only sum losses in the critical path.*
4. Select a pump that will deliver the total head loss at the required flow rate. Select a pump operating near its maximum rated efficiency as noted on the pump curve (see "Pumps and Pump Curves" section in this chapter).

Condenser Water Loop

To design a condenser water loop, the following steps are needed.

1. Locate chiller and cooling tower (or fluid cooler) and note losses through each. If it is an open cooling tower, an elevation head equal to the elevation difference between the upper tray and the basin must be included.
2. Size the water piping, noting the actual losses per unit length, and calculate the piping losses.
3. Find the total loss by adding the chiller loss, the cooler loss, and the piping loss.
4. Select the condenser water pump and evaluate the suction loop to ensure sufficient net positive suction head (NPSH) (Equation 10.11).

A PIPING DESIGN PROGRAM

This section describes a piping design/head loss calculation tool (*E-Pipealator.xls* on the CD accompanying this book) developed to follow the same logic described in the previous section for hand calculations. When properly applied, this tool offers an option that can reduce the time required to design and optimize water systems. Figure 10.11 is a view of the program input/output screen that results when Example Problem 10.1 is solved with the tool. Since it follows a procedure similar to traditional chart-hand calculation methods, the amount of time required to learn the method should be relatively short. It also offers the feature of being able to quickly correct for glycol-water mixtures, non-standard loop temperatures (up to 150°F), and pipe with other roughness values. As noted in the preface, this program is presented in an open code so that users can adjust the form to meet their individual needs.

The program will also adjust the rated losses through heat exchangers, coils, valves, and other fittings that are rated in either C_V or head loss at rated flow rate using water at a standard temperature (typically 60°F). Properties for pipe wall roughness, standard diameters

Piping Loop Head Loss Calculator - System Designer for HVAC Systems																		
Liquid	Water							Rated	Rated Δh	Inlet	Inlet		Rated					
Temp	75	°F				Coils	Flow	Flow	@ 60°F	Size	Vel	Re(in)	Vel		Re(rated)			Δh
Den	62.23	lbm/ft3					gpm	gpm	ft. water	inches	fps		fps					Ft. Liquid
Vis	6.27E-04	lbm/ft-s	0.93	cps		SX060	14	14	18	1	5.7	47335	5.7		38381			17.2
						PHE	75	90	12.9	3	3.4	84526	4.1		82246			8.9
							0	0	0	0	0.0	0	0.0		0			0.0
	HDPE Piping														**Coil sub-total**			**26.1**
Flow	Nom. Dia.	DR	I.D.	Roughness	Vel	Re	Δh/100'	Length	Fitting Selector	Leqv	Qty.	Fitting Selector	Leqv	Qty.	Fitting Selector	Leqv	Qty.	Δh
gpm	Inches	OD ÷ t	in.	feet	fps		Ft.	ft.		ft			ft			ft		Ft. Liquid
75	3	11	2.86	0.00007	3.7	88551	1.67	120	Butt90	26	4	ButtTeeRun	5	2	ButtTeeRun	5	0	4.0
63	3	11	2.86	0.00007	3.1	74383	1.22	60	ButtTeeBr	26	2	10-LoopHdrLast	34	0	Butt90	26	0	1.7
51	3	11	2.86	0.00007	2.5	60215	0.84	60	ButtTeeRun	5	2	ButtRed	10	0	Butt90	26	0	0.8
39	2	11	1.94	0.00007	4.2	67858	3.32	60	ButtTeeBr	16	2	ButtRed	7	2	SocTeeRun	3	0	3.7
25	2	11	1.94	0.00007	2.7	43499	1.51	50	ButtTeeBr	16	2	SocTeeBr	13	0	SocTeeRun	3	0	1.3
14	1.25	11	1.36	0.00007	3.1	34852	2.98	70	Butt90	10	2	ButtRed	5	6	SocTeeRun	2	0	3.7
	Steel/Brass /PVC														**HDPE sub-total**			**15.1**
Flow	Nom. Dia.	Schedule	I.D.	Roughness	Vel	Re	Δh/100'	Length	Fitting Selector	Leqv	Qty.	Fitting Selector	Leqv	Qty.	Fitting Selector	Leqv	Qty.	Δh
gpm	Inches	40 or 80	in.	feet	fps		Ft.	ft.		ft			ft			ft		Ft. Liquid
75	3	40	3.07	0.0002	3.3	82653	1.20	0	Gate Valve	4.5	4	Reducer	1.8	0	90 L	4.5	0	0.3
0	1	40	1.05	0.0002	0.0	0	0.00	0	90 L	1.3		90 L	1.3	0	90 L	1.3	0	0.0
0	1.5	40	1.61	0.0002	0.0	0	0.00	0	90 L	2.1		90 L	2.1	0	90 L	2.1	0	0.0
0	1.5	40	1.61	0.0002	0.0	0	0.00	0	90 L	2.1		90 L	2.1	0	90 L	2.1	0	0.0
0	0.75	40	0.82	0.0002	0.0	0	0.00	0	90 L	1.0		90 L	1.0	0	90 L	1.0	0	0.0
0	0.75	40	0.82	0.0002	0.0	0	0.00	0	90 L	1.0		90 L	1.0	0	90 L	1.0	0	0.0
															Fe/Br sub-total			**0.3**
						Other		Cv		Inlet	Inlet		Rated					
						Fittings	Flow	@ 60°F	Quanity	Size	Vel	Re(in)	Vel		Re(rated)			Δh
							gpm	gpm		inches	fps		fps					Ft. Liquid
						zone valve	14	13.9	1	1	5.7	47335	5.7		38107			2.2
						ball valve	14	47	2	1.25	3.7	37868	12.3		103082			0.5
							0	0	0	2	0.0	0	0.0		0			0.0
															Fitting sub-total			**2.7**
												Open Systems	Only	→	**Elevation**			**0**
															Total Loss			**43.9**

Figure 10.11 *E-Pipelator* water distribution system design program (on the CD accompanying this book).

for either schedule (40 or 80), or DR pipe are imbedded and displayed in the program. The friction factor is determined using the Churchill equation (Churchill 1977). This equation provides good agreement to the Moody diagram (Hodge and Taylor 1998) over the entire range of Reynolds (Re) numbers encountered in HVAC water piping networks. Unlike the Colebrook equation, it can be computed in a straightforward manner without iteration (Hodge and Taylor 1998).

$$f = 8\left[\left(\frac{8}{\mathrm{Re}_D}\right)^{12} + \frac{1}{(A+B)^{1.5}}\right]^{1/12}$$

$$A = \left\{2.457\ln\left[\frac{1}{(7/\mathrm{Re}_D)^{0.9} + (0.27\varepsilon/D)}\right]\right\}^{16} \qquad B = \left(\frac{37{,}530}{\mathrm{Re}_D}\right)^{16} \tag{10.8}$$

The solution to Problem 10.1 is repeated using this spreadsheet and the results are shown in Figure 10.11. The computed loss is 43.9 ft or 4% higher than the manual calculation found using the head loss charts and tables.

PUMPS AND PUMP CURVES

Figure 10.12 provides a summary of some of the basic centrifugal pump types used in HVAC applications. Additional detail and information can be found in the *2004 ASHRAE Handbook—HVAC Systems and Equipment* (ASHRAE 2004, chapter 39).

The performance of centrifugal pumps is typically provided in the form of curves. The primary curves are plots of volumetric flow rate on the horizontal axis with differential head or pressure on the vertical axis. Figures 10.13 through 10.16 demonstrate the format for a variety of pumps ranging in size from 3/4 hp to 200 hp. As noted on the box beneath the model number, these curves apply when the pumps are driven by a 1750 to 1770 rpm motor. In the US, most curves are for 1125 to 1200 rpm, 1725 to 1800 rpm, or 3450 to 3600 rpm, which represent the speeds of six-, four-, and two-pole AC motors.

Note: $$\mathrm{rpm} = \frac{2 \times f\,(\mathrm{Hz}) \times 60\ \mathrm{s/min}}{\mathrm{No.\ of\ Poles}} - \mathrm{Slip} \tag{10.9}$$

The maximum head typically occurs at shutoff (flow = 0) and the curve is relatively flat at rates up to 20% to 30% of maximum flow. Head then decreases gradually with increasing flow rate until the best efficiency point (BEP) is achieved. This typically occurs between 60% and 80% of maximum flow. In Figure 10.13 this point is 130 gpm and 43 ft of water head for the top curve (7 in.). Head decreases more rapidly from this point to the maximum flow rate.

Manufacturers typically provide multiple curves on a single sheet for a variety of pump impeller sizes that can be used in a single pump casing. For example, Figure 10.13 is for pump Model 2AC with impeller diameters of 7, 6.5, 6, 5.5, and 5 in. Pump manufacturers can machine impellers to any incremental diameter to meet the needs of an application, so values are not limited to those provided on pump curves.

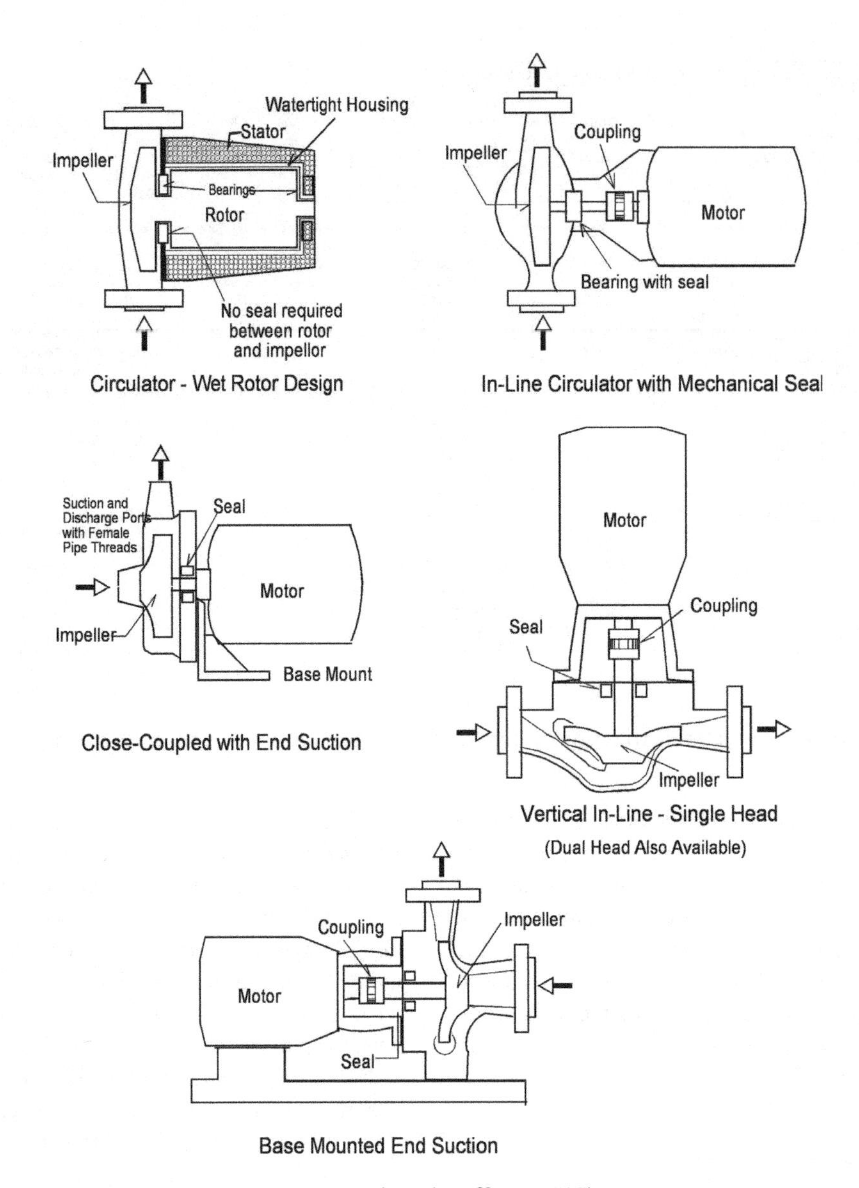

Figure 10.12 Basic HVAC pump types (Kavanaugh and Rafferty 1997).

Lines of constant efficiency can also appear on pump curves and apply to all flow vs. head curves for all impeller diameters. These lines are typically parabolically shaped but with the axis tilted toward the right of vertical. Note the highest efficiencies are attained with the largest impellers available for the pump casing.

Lines of constant power requirement are nearly linear, sloping down from left to right with increasing flow rate. It is important to select the motor to meet the requirement of the pump impeller rather than the power at the operating point. It is possible that a pump may see lower system head than at the design point, such as when two pumps are operating in parallel and one fails, leaving the second to see reduced friction loss. This will result in the operating point moving "out" (greater flow rate) on the pump curve and require greater power than the motor is designed to provide. Extended operation beyond this point may result in shortened motor life or failure. The power requirement at the operating point can also be determined from

$$w_{Req'd} = \frac{Q\Delta h}{\eta_{pump}} = \frac{Q\text{ (gpm)} \times \Delta h\text{ (ft of water)}}{3960 \times \eta_{pump}} \equiv \text{(hp)}. \quad (10.10)$$

Another important curve is net positive suction head (NPSH) required to prevent cavitation. The pump inlet must have sufficient head to prevent the static head (total head – velocity head) from dropping below the point where vaporization of the entering liquid occurs. The reduction in static pressure results from the increase of velocity head due to the high velocity of the liquid at the pump impeller entrance. The vaporization or cavita-

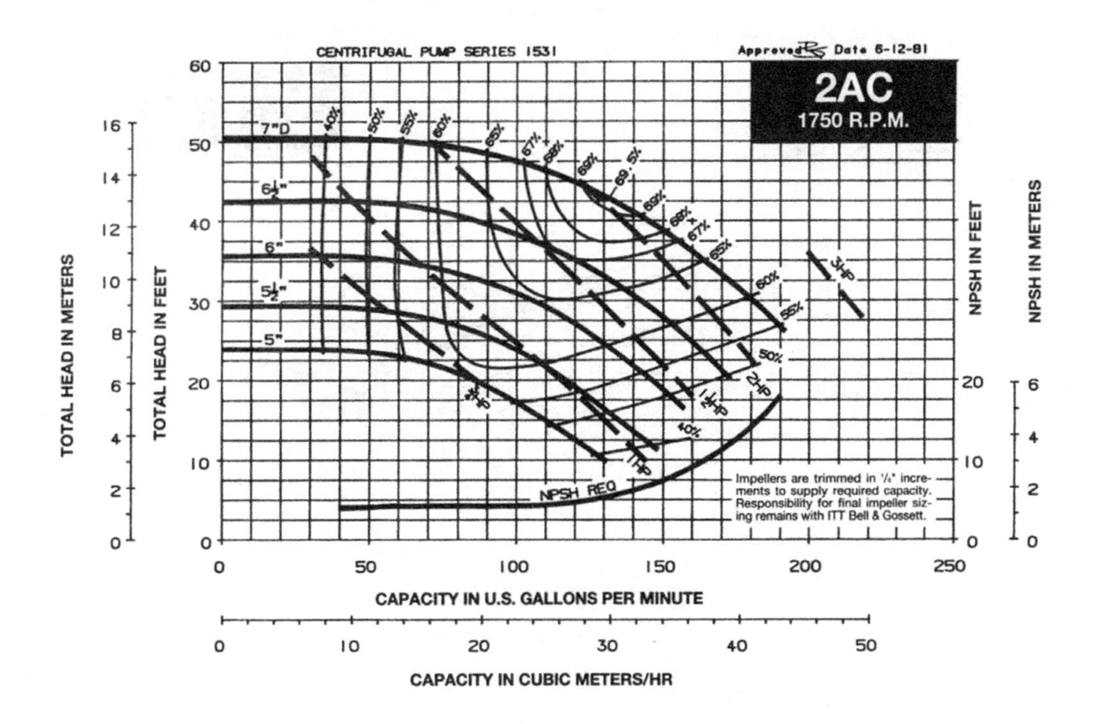

Figure 10.13 Pump curves, 1/2 to 3 hp (ITT/B&G 1996).

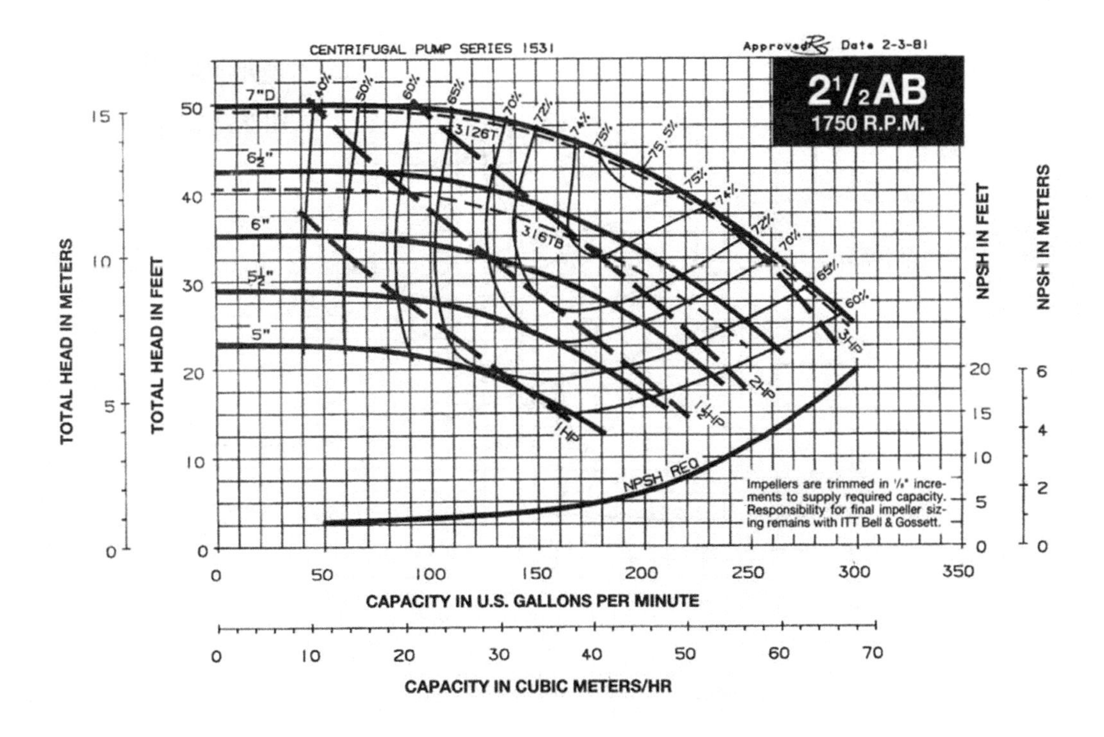

Figure 10.14 Pump curves, 1 to 3 hp (ITT/B&G 1996).

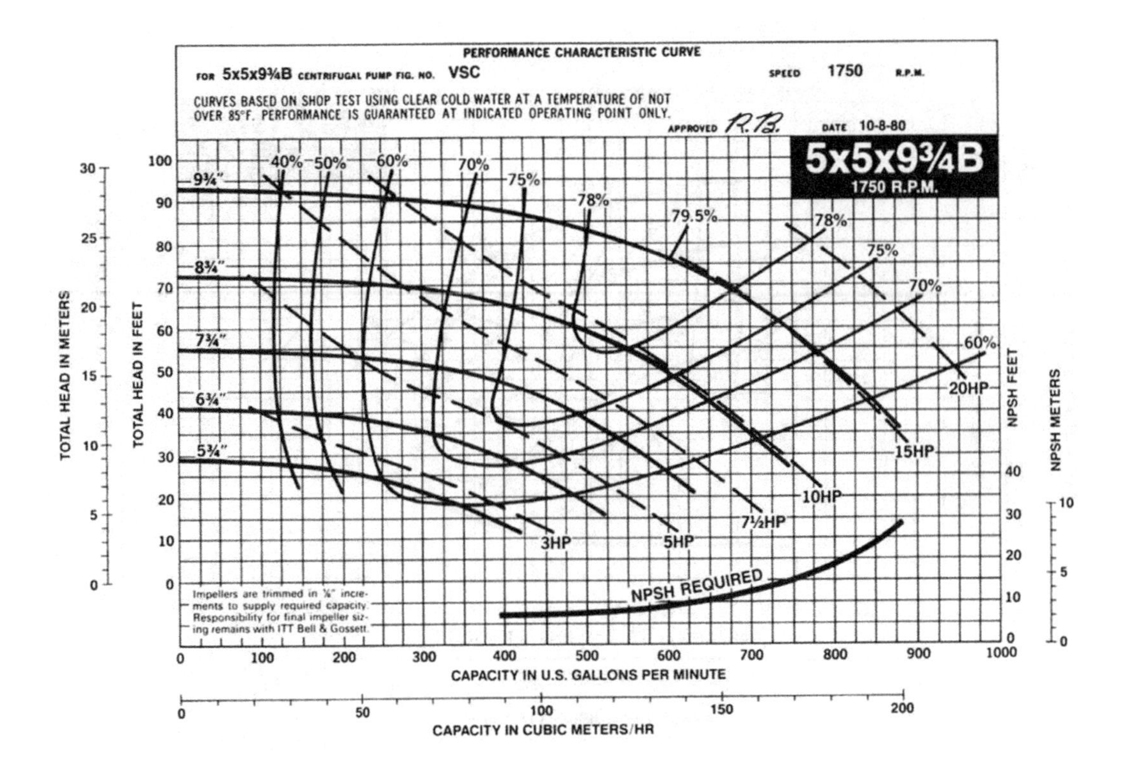

Figure 10.15 Pump curves, 3 to 20 hp (ITT/B&G 1994).

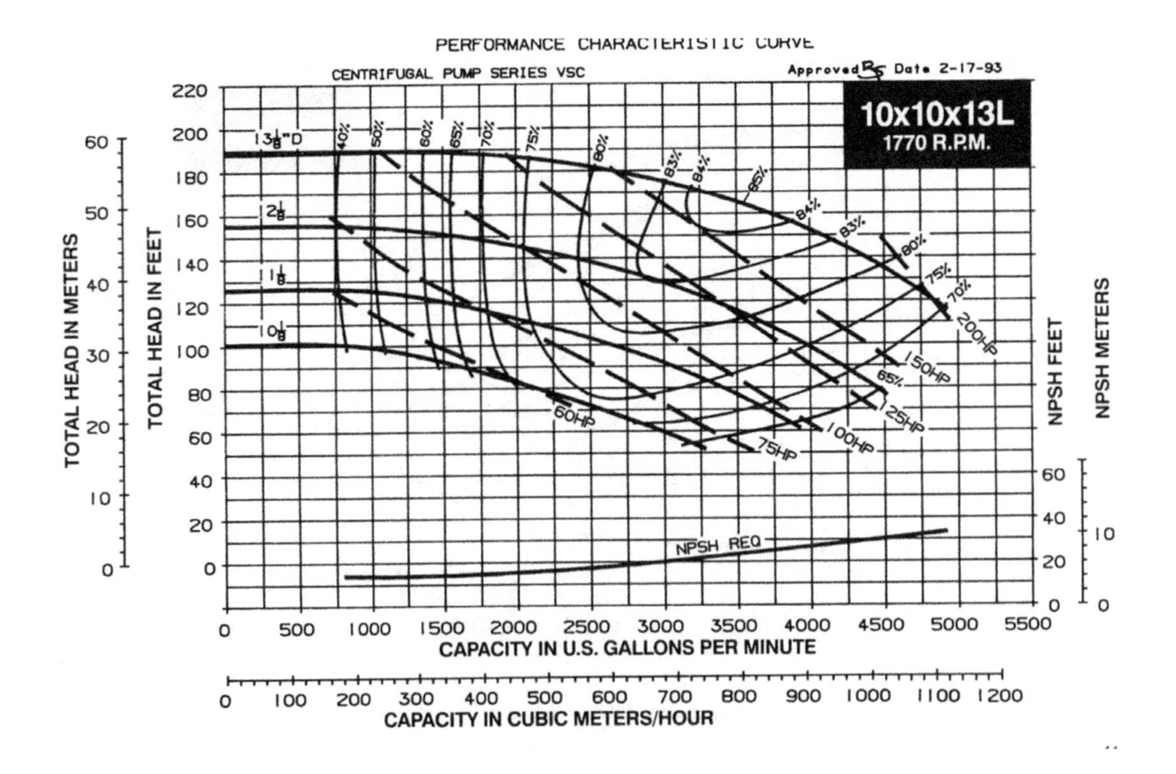

Figure 10.16 Pump curves, 60 to 200 hp (ITT/B&G 1994).

tion results in the formation of bubbles (vaporization), which can damage the pump or cause loss of circulation. The loss of suction problem is compounded if the pump must lift the fluid to an elevation (h_{Elev}) higher than the inlet and/or the friction losses in the suction piping (Δh_{Piping}) are high. At elevated temperatures, the change in the vapor pressure of the water ($P_{WaterVapor}$) must also be considered. Thus, $NPSH_{Available}$, which is calculated from Equation 10.11, must be greater than $NPSH_{Req'd}$ read from the curve.

$$NPSH_{Available} = P_{Atm.} - h_{Elev.} - \Delta h_{Piping} - P_{WaterVapor} \tag{10.11}$$

To convert pressure in psi (lb$_f$/in.^{2}) to feet of water,

$$h\ (\text{ft}) \equiv \frac{P\ (\text{lb}_f/\text{in.}^2) \times 144\ \text{in.}^2/\text{ft}^2}{\rho\ (\text{lb}_m/\text{ft}^3)} \times \frac{32.2\ (\text{ft/s}^2)}{32.2\ (\text{ft}-\text{lb}_f/\text{lb}_m-\text{s}^2)} = 2.31 \times P\ (\text{psi})@60°\text{F}. \tag{10.12}$$

In many applications, multiple pumps are required to provide either additional head (piped in series) or additional flow (piped in parallel). Figure 10.17 shows these arrangements for two pumps and the resulting combined pump curves for two pumps of identical head and flow capacities. The recommendation of manufacturers should be followed to ensure adequate separation distance between two pumps piped in series. It was previously suggested that an option was to select two pumps that, when operating in parallel, would provide rated flow and head. Figure 10.17 depicts the interaction of a system curve for a closed-loop system, the curve for a single pump, and the resulting curve for two pumps in parallel. Note that a single pump operating on the system provides approximately 70% (200 gpm) of rated flow compared to when two are operating (275 gpm).

Figure 10.18 is a series of curves for submersible pumps. These designs are intended to provide high heads, particularly in water well applications, where elevation heads cannot easily be achieved with a single pump. Impellers are stacked on a single shaft that is connected to a watertight, small-diameter electric motor. The diameter of the impeller is varied to meet flow requirements, while the number of impellers is varied to meet head requirements.

Like centrifugal fans, centrifugal pumps also follow theoretical affinity laws, often referred to as "pump laws." Equations 10.13 through 10.15 provide the base pump laws for flow rate, head, and input power as a function of angular velocity. The primary assumption, which is near valid for a narrow range of rpm ratios, is that the efficiency of the pump is constant. Observation of the efficiency of actual pump curves indicates these laws cannot be applied accurately over a wide range of operating conditions. Equations 10.16 through 10.18

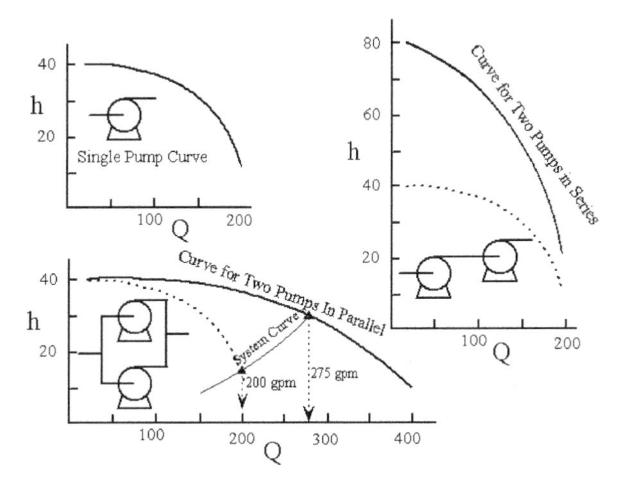

Figure 10.17 Pump curves for a single pump, two pumps in series, and two pumps in parallel.

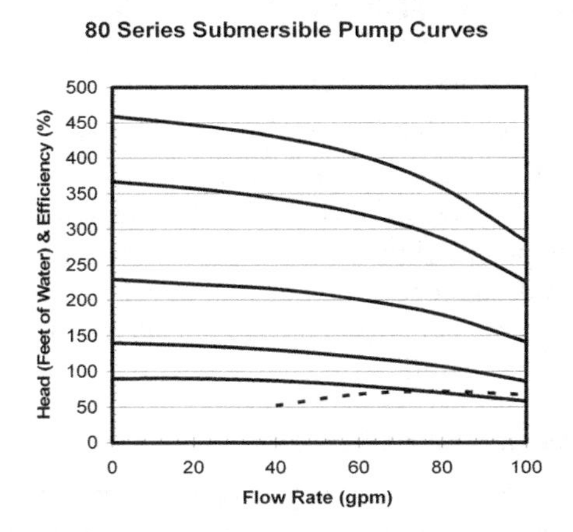

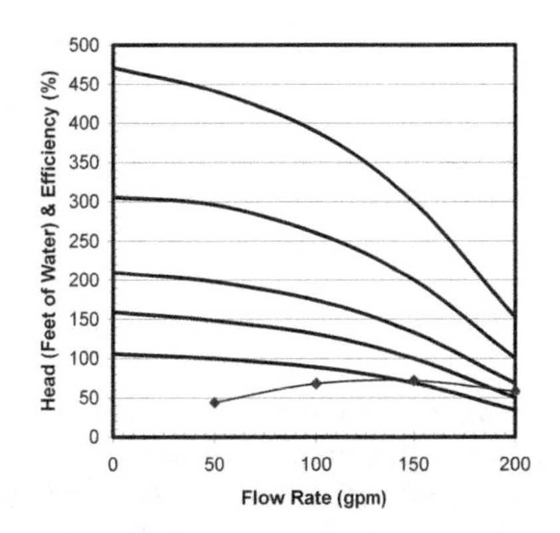

Figure 10.18 Submersible pump curves (GPC 1988).

Example Problem 10.2

Consider the $NPSH_{Available}$ for the pump in Figure 10.13 (7 in. impeller), operating at 150 gpm in an open system at sea level and 70°F, which must lift the water 10 ft from a tank with a fiction loss of 3 ft of water between the pipe inlet and the pump inlet.

Solution:

$\rho = 62.3\ lb_m/ft^3$

$P_{Atm.} = 14.7\ psi \times 144/\ 62.3 = 34.0\ ft$

$h_{Elev} = 10\ ft$

$h_{Piping} = 3\ ft$

P_v @ 70°F = 0.363 psi × 144/ 62.3 = 0.84 ft

$NPSH_{Available} = 34 - 10 - 3 - 0.84 = 20.2\ ft$

$NPSH_{Req'd}$ @ 150 gpm is 7.5 ft

Since $NPSH_{Available} > NPSH_{Req'd}$, the pump is acceptable.

demonstrate there are similar relationships with variable impeller diameter if other dimensions and conditions remain constant (ASHRAE 2004, chapter 39).

$$Q_2\ (\text{gpm}) \approx Q_1\ (\text{gpm}) \times \left(\frac{\text{rpm}_2}{\text{rpm}_1}\right) \tag{10.13}$$

$$\Delta h_2\ (\text{ft}) \approx \Delta h_1\ (\text{ft}) \times \left(\frac{\text{rpm}_2}{\text{rpm}_1}\right)^2 \tag{10.14}$$

$$w_2\ (\text{hp}) \approx w_1\ (\text{hp}) \times \left(\frac{\text{rpm}_2}{\text{rpm}_1}\right)^3 \tag{10.15}$$

$$Q_2\ (\text{gpm}) \approx \Delta Q_1\ (\text{gpm}) \times \left(\frac{D_2\ (\text{in.})}{D_1\ (\text{in.})}\right) \tag{10.16}$$

$$\Delta h_2\ (\text{ft}) \approx \Delta h_1\ (\text{ft}) \times \left(\frac{D_2\ (\text{in.})}{D_1\ (\text{in.})}\right)^2 \tag{10.17}$$

$$w_2\ (\text{hp}) \approx w_1\ (\text{hp}) \times \left(\frac{D_2\ (\text{in.})}{D_1\ (\text{in.})}\right)^3 \tag{10.18}$$

Pump Selection Steps

The steps for selection of pumps are as follows.

1. Determine the required flow rate ($gpm_{Req'd}$) to meet the demands of the HVAC condenser or evaporator.
2. Determine the system total head requirement by calculating friction losses through the piping system (pipe, fittings, valves, filters/strainers, and coils). For open systems, add the elevation head if the discharge level is above the inlet level or subtract the elevation head (but no more than $2.31 \times P_{Atm.}$) if the discharge level is below the inlet level.
3. Select a pump whose curve is on or just above the operating point determined in steps 1 (flow rate) and 2 (total head) *with* an efficiency near (±10%), the best efficiency point (BEP). If the point is not near the BEP, go to another set of pump curves.
4. Select a motor (hp and rpm) based on the pump curve (not the operating point). If the power curve intersects the pump curve at a flow rate above the expected operating point, a motor with a service factor (SF) rating greater than 1.0 is recommended (SF = maximum allowable load ÷ rated load).
5. If the operating point calculated in steps 1 and 2 is well below the pump curve, the actual operating point can be found by developing a system curve from a set of points including the original operating point and points 2 and 3 (above and below the original point).
   - Point 2: gpm_2 and $h_2 \approx \Delta h_{Piping}\ (gpm_2/gpm_{Req'd})^2 + h_{Elev}$
   - Point 3: gpm_3 and $h_3 \approx \Delta h_{Piping}\ (gpm_3/gpm_{Req'd})^2 + h_{Elev.}$

   The intersection of this curve and the pump curve is the *actual operating point*.
6. Calculate $NPSH_{Available}$ at the actual operating point to ensure it is greater than $NPSH_{Req'd}$ read from the pump curve for the actual flow rate.

PUMP AND FLOW CONTROL METHODS

Constant-Speed Pumping Systems

In a constant-speed pumping system, the pump and motor run at a constant speed. Some savings can be realized by closing valves to unused equipment. This causes the pump "riding the pump curve back" to move from the design operating point to the left and slightly up on the curve. This will reduce the required input somewhat but not significantly.

Primary-Secondary System

Primary-secondary systems use two or more pumps, which permits one or more pumps to be turned off when the requirement is lower. For example, in a

Example Problem 10.3

Select a pump and motor to meet the flow rate and head requirement of Example Problem 10.1.

Solution:

Figure 10.13 contains the curves for pumps within the 75 gpm flow rate required. A pump with a 6.5 in. diameter impeller will deliver 41 ft of head at the specified motor speed of 1750 rpm, which is slightly below the 42.2 ft requirement. A 7 in. impeller can be selected that will provide 50 ft of head. Although the power requirement is 1.5 hp at the operating point, a 3.0 hp motor is recommended since conditions in the system might cause the pump to operate at higher flow rates beyond the point that can be handled by even a 2.0 hp ($Q > 125$ gpm). A more prudent option would be to specify the impeller be trimmed to 6 in., which would allow the pump to deliver slightly more than the 42.2 ft requirement and only require a 2.0 hp motor. Another option is to analyze other pump curves to see if a more appropriate option is available that can meet the system requirements at a higher efficiency than the rather modest value of 61%. In fact, this particular manufacturer lists a slightly smaller pump model that meets the system requirements with a 65% efficiency and lower cost.

Example Problem 10.4

Determine the head of the pump shown in Figure 10.13 with a 7.0 in. diameter impeller if the speed is reduced to 1725 rpm.

Solution:

If the speed is reduced to 1725 rpm, flow and head will be lower. In order to determine operating points, a new pump curve is necessary for this lower speed. However, only two points are necessary since the range of interest is near the original speed of 1750 rpm. Thus, a point for a slightly lower flow (70 gpm) and higher flow (80 gpm) is generated. For the original pump speed, h @ 70 gpm = 50 ft and h @ 80 gpm = 49 ft. Equations 10.13 and 10.14 are applied at both points.

$$Q_1 \approx 70 \times (1725/1750) \approx 69 \text{ gpm}, \Delta h_1 \approx 50 \times (1725/1750)^2 \approx 48.6 \text{ ft}$$

$$Q_2 \approx 80 \times (1725/1750) \approx 79 \text{ gpm}, \Delta h_1 \approx 49 \times (1725/1750)^2 \approx 47.6 \text{ ft}$$

By interpolation between points 1 and 2 for the 1725 rpm curve, a value of 48.0 ft is determined for a flow rate of 75 gpm. This is sufficient to overcome the requirement of the piping system (43.9 ft).

ground-source heat pump system that requires two 7.5 hp pumps, a 7.5 hp pump could provide flow continuously through the building. A second 7.5 hp pump delivers flow through the ground loop and is cycled as a function of building load. Other options for primary-secondary arrangements are discussed in the "Piping Loop Systems" section earlier in this chapter.

Variable-Speed System

As discussed in the "Piping Loop Systems" section, pumps driven by motors with variable-speed drives (VSDs) are effective tools in reducing energy consumption. When the flow rate requirement goes down, the pump speed is reduced to match load. Considerable energy savings are possible. Figure 10.19 shows the system and pump curves for the central piping system in a school. A theoretical system curve is also plotted on the graph. This curve assumes that no valves are closed and that no minimum differential pressure is required. In reality, the system with the VSD operates on the concept that flow is reduced by closing two-way valves on the heat pumps that are not operating. Additionally, a minimum differential pressure is maintained across the building supply and return headers to ensure that adequate flow will be maintained in each unit. The resulting system curve will resemble the actual curve shown in the figure, which assumes a 20 ft head will be maintained across the supply and return headers.

The manufacturer of the pump represented in Figure 10.19 provides curves for 1150 and 1750 rpm (ITT/B&G 1995). Points of equal efficiency (50%, 60%, 70%, 80%, and 83%) are located on both curves and connected. This permits the computation of the required input power for any operating point between the two curves. For example, a flow of 500 gpm and a required head of 58 ft is found on the system curve. At this point the pump efficiency is 82%. The required input power to the pump (or motor output) is found from Equation 10.10.

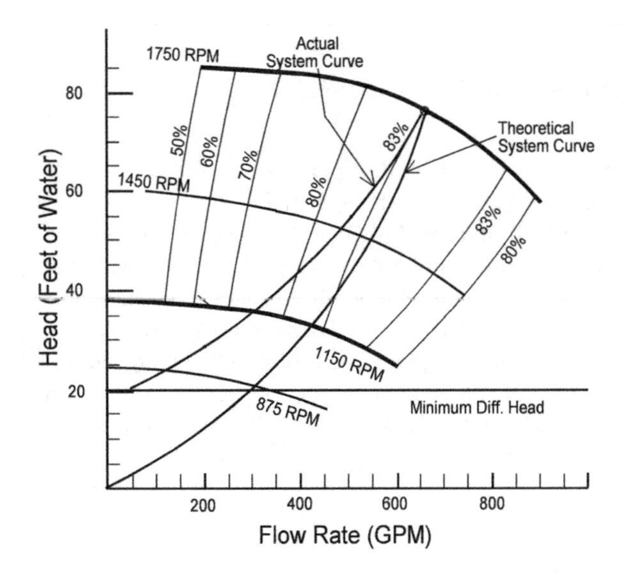

Figure 10.19 Pump and system curves for VSD 15 hp application (ITTC 1995).

$$W_{Pump} = \frac{Q\Delta h}{\eta_{pump}} = \frac{500 \text{ gpm} \times 60 \text{ ft}}{3960 \times 0.82} = 8.9 \text{ hp} = 6.7 \text{ kW} \tag{10.19}$$

In order to determine the electrical energy consumption using weather data and load data, a relationship between water flow rate and pump electrical demand must be developed. For a pump with a VSD, both the drive and motor efficiencies influence demand.

$$W_{Pump\ Out}\ (\text{hp}) = \Delta h\ (\text{ft of water}) \times Q\ (\text{gpm})/3960 \tag{10.20}$$

$$W_{Pump\ In}\ (\text{hp}) = W_{MotorOut} = \Delta h\ (\text{ft of water}) \times Q\ (\text{gpm})/\eta_{Pump} \times 3960 \tag{10.21}$$

$$W_{Motor\ In}\ (\text{kW}) = \frac{0.746 \text{ kW/hp} \times W_{Pump\ In}\ (\text{hp})}{\eta_{Motor} \times \eta_{VS\ Drive}} \tag{10.22}$$

$\eta_{VS\ Drive} = 1.0\ (100\%)$ for constant-speed motors.

Values for motor and VSD efficiencies can be found in chapter 11.

CHAPTER 10 PROBLEMS

10.1 Size the piping (Schedule 40 steel) and compute the system head loss for the (direct-return) chilled water system shown in Figure 10.1 (and Figure 5.10). The distance between each FCU is 50 ft, and the distance between the last FCU and the headers is 30 ft. The distance between the first FCU and the chiller is 120 ft. The chiller head loss is found using the specifications of a Model 060 scroll compressor design (Table 5.10). The fan-coils units are Model 60-HW-4 (Table 5.12) with a flow rate of 20 gpm each. Use ball valves for all valves 2 in. and smaller and gate valves for larger valves.

10.2 Select a chilled water pump and corresponding motor for the system described in Problem 10.1.

10.3 Size the piping, compute the required head, and select a pump and motor for the condenser water loop shown in Figure 10.7. Use the Model 060 chiller from Problem 10.1 with a flow rate of 200 gpm. Use SDR 11 high-density polyethylene to eliminate corrosion in this open loop. The distance between the basin and upper tray in the cooling tower is 12 ft and the distance from the chiller to the cooling tower is 200 ft.

10.4 Design the chilled water piping loop for the system shown below using Schedule 40 steel pipe and gate valves on the main piping, with two-ball valves and one motorized zone valve (C_v = 18) on the fan-coil loops.

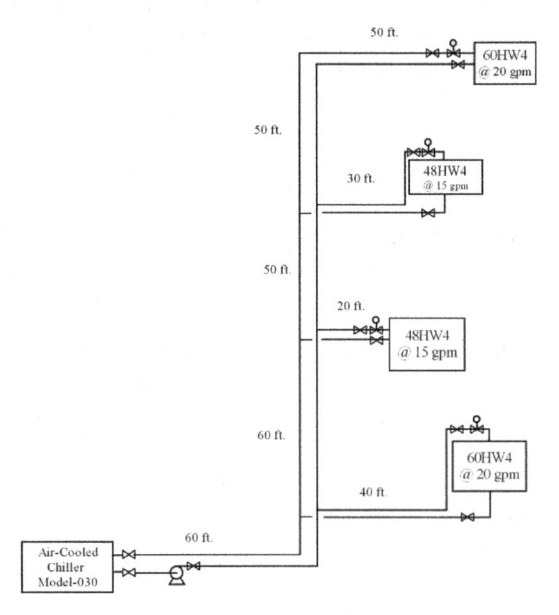

10.5 Compute the required head loss and select a chilled water pump for Problem 10.4.

REFERENCES

ASHRAE. 2000. *ASHRAE Guideline 12, Minimizing the Risks of Legionellosis Associated with Building Water Systems*. Atlanta: American Society of Heating, Refrigerating and Air-Conditioning Engineers, Inc.

ASHRAE. 2004. 2004 *ASHRAE Handbook—HVAC Systems and Equipment,* chapter 12, Hydronic heating and cooling system design; chapter 39, Centrifugal pumps; chapter 42, Valves. Atlanta: American Society of Heating, Refrigerating and Air-Conditioning Engineers, Inc.

ASHRAE. 2005. *2005 ASHRAE Handbook—Fundamentals,* chapter 36, Pipe sizing. Atlanta: American Society of Heating, Refrigerating and Air-Conditioning Engineers, Inc.

Broderick, J., and D. Westphalen. 2001. Uncovering auxiliary energy use. *ASHRAE Journal* 43(2):58–61.

Churchill, S.W. 1997. Friction-factor equation spans all fluid-flow regimes. *Chemical Engineering* 7(Nov.):91–92.

GPC. 1988. Stainless steel submersible pumps. Publication No. L-SP-TL-051. Grundfos Pumps Corporation, Clovis, CA.

HDR Design Guide. 1981. Omaha, NE: Hennington, Durham, and Richardson.

Hodge, B.K., and R.P. Taylor. 1998. *Analysis and Design of Engineering Systems,* 3d ed. Upper Saddle River, NJ: Prentice-Hall.

ITT/B&G. 1994. Base mounted-double suction pump performance curves and dimensions. Curve Booklet B-460F. Morton Grove, IL: ITT Industries-Bell & Gossett.

ITT/B&G. 1996. Close coupled centrifugal pump performance curves. Curve Booklet B-360C. Morton Grove, IL: ITT Industries-Bell & Gossett.

Kavanaugh, S.P., and K.D. Rafferty. 1997. *Ground-Source Heat Pumps: Design of Geothermal Heating and Cooling System for Commercial and Industrial Buildings.* Atlanta: American Society of Heating, Refrigerating and Air-Conditioning Engineers, Inc.

PPI. 1971. *Water Flow Characteristics of Thermoplastic Pipe.* New York: Plastic Pipe Institute.

Rishel, J.B. 2005. Simplifying contemporary HVAC piping. *ASHRAE Journal* 47(2):16–22.

Smith, T. 1983. Reducing the corrosion in heating plants with special reference to design considerations. *Anti-Corrosion Methods and Materials* 30(October):4.

Stein, J., and S.T. Taylor. 2005. It's in the details. *ASHRAE Journal* 47(10):50–53.

11 Motors, Lighting, and Controls

ELECTRIC CIRCUITS, MOTORS, AND DRIVES

Electric motors are the primary driver of almost all HVAC equipment. This chapter attempts to provide an overview of a variety of motors and the electric circuits that power and control these devices. Alternating current (AC) motors are the most common type. Direct current (DC) motors are used only to a limited degree. However, a discussion of electrical circuits is first provided as a background to motor types and characteristics.

Figure 11.1a is a diagram of a single-phase circuit breaker panel with typical wiring connections to commonly used receptacles. Power is connected from the voltage transformer on a nearby power pole or underground connection. The supply cable contains three wires—two line voltage (or hot wires) and a neutral. In the circuit breaker panel, the lines L1 and L2 are connected to two interlaced terminal bars. The neutral wire is connected to a neutral or common terminal bar. A fourth terminal bar is connected to a rod driven into the ground near the panel to a depth of approximately 5 ft. Multiple rods may be required to attain grounding circuit resistance of 5 to 10 ohms. The interlaced line voltage terminal bars are arranged so that a double-pole circuit breaker can be inserted into two adjacent connectors on each bar. The over-current breakers have mating connectors on one end and screw terminals on the other to connect the "hot" wires. The two "hot" wires and a "common" wire are routed from the circuit breaker and common terminal in a multi-wire cable or through an electrical conduit to a 230 voltage-alternating current (VAC) receptacle, as shown in the figure. Some circuits also require a fourth wire for a ground wire. Connections to 115 VAC receptacles are accomplished by using a single-pole circuit breaker connected to only one hot leg (L1 or L2). The connecting cable also has two wires and a ground.

The second wire is connected to the neutral bar and the ground wire is connected to the ground terminal bar. Note the hot wire connection in the 115 VAC receptacle is to the narrow vertical slot and the common wire is to the wider vertical slot. Figure 11.1b depicts a three-phase circuit breaker panel. The panel contains three interlaced line voltage terminal bars to which three-pole circuit breakers are connected. Equal voltage appears across all three legs, creating the three phases (L1 to L2, L1 to L3, and L2 to L3). A fourth wire is connected to the common terminal bar, and in some cases a ground wire is also required.

Motor Characteristics

The nominal rating for single-phase circuits is 240 VAC. Motors and receptacles are typically labeled 230 (or 220) VAC, which corresponds to the slightly higher circuit rating. Line voltages for three-phase circuits include 208, 240, 440, and 480 VAC. Single-phase circuits can be created from 208 or 240 VAC three-phase panels by using only two of the three line voltage legs. Lower voltage (120 VAC) circuits are also available from one of the legs and a wire to the neutral connection. Higher voltage (440 or 480 VAC) three-phase circuits often have a 277 VAC leg-to-neutral circuit that is used for lighting and smaller refrigerating and air-conditioning units.

A variety of AC motor types are used for HVAC applications. Table 11.1 lists the common types used with their operating characteristics and typical applications. The choice of motor is dictated by many factors, not the least of which is cost. Relatively low-cost and low-efficiency single-phase motors dominate the market in the smaller (fractional horsepower) range. Another important factor in motor choice is starting torque. Compressors, pumps, and larger blowers must start against larger inertial and differential pressure requirements. Stored charge in "start" capacitors is used to momentarily boost current through the motor to provide additional torque for starting. Above the one horsepower size, efficiency is improved since minimum efficiencies are mandated (NEMA 2003). Three-phase motors are almost always a wise choice if these circuits are available since the motor size is often reduced compared to a single-phase motor of equivalent power. A more recent innovation for smaller fan applications is the electronically commutated motor (ECM) that can be driven by single-phase power, which is converted to a three-phase signal with variable-speed capability and high efficiency (GE 2000).

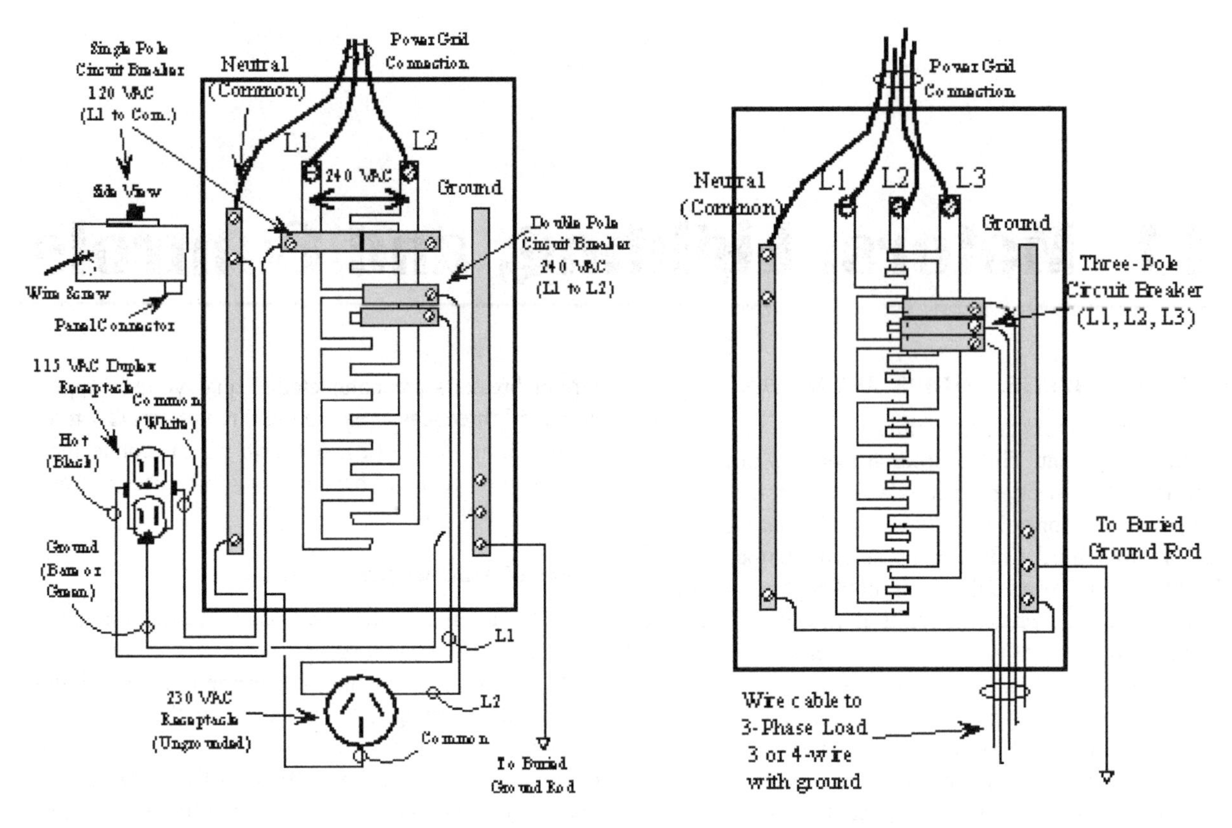

(a) Single-phase panel.

(b) Three-phase panel.

Figure 11.1 Circuit breaker panels with wiring connections.

Table 11.1 AC Motor Types, Characteristics, and Applications (Grainger 2005; GE 2000)

Phase	Type	Typical rpm	Start Torque, % Full Load	Relative Efficiency	Typical Uses
1	Shaded pole	1050, 1550, 3000	50%–100%	Low	Small direct-drive fans and blowers
1	Perm. split capacitor	825, 1075, 1625	75%–150%	Moderate	Direct-drive fans and blowers
1	Split-phase	1140, 1725, 3450	130%–170%	Moderate	Fans, blowers, small tools, pumps, appliances
1	Capacitor-start	1140, 1725, 3450	200%–400%	Moderate to High	Pumps, compressors, tools, conveyors, farm equipment
1/3	Electronically commutated	300-1200		High	Fans, blowers
3	Three-phase	1140, 1725, 3450	200%–300%	High	Applications where three-phase power is available

Figure 11.2 shows several mounting options for motors. Larger motors are typically base mounted. Another base-mounting option used with smaller motors includes a cradle base, which is attached to a raised surface on both ends of the motor. As discussed in chapter 2, compressor motors are often mounted inside a pressurized casing with the compressors to minimize shaft-seal refrigerant leakage. Direct-drive pumps or fans attach directly to the face of the motor. The pump or fan housing is then mounted to a base. Dual-shaft fans are available in face mount and cradle-base mount styles. These motors are used to drive both the condenser and evaporator fans of unitary equipment, as shown in Figures 5.1 and 5.2.

Table 11.2 lists the *minimum* full-load efficiency requirements for typical motors for three different nomi-

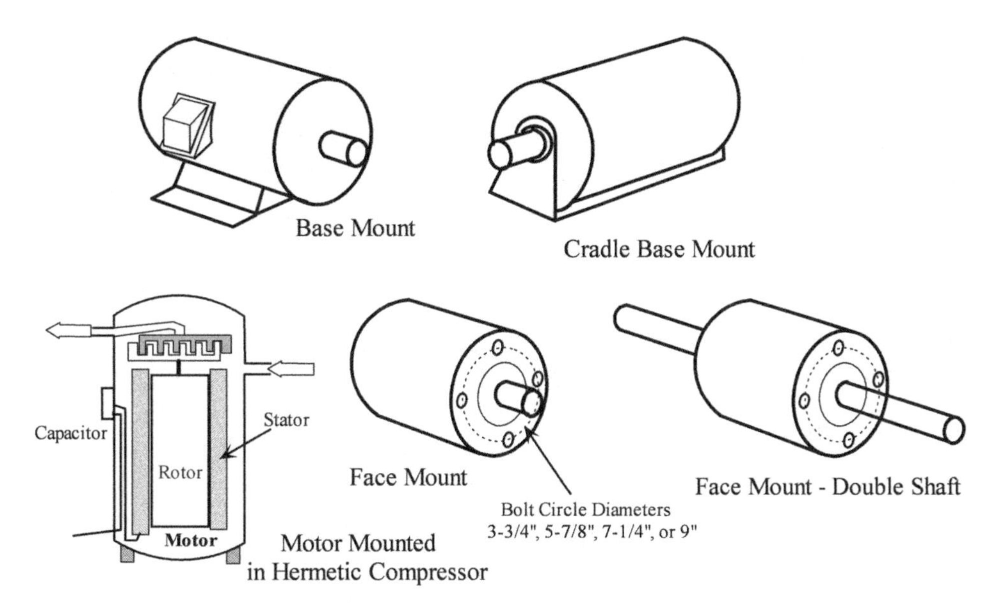

Figure 11.2 HVAC motor mounting options.

Table 11.2 Minimum Motor Full-Load Efficiencies (η_{FL}) in Percent (DOE 1999)*

Output Power (hp)	Nominal Speed ~3600 rpm (2-Pole)	~1800 rpm (4-Pole)	~1200 rpm (6-Pole)
1	–	82.5	80.0
2	84.0	84.0	85.5
3	84.0	86.5	86.5
5	85.5	87.5	87.5
10	88.5	89.5	90.2
15	89.5	91.0	90.2
20	90.2	91.0	91.0
30	91.0	92.4	92.4
50	92.4	93.0	93.0
75	93.0	94.1	93.6
100	93.0	94.1	94.1
200	94.5	95.0	94.5
300	95.0	95.4	95.4
400	95.4	95.4	–
500	95.8	95.8	–

* Additional Reference: NEMA Standards MG 1-12.59, Table 12-10 (see Table 12-11 for other designs).

nal speeds for NEMA A, B, C, and open motors. In many cases, the economic value of premium efficiency motors is easily justified, especially in applications in which the motors operate for extended periods.

Table 11.3 is a listing of correction factor multipliers that are used to compute the efficiency of a motor at part-load operation as a function of full-load efficiency (DOE 1996). Table 11.4 is a similar presentation that provides the power factor of motors from 50% to full load. These values can be used to compute amperage and reactive power (kVAR).

Table 11.3 Motor Efficiency Multipliers for Part-Load Operation ($\eta_{PL} = f_{PL} \times \eta_{FL}$)

Power (hp)	Percentage of Full Load 20%	40%	60%	80%	100%	120%
0.1-1	0.59	0.82	0.90	0.96	1.0	0.95
1.5-5	0.66	0.93	1.0	1.0	1.0	0.96
7.5-10	0.80	0.96	1.0	1.0	1.0	0.98
15-25	0.87	0.98	1.0	1.0	1.0	0.98
30-60	0.92	0.99	1.0	1.0	1.0	0.99
75-100	0.96	0.99	1.0	1.0	1.0	0.99

As noted in previous chapters (5, 9, and 10), the required power output and speed of a motor is often dictated by the requirements from the fan or pump manufacturer's performance table or curve. In other cases, the power requirement of a motor can be computed from

$$W_{Req'd}\,(hp) = \frac{2\pi \times F\,(lb_f) \times L\,(ft) \times rpm}{33000(ft{\cdot}lb_f/hp{\cdot}min)} = \frac{2\pi \times T\,(ft{\cdot}lb_f) \times rpm}{33000\,(ft{\cdot}lb_f/hp{\cdot}min)} \qquad (11.1)$$

The required input power into an electric motor can be found from Equation 11.2, provided efficiency values and part-load corrections similar to those in Tables 11.2 and 11.3 are available.

$$W_{MotorIn}(kW) = \frac{0.746\ kW/hp \times W_{Req'd}\,(hp)}{\eta_{Motor}}$$

$$\text{or} \quad W_{MotorIn}(W) = \frac{746\ W/hp \times W_{Req'd}\,(hp)}{\eta_{Motor}} \qquad (11.2)$$

In many cases, only the voltage and current are known, so power input must be estimated from

$$W\text{ (watts)} = E\text{ (volts)} \times I\text{ (amps)} \quad \text{(for DC motors)}. \tag{11.3}$$

For AC loads, Equation 11.3 is adjusted to account for the fact that nonresistive loads result in the voltage and current sinusoidal waveforms being "out-of-phase." Inductive loads such as motors (especially those operating with low loads) and fluorescent lights are primary contributors to this phenomenon. The angle (θ) designates the degree to which the signals are out of phase, and the resulting power is corrected by the power factor (Power Factor = cosign of θ).

$$W\text{ (watts)} = E\text{ (volts)} \times I\text{ (amps)} \times \text{PowerFactor} \quad \text{(for single-phase AC motors)} \tag{11.4}$$

Recall that single-phase motors have two power wires (L1 and L2 for 230 VAC motors), a common wire, and/or a ground wire. In 115 VAC motors are a single power wire (hot), a common wire, and a ground wire.

$$W\text{ (watts)} = E\text{ (volts)} \times I\text{ (amps)} \times \text{PowerFactor} \times \sqrt{3} \quad \text{(for three-phase AC)} \tag{11.5}$$

Three-phase motors have three power wires (L1, L2, and L3), a common wire, and/or a ground wire.

However, the angular velocity of a motor (a.k.a. motor speed) is typically expressed in revolutions per minute (rpm). For a DC motor, speed can be controlled by varying the DC voltage. Load will also affect motor speed. The rotational direction (clockwise [CW] or counterclockwise [CCW]) of a DC motor can be reversed by changing polarity (switching + and – leads).

The angular velocity of an AC motor is a function of the number of motor poles and the frequency of the AC power source, which is typically 60 Hz (3600 cycles

Table 11.4 Motor Power Factors (cos θ) (Turner 1992)

Motor Output (hp)	~3600 rpm % of Full Load			~1800 rpm % of Full Load			~1200 rpm % of Full Load		
	50%	75%	100%	50%	75%	100%	50%	75%	100%
1	0.68	0.79	0.85	0.54	0.66	0.74	0.37	0.47	0.56
1.5–5	0.65	0.77	0.83	0.63	0.75	0.83	0.45	0.57	0.65
7.5–10	0.74	0.83	0.86	0.77	0.83	0.86	0.71	0.79	0.84
15–25	0.84	0.88	0.89	0.78	0.85	0.87	0.74	0.81	0.84
30–60	0.85	0.88	0.88	0.84	0.89	0.90	0.77	0.84	0.87
75–100	0.87	0.90	0.90	0.79	0.85	0.87	0.77	0.84	0.86

Example Problem 11.1

Select a 460 VAC, three-phase, 3450 rpm motor to provide 20,000 cfm and a total static pressure of 3.5 in. of water from Table 9.13. Compute the percent of rated load, efficiency, demand (kW), power factor, and current of the selected motor.

Solution:
From Table 9.13, a 30 DE-25 in. diameter fan operating at 1603 rpm is selected to deliver the required flow and TSP. Equation 9.20 is used to compute the required pulley sizes for the 3450 rpm motor. The required motor output power (bhp) is 21.1 hp. The next largest incremental motor size is 25 hp. Thus,

$$\%\text{ rated load} = 21.1 \div 25 = 84\%.$$

By interpolation from Tables 11.2 and 11.3, the full-load efficiency is 90.6% and the part-load correction factor is 1.0. Thus,

$$\eta_{PL} = f_{PL} \times \eta_{FL} = 1.0 \times 90.6 = 90.6\%.$$

From Equation 11.2,

$$w_{MotorIn}\text{ (kW)} = \frac{0.746\text{ kW/hp} \times w_{Req'd}\text{ (hp)}}{\eta_{Motor}} = \frac{0.746\text{ kW/hp} \times 21.1\text{ hp}}{0.906} = 17.4\text{ kW}.$$

From Table 11.4 the power factor for a 25 hp, 84% loaded motor is 0.88. By rearranging Equation 11.5, the current is

$$I\text{ (amps)} = W\text{ (watts)}/[E\text{ (volts)} \times \cos\theta(\text{PowerFactor}) \div \sqrt{3}\,] = 17{,}400/(460 \times 0.88 \times 1.732) = 24.8\text{ amps}.$$

Table 11.5 Typical AC Motor Performance Data (Grainger 2005)

		Performance Data for Three-Phase Energy-Efficient Motors			
hp	rpm	Volts	Full-Load Amps	Service Factor	NEMA Eff. %
1/4	1725	208–230/460	1.3–1.3/0.65	1.35	62.0
1/4	1725	230/460	0.8/0.40	1.35	77.0
1/3	3450	208–230/460	1.4–1.3/0.70	1.35	64.0
1/3	1725	208–230/460	1.7–1.5/0.80	1.35	62.0
1/2	3450	208–230/460	1.8–1.8/0.90	1.25	70.0
1/2	1725	230/460	1.6/0.80	1.25	81.5
3/4	3450	230/460	2.4/1.2	1.25	74.0
3/4	1725	208–230/460	2.4–2.3/1.1	1.25	81.5
3/4	1140	208–230/460	3.5–3.5/1.7	1.15	75.5
1	3450	208–230/460	3.8–3.8/1.9	1.25	77.0
1	1725	208–230/460	4.0–3.8/1.9	1.15	82.5
1	1140	208–230/460	4.3–4.2/1.7	1.15	75.5
1½	3450	208–230/460	4.6–4.3/2.1	1.15	82.5
1½	1725	208–230/460	5.0–4.9/2.5	1.15	84.0
1½	1140	208–230/460	6.3–6.2/3.1	1.15	80.0
2	3450	208–230/460	6.0–5.5/2.8	1.15	84.0
2	1725	208–230/460	6.6–6.6/3.3	1.15	85.5
2	1170	208–230/460	7.2–7.0/3.5	1.15	85.5
3	3450	208–230/460	9.2–8.7/4.3	1.15	84.0
3	1725	208–230/460	9.7–9.4/4.6	1.15	86.5
3	1175	208–230/460	9.7–9.2/4.6	1.15	86.5
5	3450	208–230/460	9.2–8.7/4.3	1.15	84.0
5	1725	208–230/460	9.7–9.4/4.6	1.15	86.5
5	1175	208–230/460	9.7–9.2/4.6	1.15	86.5
7½	3490	208–230/460	20.3–18.3/9.1	1.15	87.5
7½	1760	208–230/460	20.9–19.2/9.6	1.15	88.5
7½	1165	208–230/460	20.7–18.8/9.4	1.15	88.5
10	3500	208–230/460	27.4–25.4/12.7	1.15	88.5
10	1760	208–230/460	27.2–24.6/12.3	1.15	89.5
10	1165	208–230/460	27.0–24.4/12.2	1.15	90.2
15	3485	208–230/460	40.3–36.4/18.2	1.15	89.5
15	1770	208–230/460	40.5–38.0/19.0	1.15	91.0
15	1180	208–230/460	41.3–37.6/18.8	1.15	90.2
20	3530	208–230/460	54.0–47.2/23.6	1.15	90.2
20	1765	208–230/460	55.0–49.8/24.9	1.15	91.0
20	1180	208–230/460	55.0–50.0/25.0	1.15	91.0
25	3530	208–230/460	66.0–58.0/29	1.15	91.0
25	1775	208–230/460	63.0–61.0/30.5	1.15	91.7
25	1185	208–230/460	66.3–59.4/29.7	1.15	91.7

per minute) in the US and 50 Hz in some other countries. Common AC motors have two, four, and six poles. The synchronous speed (angular velocity of a motor operating at an exact multiple of the AC frequency) of an AC motor can be computed.

$$rpm_{Syn} = \frac{7200}{\text{No. of poles}} \qquad (11.6)$$

When loads are applied, motor speeds will be reduced (slip) compared to the synchronous speeds computed by Equation 11.6. Rated speeds are typically between 3400 and 3500 rpm for two-pole motors, 1725 and 1775 rpm for four-pole motors, and 1050 and 1175 rpm for six-pole motors. Table 11.5 is a typical presentation of motor characteristics for a product line of three-phase motors from 1/4 through 25 horsepower. The motors in this product line can be wired in either 230 or 460 VAC three-phase circuits. Note that most of the motors can also be used in 208 VAC circuits. However, care should be exercised because some motors are not rated for 208 VAC applications. Small eight-pole motors (900 rpm) are also available but are not included in the table. Three values of current are provided, which correspond to the values of rated voltage (first value is for 208 VAC, second for 230 VAC, and third for 460 VAC). If the actual voltage available in the field is different from the rated value, the current will change accordingly (if voltage is lower than rated value, current will increase). Table 11.5 also includes service factor values, which indicate the amount of continuous overload the motors can survive (1.15 service factor = 115% of rated load). It is also important to note that motors that are not fully loaded will draw less power and current than the rated values. To predict the actual power and current at part-load, Equation 11.2 is modified to incorporate the part-load efficiency correction values (f_{PL}) found in Table 11.3.

$$w_{MotorIn(PL)}\ (W) = \frac{746\ W/hp \times w_{PL}\ (hp)}{\eta_{Motor} \times f_{PL}} \quad (11.7)$$

Equation 11.5 is rearranged and part-load power factors (Table 11.4) are used to find the current.

$$I\ (\text{amps}) = W\ (\text{watts})/[E\ (\text{volts}) \times \cos\theta(\text{PowerFactor}) \div \sqrt{3}\,] \quad (11.8)$$

Often single-phase AC motors will turn in only one direction (CW or CCW), but many can be wired to different connections to reverse direction. Three-phase AC motors can often be reversed by changing two of the line connections. The speed in *some* small AC series-wound motors can be varied with voltage, but changing the voltage in most AC motors will result in burnout. Speed is adjusted in most AC motors with a variable-frequency drive (a.k.a. an inverter drive) that converts the AC signal to DC and then electronically pulses the DC signal in a manner that mimics an AC signal with varying frequencies.

Variable-Speed Drives

As discussed in previous chapters, fan and pump laws demonstrate the significant energy-saving potential when rotating equipment can be operated at reduced speed. This phenomenon is also present in compressors. However, the speed of most AC motors is primarily a function of signal frequency and the number of poles. Neither of these is easily adjusted.

Variable-speed drives (VSDs) for AC motors involve rectifying and filtering the input signal to a DC output coupled with a series of electronic pulses that are modulated to mimic an alternating current pattern. VSDs employing this pulse width modulation (PWM) technology have become more popular since costs have declined and signal quality has improved. However, careful application of this technology is required to prevent motor winding failure, high levels of audible noise, electrical noise, vibration, and feedback in nearby sensitive electronic equipment (ASHRAE 2004b, chapter 40). Inverter duty motors, specification of equipment with lower noise (electronic and audible) levels, and paying close attention to the manufacturer's installation practices will mitigate these concerns.

Table 11.6 is a listing of typical efficiencies of PWM VSDs over the range from 25% to 100% full load. In order to obtain overall motor efficiency ($\eta_{VSD\ Motor}$) when a VSD is applied, the drive efficiency must be multiplied by previously mentioned motor efficiency and part-load correction factors. There is a marked loss in efficiency when the VSD speed falls below 50% of the rated speed. This phenomenon must be considered when computing the savings of VSDs. If designers choose to oversize systems because of the reduced full-load demand penalty offered by variable-speed drives, the equipment will operate many hours at a much lower than necessary speed during part-load conditions. At these much lower speeds, system efficiency will be poor and much lower-than-expected savings will result. Furthermore, efficiency loss is typically converted to unwanted heat generation that may have an impact on equipment life since cooling airflow is reduced at lower speeds. For this reason, some manufacturers provide a lower limit of operating speed.

LIGHTING

Lighting is a major source of energy consumption as well as a primary heat source that must be removed by the building cooling system. Thus, optimization is a major concern when attempting to minimize the required size of the cooling equipment and reduce both lighting and HVAC energy consumption. Required lighting levels vary with application and type of occupant. Rooms in which occupants perform tasks with

Table 11.6 PWM Variable-Speed Drive Efficiencies (η_{VSD}) in Percent (DOE 2002) ($\eta_{VSD\ Motor} = f_{PL} \times \eta_{FL} \times \eta_{VSD}$)

VSD	% of Full-Load Speed			
Rated hp	25%	50%	75%	100%
1	9.4	44.2	70.5	82.5
5	29.6	74.7	88.3	92.4
10	35.3	79.0	90.3	93.5
25	35.6	79.4	90.6	93.8
50	43.3	83.5	92.1	94.4
100	54.8	89.1	95.0	96.6
250	61.2	91.3	96.1	97.3

Example Problem 11.2

Compare the demand of a 10 hp, four-pole motor at full load with and without a VSD. Repeat the calculation if the motor is operating at 600 rpm with a 1.4 hp load.

Solution:

From Table 11.5, a four-pole motor (~1800 rpm - slip) operates at 1760 rpm with an efficiency of 89.5%. From Equation 11.2,

$$w_{MotorIn}\ (\text{kW}) = \frac{0.746\ \text{kW/hp} \times w_{Req'd}\ (\text{hp})}{\eta_{Motor}} = \frac{0.746\ \text{kW/hp} \times 10\ \text{hp}}{0.895} = 8.3\ \text{kW}.$$

If a VSD is used, the drive efficiency from Table 11.6 must be included.

$$w_{MotorIn}\ (\text{kW}) = \frac{0.746\ \text{kW/hp} \times w_{Req'd}\ (\text{hp})}{\eta_{Motor} \times \eta_{VSD}} = \frac{0.746\ \text{kW/hp} \times 10\ \text{hp}}{0.895 \times 0.935} = 8.9\ \text{kW}$$

When the motor operates at reduced speed and part load, the percentage of full-load speed is computed and the efficiency is determined from Table 11.6. The percent of rated speed is

$$\%\ \text{Rated Speed} = 600 \div 1760 = 34\%.$$

Interpolating between 25% and 50% of rated speed for a 10 hp motor in Table 11.6, the VSD efficiency is 51%. Thus,

$$w_{MotorIn}\ (\text{kW}) = \frac{0.746\ \text{kW/hp} \times w_{Req'd}\ (\text{hp})}{\eta_{Motor} \times \eta_{VSD}} = \frac{0.746\ \text{kW/hp} \times 1.4\ \text{hp}}{0.895 \times 0.51} = 2.4\ \text{kW}.$$

Table 11.7 Lighting Illuminance Categories and Values for Generic Activities (IESNA 1993)

Type of Activity	Illum. Category	Range of Illuminances Lux (Lumen/m^2)	Range of Illuminances Foot-candles	Work Plane
Public spaces with dark surroundings	A	20-30-50	2-3-5	General lighting throughout space
Simple orientation for short, temporary visits	B	50-75-100	5-7.5-10	
Work spaces with few visual tasks	C	100-150-200	10-15-20	
Work spaces with high-contrast visual tasks	D	200-300-500	20-30-50	Illuminance on task
Work spaces with medium-contrast visual tasks	E	500-750-1000	50-75-100	
Work spaces with low-contrast visual tasks	F	1000-1500-2000	100-150-200	
Extended work periods of small, low-contrast tasks	G	2000-3000-5000	200-300-500	Illuminance on task + general lighting throughout space
Prolonged work periods of small, low-contrast tasks	H	5000-7500-10000	500-750-1000	
Performance of very specialized, very small, low-contrast tasks	I	10000-15000-20000	1000-1500-2000	

very small, low-contrast components over extended periods will need very high lighting levels. In areas requiring high lighting levels, designers should consider small "task" lights on the work areas in conjunction with low- to medium-level lighting throughout the space. Lighting levels also increase if the space is populated with older occupants who typically have some level of diminished eyesight. Table 11.7 is a sample of recommendations from the Illuminating Engineering Society of North America (IESNA) that lists illumination categories of A (lowest lighting level) through I (highest level) with a range of requirements (low-medium-high) that considers the age of the occupants. The units of illuminance are lux (lumens/m^2) or foot-candle (lumens/ft^2). Table 11.8 provides a list of tasks and related building types with the suggested illumination categories (IESNA 1993).

ASHRAE Standard 90.1-2004 (ASHRAE 2004a) also mandates lighting control. Buildings larger than 5000 ft^2 must have automatic control devices to shut off interior lighting during unoccupied periods. The devices can be either programmed time-of-day controllers or occupancy sensors. This standard also places limits on the amount of area (i.e., 2500 ft^2 for buildings less than 10,000 ft^2) and the number of rooms or lamps that can be placed on a single switch (e.g., motel rooms must have a switch at the main entry for all permanent luminaries). Exterior lighting must be controlled by a photosensor or time clock, unless it is required for emergencies, to provide direction or an advertisement, or if it is on a historic building.

ASHRAE Standard 90.1-2004 (ASHRAE 2004a) prescribes maximum allowable lighting power density (LPD in W/ft^2) for a variety of building types, as shown in Table 11.9. Compliance with the Building Area Method requires that the total building installed lighting wattage divided by the lighted building area be less than the values in the table. Additionally, buildings larger than 5000 ft^2 must have either occupant sensor or time-of-day controls that turn lights off during unoccupied periods.

In some cases, a building may contain a higher than normal percentage of spaces requiring high lighting levels. Compliance with the Building Area Method prescriptions may be an unreasonable expectation. An alternate method is the Space-by-Space Method, which allows compliance for individual spaces rather than for the entire building. However, the limits are set by both space and building type, as shown in Table 11.10. This table provides values for only four building types, but the standard lists individual space values for all building types that are shown in Table 11.9.

In order to meet these standards, it is often necessary to use types of lighting with higher efficacy (efficacy = lumens/watt) such as fluorescent or high-pressure sodium fixtures rather than low-efficacy fixtures like incandescent bulbs. Table 11.11 provides information for typical lighting options. This information includes the size, lamp or bulb wattage, total wattage (lamp + ballast), mean lamp output in lumens over the life of the fixture, and average hours of life. In addition to the reduced energy consumption and waste heat generation, designers should note the much greater life of high-efficacy lighting compared to lower-efficacy incandescent bulbs.

The amount of illumination required in a room is affected by the reflectance of the surfaces, the presence of natural lighting, and other factors. These factors can be considered in detail by following practices recommended by the Illuminating Engineering Society of North America (IESNA 1993). However, a simplified approach can compute the approximate number of luminaries (a.k.a. lightbulbs, lamps, or light fixtures) required.

$$\text{Luminaries}_{Req'd} = \text{Illum. (foot-candles)} \times \text{Floor Area (ft}^2) \div \text{Lumens/bulb} = \text{Illum. (lux)} \times \text{Floor Area (m}^2) \div \text{Lumens/bulb} \quad (11.9)$$

Compliance with ASHRAE Standard 90.1-2004 lighting power density can be estimated from

$$(W/A)_{max}\ (\text{W/ft}^2) = \text{No. of fixtures} \times \text{watts/fixture} \div Area\ (1\ \text{ft}^2 = 0.0929\ \text{m}^2)\,. \quad (11.10)$$

The financial impact of bulb life (including labor cost to install or replace) should be included in addition to energy and demand costs, as shown in Equation 11.11.

$$\text{Annual Cost (\$)} = \frac{\text{Hours}}{\text{Year}} \times \left[\frac{\text{Watts} \times \text{\$/kWh}}{1000} + \frac{\text{\# Bulbs} \times \text{\$/Bulb}}{\text{Bulb Life (Hours)}} + \frac{\text{Replacement Labor (\$)}}{\text{Bulb Life (Hours)}}\right] \quad (11.11)$$

Table 11.8 Lighting Recommendations for Specific Tasks (IESNA 1993)

Bakeries	D	Inspection, difficult	F	Building entrances	A
Classrooms	D or E	Inspection, exacting	H	Bulletin boards, dark	E or F
Conference rooms	D	Machine shops	D to H	Bulletin boards, bright	C or D
Drafting rooms	E to F	Material handling	C to D	Boiler areas	A
Hotel lobbies	C to D	Storage, inactive	B	Parking areas	A
Home kitchens	D	Storage, bulky items	C		
Inspection, simple	D	Storage, small items	D		

Table 11.9 Lighting Power Densities for Building Area Method (ASHRAE 2004a)

Building Area Type	W/ft^2	Building Area Type	W/ft^2	Building Area Type	W/ft^2
Automotive facility	0.9	Hospital	1.2	Performing arts theater	1.6
Convention center	1.2	Hotel	1.0	Police/fire station	1.0
Courthouse	1.2	Library	1.3	Post office	1.1
Dining: bar/lounge	1.3	Manufacturing	1.3	Religious building	1.3
Dining: cafeteria	1.4	Motel	1.0	Retail	1.5
Dining: fast food	1.6	Picture theater	1.2	School/university	1.2
Dining: family	1.9	Multi-family	0.7	Sports arena	1.1
Dormitory	1.0	Museum	1.1	Town hall	1.1
Exercise center	1.0	Office	1.0	Transportation	1.0
Gymnasium	1.1	Parking garage	0.3	Warehouse	0.8
Health care—clinic	1.0	Penitentiary	1.0	Workshop	1.4

Table 11.10 Lighting Power Densities (W/ft²) for Space-by-Space Method (ASHRAE 2004a)

Common Space Types[1]	LPD (W/ft²)	Building Specific Space Types	LPD (W/ft²)
Office—Enclosed	1.1	**Gymnasium/Exercise Center**	
Office—Open Plan	1.1	Playing area	1.4
Conference/meeting/multipurpose	1.3	Exercise area	0.9
Classroom/lecture/training	1.4	**Courthouse/Police Station/Penitentiary**	
For penitentiary	1.3	Courtroom	1.9
Lobby	1.3	Confinement cells	0.9
For hotel	1.1	Judges chambers	1.3
For performing arts theater	3.3	**Fire Stations**	
For motion picture theater	1.1	Fire station engine room	0.8
Audience/seating area	0.9	Sleeping quarters	0.3
For gymnasium	0.4	Post office—Sorting area	1.2
For exercise center	0.3	Convention center—Exhibit space	1.3
For convention center	0.7	**Library**	
For penitentiary	0.7	Card file and cataloging	1.1
For religious buildings	1.7	Stacks	1.7
For sports arena	0.4	Reading area	1.2
For performing arts theater	2.6	**Hospital**	
For motion picture theater	1.2	Emergency	2.7
For transportation	0.5	Recovery	0.8
Atrium—First three floors	0.6	Nurse station	1.0
Atrium—Each additional floor	0.2	Exam/treatment	1.5
Lounge/recreation	1.2	Pharmacy	1.2
For hospital	0.8	Patient room	0.7
Dining area	0.9	Operating room	2.2
For penitentiary	1.3	Nursery	0.6
For hotel	1.3	Medical supply	1.4
For motel	1.2	Physical therapy	0.9
For bar lounge/leisure dining	1.4	Radiology	0.4
For family dining	2.1	Laundry—Washing	0.6
Food preparation	1.2	Automotive—Service/repair	0.7
Laboratory	1.4	**Manufacturing**	
Restrooms	0.9	Low bay (< 25 ft floor to ceiling height)	1.2
Dressing/locker/fitting room	0.6	High bay (≥ 25 ft floor to ceiling height)	1.7
Corridor/transition	0.5	Detailed manufacturing	2.1
For hospital	1.0	Equipment room	1.2
For manufacturing facility	0.5	Control room	0.5
Stairs—Active	0.6	Hotel/motel guest rooms	1.1
Active storage	0.8	Dormitory—Living quarters	1.1
For hospital	0.9	**Museum**	
Inactive storage	0.3	General exhibition	1.0
For museum	0.8	Restoration	1.7
Electrical/mechanical	1.5	Bank/office—Banking activity area	1.5
Workshop	1.9	**Religious Buildings**	
		Worship pulpit, choir	2.4
		Fellowship hall	0.9
		Retail [For accent lighting, see 9.3.1.2.1(c).]	
		Sales area	1.7
		Mall concourse	1.7
		Sports Arena	
		Ring sports area	2.7
		Court sports area	2.3
		Indoor playing field area	1.4
		Warehouse	
		Fine material storage	1.4
		Medium/bulky material storage	0.9
		Parking garage—Garage area	0.2
		Transportation	
		Airport—Concourse	0.6
		Air/train/bus—Baggage area	1.0
		Terminal—Ticket counter	1.5

[1] In cases where both a common space type and a building specific type are listed, the building specific space type shall apply.

Table 11.11 Wattage, Output, and Bulb Life of Typical Lighting Types (Grainger 2005)

Fluorescent Bulbs and Ballasts

Length (in.)	Dia.	Lamp Wattage	Wattage w/Mag. Ballast	Wattage w/Elect. Ballast	Mean Lumens	Lamp Life Hours
18	T8	15	19	–	725	750
18	T12	15	19	–	685	9000
24	T8	17	24	16	1260	20000
24	T12	20	28	–	1150	9000
36	T8	25		24	1970	18000
36	T12	30	40	–	1910	18000
48	T8	32	35	31	2710	20000
48	T12	34	40	31	2280	20000
60	T8	40	–	36	3240	20000
60	T12	50	63	45	3310	12000
72	T12	55	66	59	4140	12000
96	T8	59	–	55	5310	15000
96	T12	60	70	55	5240	12000

Other Nonincandescent Lighting

Compact Fluorescent			High Pressure Sodium			Metal Halide		
watts rated/act.	Output Lumens	Life Hours	watts rated/act.	Output Lumens	Life Hours	watts rated/act.	Output Lumens	Life Hours
13/17	700	9000	50/66	3600	24000	50/72	1900	5000
18/23	1100	15000	70/95	5450	24000	100/128	6200	15000
22/24	1520	12000	100/138	8550	24000	175/215	8800	10000
26/33		12000	200/250	19800	24000	250/295	13500	10000
			400/465	48600	24000	400/458	25500	20000

Incandescent Lighting

Watts	Type	Lumens	Life Hours
40	Standard	500	1000
40	PAR-Halogen	510	2500
60	Standard	865	1000
60	PAR-Halogen	800	3000
75	Standard	1190	750
75	PAR-Halogen	1030	2500
100	Standard	1710	750
100	PAR-Halogen	1400	2000
150	Standard	2850	750
150	PAR-Halogen	1690	2000
250	PAR-Halogen	3600	4200

Example Problem 11.3

Design a lighting system for a 1000 ft^2 training room that may include adults using fixtures with two 48-inch T8 fluorescent bulbs and electronic ballast. Check for ASHRAE Standard 90.1-2004 compliance.

Solution:

Classrooms must meet level D or E (Table 11.8). Use the upper level of 100 foot-candles (1000 lux) for level E since adults may be present. A 48-inch T8 bulb produces 2710 lumens and draws 31 watts with an electronic ballast (Table 11.11). From Equation 11.9,

$$\text{Luminaires} = 100\ \text{lumens/ft}^2 \times 1000\ \text{ft}^2 \div 2710\ \text{lumens} = 37\ \text{bulbs (19 two-bulb fixtures).}$$

From Equation 11.10, the ASHRAE Standard 90.1-2004 required power density of 1.4 W/ft^2 (Table 11.10) is met.

$$(W/A)_{max} = 19 \times 2\ \text{bulbs/fixture} \times 31\ \text{watts/bulb} \div Area = 19 \times 2 \times 31 \div 1000 = 1.18\ \text{W/ft}^2$$

HVAC CONTROL COMPONENTS AND CIRCUITS

A detailed discussion of HVAC controls is beyond the scope of an introductory text intended primarily for mechanical engineers and related disciplines. Modern commercial applications typically have direct digital controls (DDC) with electronic hardware circuits and software interfaces. Some buildings require high levels of sophisticated controls to satisfy a variety of occupants and applications. There are needs for well planned, specified, installed, and documented building automation systems (BAS). The level of sophistication should match the needs of the building owners and occupants. Designers should consider installation costs, level of difficulty to operate and service, and compatibility with "open protocol" devices that can be interfaced with devices from other manufacturers. Cost considerations should also include potential expenditures for replacements or interfacing with future expansion, especially if the control protocols are proprietary or not clearly documented.

This section will introduce the topic of controls for unitary equipment, which are primarily ON-OFF and are suited to either open protocol wiring diagrams with electromechanical (EM) components or proprietary integrated circuits with similar EM devices. This discussion will be followed by an overview of control systems for larger equipment and systems.

Unitary HVAC Controls—Components and Circuits

Figure 11.3 is a diagram of two manual switches, a basic but widely used method of HVAC control. The upper switch in the figure is single-pole (single connection), single-throw (ON-OFF switch) (SPST). Connections are made via screw terminals for the type shown. SPST switches can also be ON-ON designs if a third terminal is used. Single-pole switches are typically used in circuits with a single hot wire, such as 115 VAC or DC wiring. The lower switch in the figure is a double-pole, double-throw (DPDT) switch. Two pole switches are applied when two line voltage connections are used (such as single-phase 240 VAC). The double-throw capability is often necessary when an OFF position and two functions of a control circuit are needed. An example would be a reversible motor with two poles (clockwise-OFF-counterclockwise).

Several other types of manual switches and circuit protection switches are shown in Figure 11.4. Slide and rocker switches are available in single-throw, double-throw, single-pole, and multiple-pole designs. Button and thumb wheel switches are typically single throw. Many of these types of switches are all available with spring-loaded action (momentary switches) that requires the user to maintain contact to hold the switch in the closed position. The snap switch is one such design and may be equipped with a lever, as shown. This arrangement is known as a *limit switch* that will break a circuit (NC limit switch) or activate a circuit (NO) if motion moves the lever.

Figure 11.4 also provides three types of switches that are typical of devices that provide either protection or control of HVAC equipment. A pressure switch is shown in a package that is commonly used to switch on or off with changes in pressure. Examples are a nor-

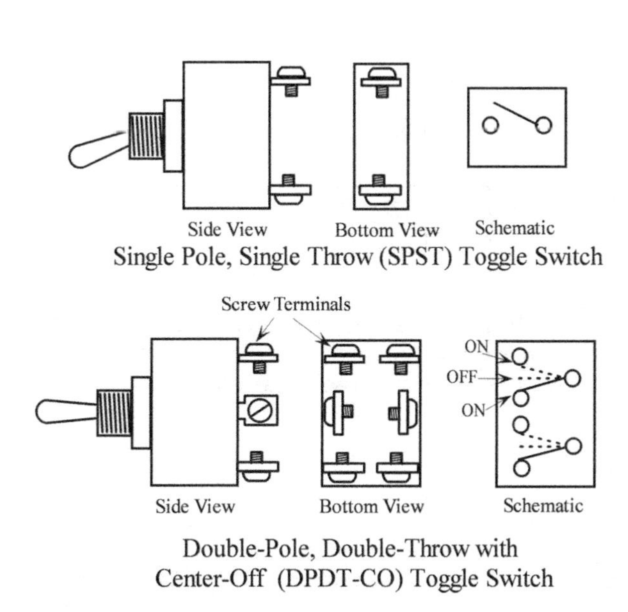

Figure 11.3 Two types of toggle switches.

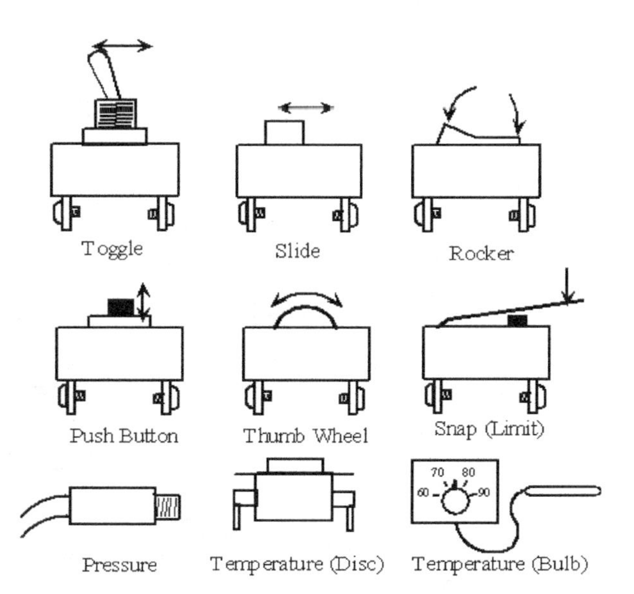

Figure 11.4 Manual and circuit protection switches for HVAC applications.

mally closed switch that will open and deactivate the compressor when the discharge pressure exceeds acceptable levels because of a component malfunction. A similar low-pressure switch is common on the suction side, but it opens when the pressure falls below an acceptable value. Normally open switches are also used, such as ones that activate additional fans when the refrigerant discharge pressure rises. In many cases, a belt-and-suspender approach to compressor protection might be taken by putting a normally closed disc type switch (shown in the figure) that will deactivate a circuit when the temperature is excessive. This type of switch is also commonly used in electric furnaces. The final item shown in Figure 11.4 is an adjustable bulb temperature switch. A capillary tube connects an ideal gas bulb to a diaphragm-activated switch. The adjustment knob shown sets the diaphragm counterforce with a spring, which is selected so that the indicator position represents the temperature surrounding the bulb.

Figure 11.5 provides two views and two schematic diagrams of a relay, which is simply an "electrically activated switch." The activating signal is typically a low-voltage (and low-power) control signal. Many electromechanical HVAC devices use a 24 VAC signal that is provided by on-board transformers. Typically a thermostat, timer, or some other control device will route the low-voltage signal to solenoid coil terminals in the relay. The solenoid will close the switch contacts, and current from the higher line voltage source will flow to the HVAC equipment (compressor, pump, fan, etc.). The relay shown in Figure 11.5 is used as an example. The clear cover provides a good view of the components. In addition to the actual physical representation of a relay, Figure 11.6 provides two schematic representations. The left schematic more closely resembles the actual contact movement. When there is no voltage to the coil, the conducting lever is connected from the common terminal to the NC terminal. When voltage is applied to the coil, the

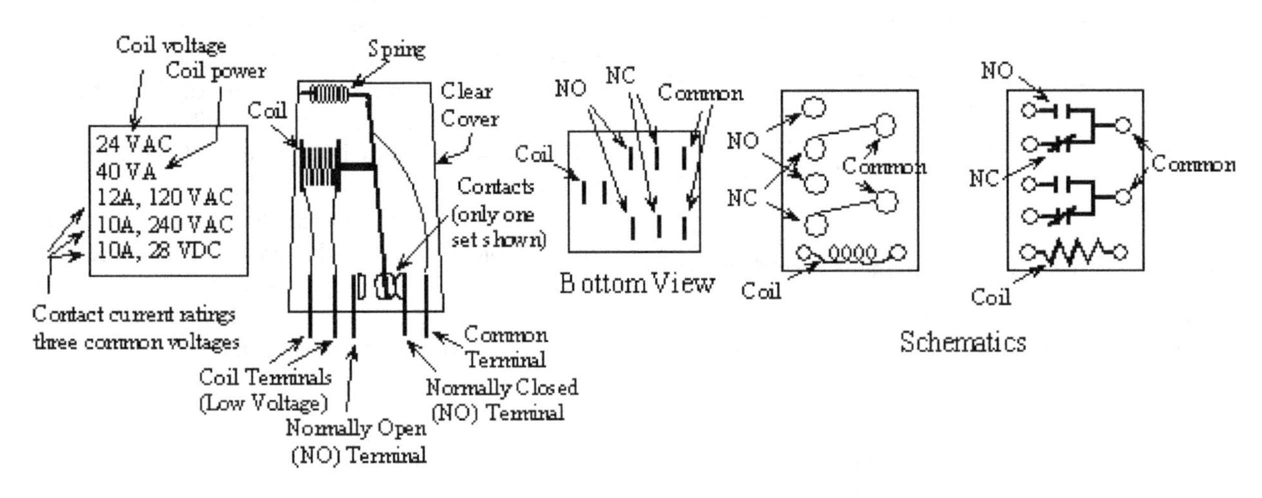

Figure 11.5 Side and bottom specifications and schematic views of DPDT relay.

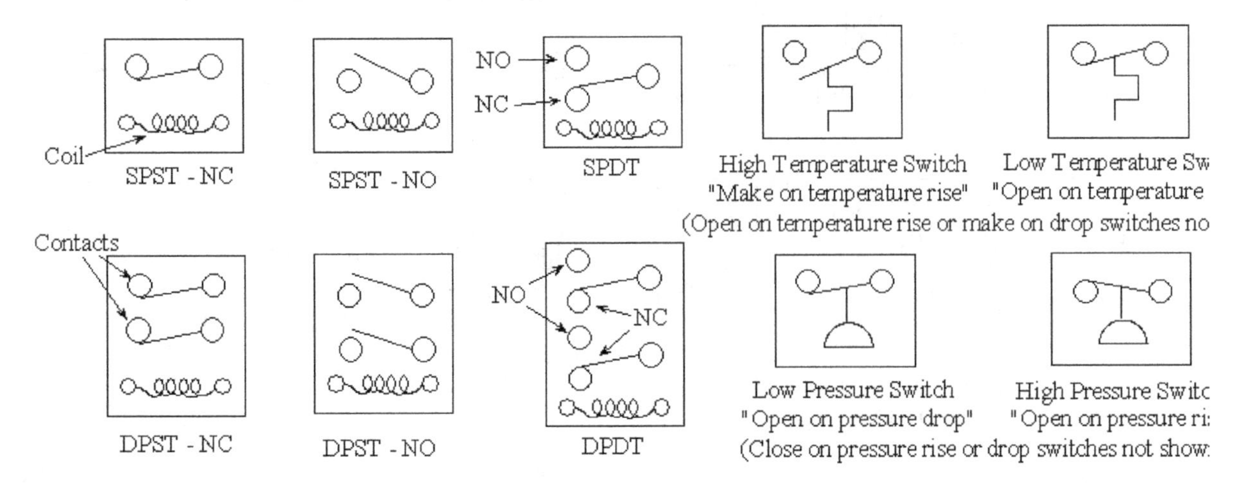

Figure 11.6 Symbols for typical unitary control components—relays and switches.

lever connects the common and NO terminals. A commonly used alternate schematic format is shown on the right. The normally open contacts are represented by two parallel lines and the normally closed contacts are represented by two parallel lines with a diagonal connecting line. Figure 11.5 also includes the relay specifications that are printed on the cover. Some relays used in control circuits are sealed with only the terminals exposed. One common type of relay used with compressors and larger fans and pumps is often called a contactor. These relays are normally open with exposed coils and contacts. They are either single- or multi-pole and capable of handling large, high voltage currents with little coil power consumption.

Figure 11.6 provides a summary of the schematic view of some typical relay and circuit protection switches. The symbols attempt to indicate graphically the arrangement of the switches in their normal position (no voltage applied to the coil). For example, consider a low-temperature switch used in a freeze-protection application. The crooked line in the low-temperature switch schematic indicates an element that will contract when its temperature declines, thereby opening the contacts and deactivating the device that is responsible for creating a temperature below the freeze point.

Figure 11.7 shows circuit diagrams for two unitary heating and cooling systems. The schematic diagram in Figure 11.7a represents a water-to-air heat pump and Figure 11.7b depicts an air-cooled direct expansion air conditioner with a natural gas furnace heating system. The upper portions of the diagrams are the higher line voltage circuit devices, which are separated from the low-voltage portion of the diagrams by the voltage transformer. This type of ladder diagram simplifies the complexity of the actual wiring patterns, but it requires that the line voltage contacts of the relays be separated from the low-voltage coils on the diagrams. The symbols for the relays are noted near the diagrams for the contact and coil symbols to assist with the interpretation of the circuit.

Power to the water-to-air heat pump circuit shown in Figure 11.7a is single-phase high voltage. The ground and common wires are not shown. The input lines (L1 and L2) pass through a manual disconnect switch. The first component in the circuit is the compressor motor. A DPST compressor contactor (CC) is used to control a capacitor-start motor. One leg connects the capacitor to the start connection (S) and the run (R) connection. The second leg connects to the common connection, which also has a thermal switch to protect the motor from overload. The indoor fan motor has a similar connection except that the motor is three-speed, which allows fan speed to be changed in the field to match indoor airflow requirements. The pumps in water-to-air heat pumps are almost always field-installed since size requirements are rarely known by the manufacturer. Thus, the relay connections are provided but the pump is field-installed as shown.

The next component shown on the diagram is the voltage transformer. Often it will have multiple taps on the line voltage (input) side in order to produce the 24 VAC on the low-voltage connections (output) for a variety of line voltages. The installer must select the tap for the appropriate line voltage rating (208 or 230 VAC in this case) that will result in a value near 24 VAC on the secondary circuit. The low-voltage circuit has either a fuse or a circuit breaker to protect the transformer. Power is fed to the thermostat connection terminal via a red wire (R). The green (G) terminal is connected to the indoor fan relay (FR) coil. Thus, if wire from a switch or thermostat relay connects the R terminal to the G terminal, the coil will energize, close the fan relay contacts, and start the indoor fan. Again, note that in the ladder diagram format of Figure 11.7, the relays are separated with the coil being shown as part of the low-voltage circuit and the corresponding contacts shown on the line voltage portion of the diagram.

The control circuit to protect the compressor is more elaborate than the fan motor. In addition to the line voltage high-temperature switch, the Y terminal for the compressor is connected to refrigerant high-pressure and low-pressure switches and a low-temperature switch located on the water line to prevent coil freezing. All three of these protection switches are wired in parallel with a lockout relay (LOR) to prevent the compressor from restarting until an operator switches the power to the low-voltage circuit OFF. Additionally, a time delay relay (TDR) will open when the compressor is cycled OFF for a brief period of time. This allows the refrigerant compressor inlet and outlet pressures to equalize, thus reducing the required compressor starting torque and current. When all four protection switches are closed and low voltage is routed from the R to the Y terminals, current flows through the compressor and pump relay coils, which results in the corresponding two sets of contacts closing. The final control terminal shown in this circuit is the reversing valve relay coil. Typically, heat pumps are piped so that the compressor discharge is routed through the four-way valve (see Figure 2.11, 5.6, or 5.7) to the indoor coil. This is the heating mode of the heat pump, which is often more critical. Thus, to switch the reversing valve to route the compressor discharge to the condenser and bring the unit into the cooling mode, both the Y terminal and the O terminal must be connected to the R terminal. However, some manufacturers of heat pumps for cooling-dominated applications may choose to route the reversing valve piping in a manner that would make cooling the default mode when the reversing valve relay is not activated.

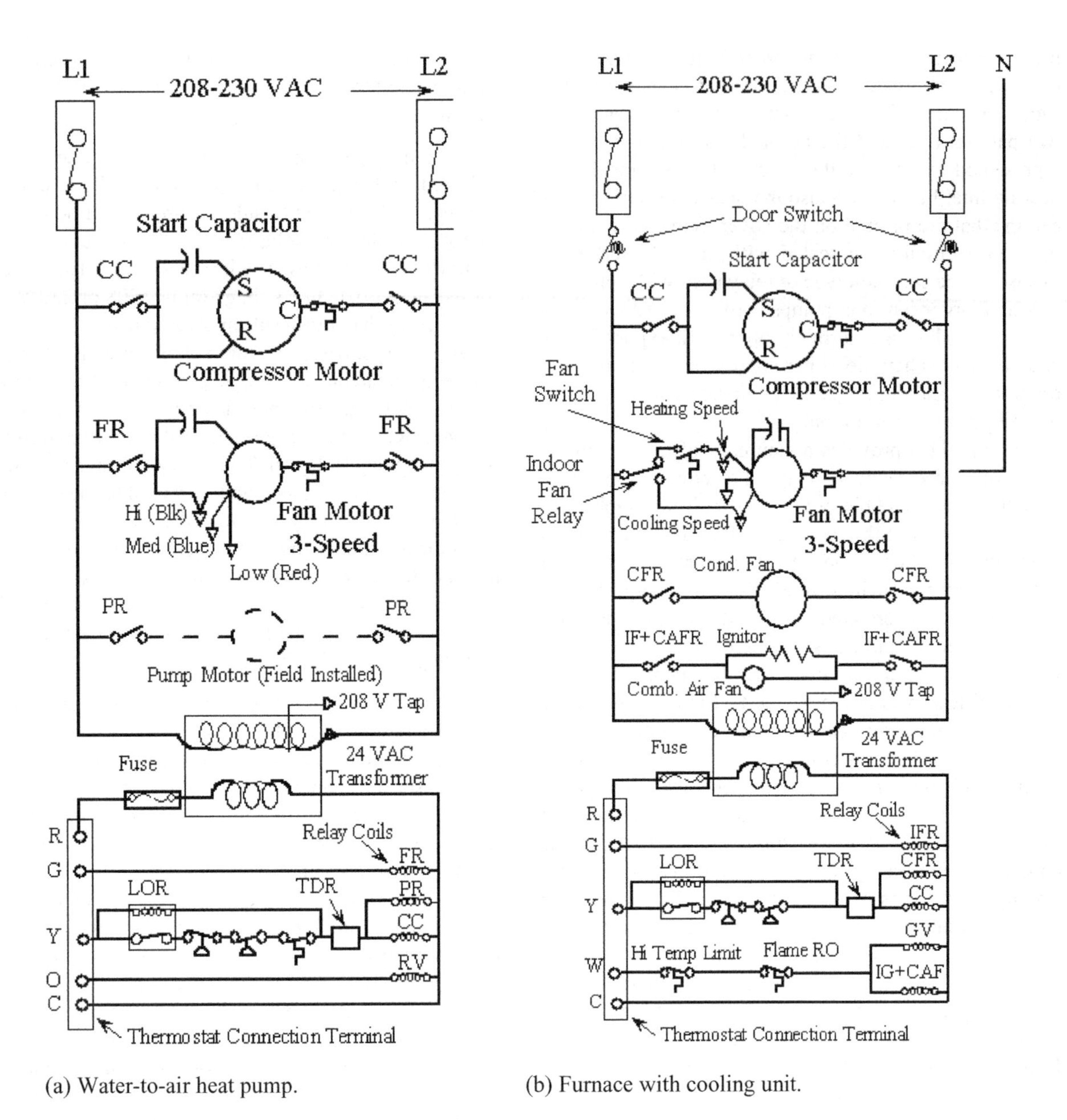

(a) Water-to-air heat pump.

(b) Furnace with cooling unit.

Figure 11.7 Circuit diagrams for unitary equipment.

Figure 11.7a does not show the auxiliary heat or emergency control circuits, which are in a separate cabinet but typically in the same thermostat control circuit. The auxiliary heat is activated by a second-stage temperature switch that is set to close several degrees below the first-stage heating switch for the compressor. The emergency heat is activated by the occupant when the heat pump is not functioning. Typically, the auxiliary and emergency heat is provided by electric furnaces.

Figure 11.7b is a typical wiring diagram for a natural gas furnace and electric cooling system. In many cases, these are split systems with the furnace, evaporator coil, and indoor fan located in one package and the compressor and condenser fan located in a separate outdoor package (see Figure 5.5). However, the Figure 11.7b diagram is for a single packaged system located outdoors with supply and return ducts routed to and from the indoor space (see Figure 5.4).

Line voltage is connected to the unit via a door switch that prevents opening the access door with the furnace operational. The compressor is typically circuited as described for the water-to-air heat pump. The condenser (outdoor) fan is activated simultaneously with the compressor. In the cooling mode, the indoor fan can be activated continuously or when the compressor is operating. However, in the heating mode, the fan is activated with a temperature switch located near the air outlet that closes the contact(s) soon after the furnace ignites. This arrangement prevents cold air from being delivered at start-up before the air is warm enough for comfort and also continues air circulation for a short period after the furnace has shut off to prevent system overheating.

In order to reduce fuel consumption, furnace manufacturers have replaced standing pilot lights with igniters. An igniter is activated in parallel with the opening of the gas valve (GV) when the control system calls for heat by connecting the R low-voltage terminal to the W heating terminal. In the diagram, a high-temperature limit switch and a flame roll-out (RO) switch protect the system from unsafe combustion conditions by preventing igniter or gas valve operation. The cooling mode circuit is similar to the water-to-air heat pump, except the outdoor condenser fan is activated rather than a pump and no water freeze-protection switch is required.

Both Figures 11.7a and 11.7b are single-stage devices. Multi-stage (or variable capacity) cooling and heating units are becoming increasingly popular. They require multiple temperature switches, multi-speed motors, and/or multiple furnace burners. Control circuits become increasingly complex and integrated printed circuit boards are the norm.

Figure 11.8 depicts a simple thermostat that could be used with a control system such as one shown in Figure 11.7b. The low-voltage input is provided to terminal R. Power is circuited to the input of the system (operating mode) switch and to the continuously ON terminal of the fan switch. If this switch is in the ON position, the indoor fan will operate with or without the cooling unit or furnace being on since the G terminal is connected directly to R. If the system switch is set to the "Cool" position and the room temperature rises enough to close the "make on temperature rise" switch, power will be fed to the Y terminal (which activates the compressor and the outdoor fan) and to the "Auto" terminal of the fan switch, which connects the G terminal (thus activating the indoor fan).

If the system switch is set to the "Heat" position and room temperature falls enough to activate the "make on temperature drop" switch, power is routed to the W terminal, which will activate the furnace igniter and open the gas valve. However, power is not fed to the G terminal through the thermostat since the indoor fan will be activated by the line voltage fan temperature switch located near the furnace outlet.

The thermostat circuitry required for a single-stage heat pump is somewhat more complicated. As previously mentioned, the system typically has two supplemental heating sources. The auxiliary heating source is used during those periods of higher heating loads when the heat pump is not able to meet the building requirement. An emergency heater is normally used if the heat pump malfunctions or for air-source heat pumps (Figure 5.6) during periods of very low outdoor temperatures when this particular type of unit is ineffective.

Figure 11.9 is a block diagram of the thermostat connection terminals and associated relays for heat pump operation. In addition to the need for auxiliary and emergency heating, the circuitry must deal with the fact that the compressor must be activated in both the heating and cooling modes. Therefore, the circuit must include a means of preventing the reversing valve relays from being activated in the heating mode via feedback from the signals that activate the compressor or the fan relays. Figure 11.9 also brings attention to the lack of convention for the terminal labeling among heat pump manufacturers. In many cases, this will limit the choice to compatible and proprietary thermostats available only

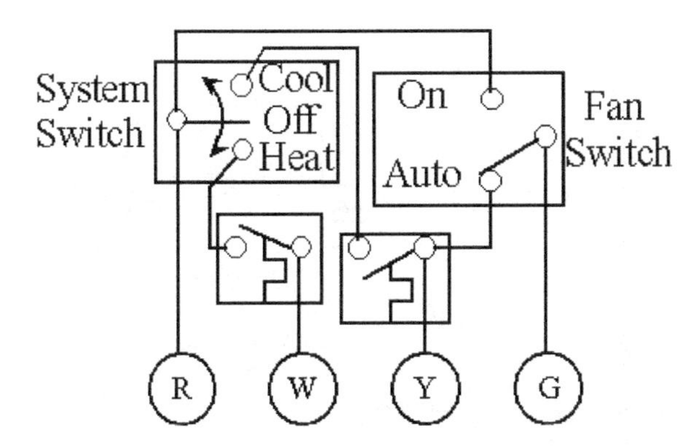

Figure 11.8 Simple single-stage heat and cool thermostat.

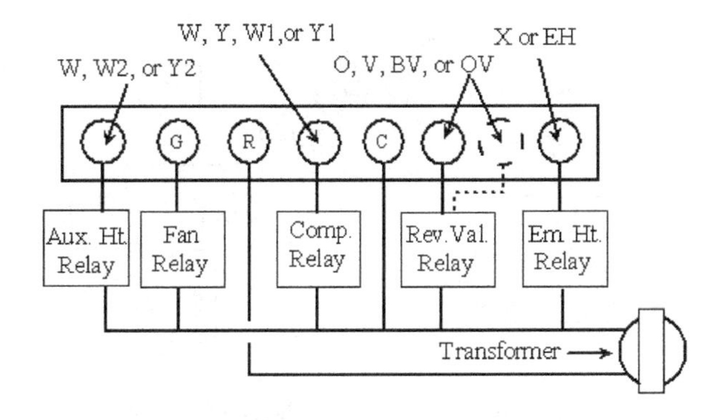

Figure 11.9 Block diagram of heat pump thermostat with various in-terminal designations.

from the heat pump manufacturer. This situation is compounded with the use of integrated circuits with embedded software logic control systems that are available on some unitary equipment and nearly all central HVAC systems in nonresidential applications.

Central HVAC System Controls

Simple ON-OFF control in most cases is inadequate for the needs of larger HVAC systems. The *1995 ASHRAE Handbook—HVAC Applications* (ASHRAE 1995, chapter 42) provides an extensive discussion of the components, systems, and logic of controls for more complex systems. More recent versions of this publication repackaged the presentation of this material by decreasing the amount of discussion on the hardware and adding a chapter on control strategies and optimization (ASHRAE 2003, chapter 41). The information provided in this brief section serves only as a sample of the many components and control strategies currently available.

This discussion will include water loop and air system distribution control. Refrigerant flow control is not included other than to mention capacity variation of chillers. Figure 11.10 provides a diagram of basic two-way and three-way water valves. The three-way valve shown is a mixing valve in that flows enter two ports and exit the third. A three-way diverting valve (not shown) would have a single inlet port and two outlets. The designs shown in the figure are referred to as *globe* or *plug* valves, which have relatively high friction losses compared to ball valves, gate valves, or butterfly valves. As discussed in chapter 10, the flow coefficients (C_v = flow rate in gpm that generates 1 psi friction loss) of these types of valves are greater so that system head requirements will be lower than with more restrictive valve designs. The figure also indicates the type of actuators available. Solenoid and motorized valves have simple ON-OFF motion with the solenoid having a "quick" and relatively noisy movement and the motorized valve having a slower, quieter opening and closing. An electric motor can also be used as a modulating valve that proportions the flow rate through the ports based on a control signal. Although pneumatic diaphragms are used less frequently in modern control systems, they continue to be used in some applications since they offer dependable actuation when a large force is needed to close valves and dampers. Diaphragm actuators are used in HVAC systems with fluids other than air, such as refrigerants. Additional details and valve sizing recommendations are available in the *2004 ASHRAE Handbook—HVAC Systems and Equipment* (ASHRAE 2004b, chapter 42).

Figure 11.11 is a side view diagram of three types of airflow control devices that can be connected to the same type of actuators listed with the water valves in Figure 11.10. A wide variety of blade dampers are available for both rectangular and round duct. In addition to automated flow control, manually actuated dampers with set screws or nuts are used to balance flow through duct mains, at take-offs, and at diffuser inlets. The variable air volume (VAV) terminal air valve shown in the figure is an alternative method of regulating flow. However, VAV terminals with blade type dampers are also available.

The variable-speed drive for fan and pump motors is a valuable control system tool in reducing energy consumption and improving acoustics when properly applied. Closing or modulating flow rates with valves and dampers alone only reduces electrical demand slightly, and noise can be generated when friction losses are high. Reducing pump and fan static pressure via the combination of closing/modulating valves with motor speed reduction saves much more energy and reduces friction-related noise. However, it should be pointed out

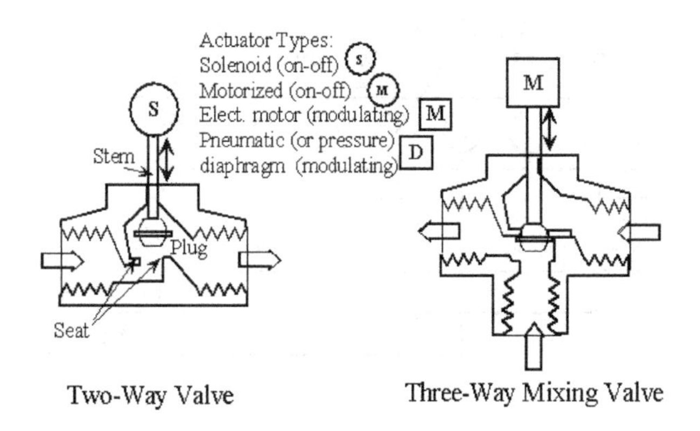

Figure 11.10 Water flow control valves with female pipe thread connections.

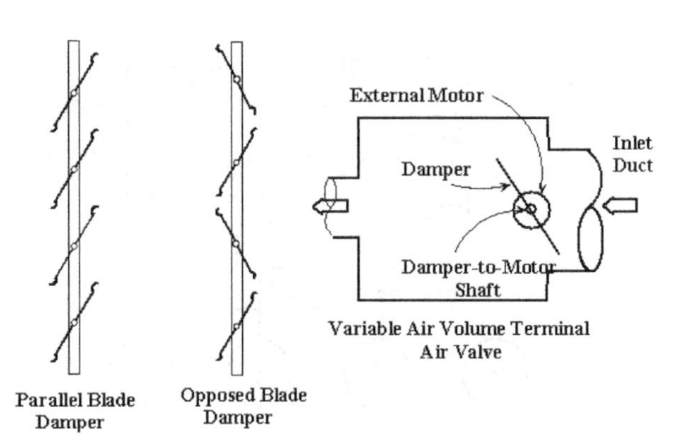

Figure 11.11 Airflow control devices.

that three-way diverting valves should be used sparingly (in favor of two-way valves) in variable-speed applications since they do not provide energy savings gained from flow reductions.

Figure 11.12 is included to demonstrate system flow control technique options used in a variable-flow chilled water system that incorporates previously mentioned devices. Other piping and control options are described in the *2003 ASHRAE Handbook—HVAC Applications* (ASHRAE 2003, chapter 46). The system in the figure is categorized as a primary-secondary loop in that the chillers have dedicated pumps (primary) and the fan-coil loop also has a pump (secondary). This arrangement allows flow in one chiller to operate more frequently near rated capacity and flow in the other chiller to be deactivated during the many hours when the load is below 50% of design.

Variations on this system include primary-only pumping in which the same set of pumps provides flow through the chillers and fan coils, constant-flow chillers that divert water through three valves on the FCUs, and chilled water systems with two or more chillers of different size to improve the load-matching capabilities of the system.

Figure 11.13 is an example of some options for airflow control of a central VAV system. A two-pipe chilled water air handler that has both variable water and air flow provides the primary chilled air. Air temperature can be controlled in response to cooling and heating load (and in some cases dehumidification requirement). The ventilation air is provided with the primary air so the control system must be programmed to comply with the algorithms of ASHRAE Standard 62.1 for multiple-zone systems (ASHRAE 2004c). This requires that the airflow rates be monitored at the outdoor air inlet and at each zone. Note the VAV terminals also include sensors for this purpose. In this system, heating is accomplished at the VAV terminal. This also serves as a convenient source of reheat when the minimum ventilation air requirement results in overcooling of the room during periods of low load.

Digital Controls

The complexity of many larger HVAC systems dictates the use of microprocessors. Digital controls use

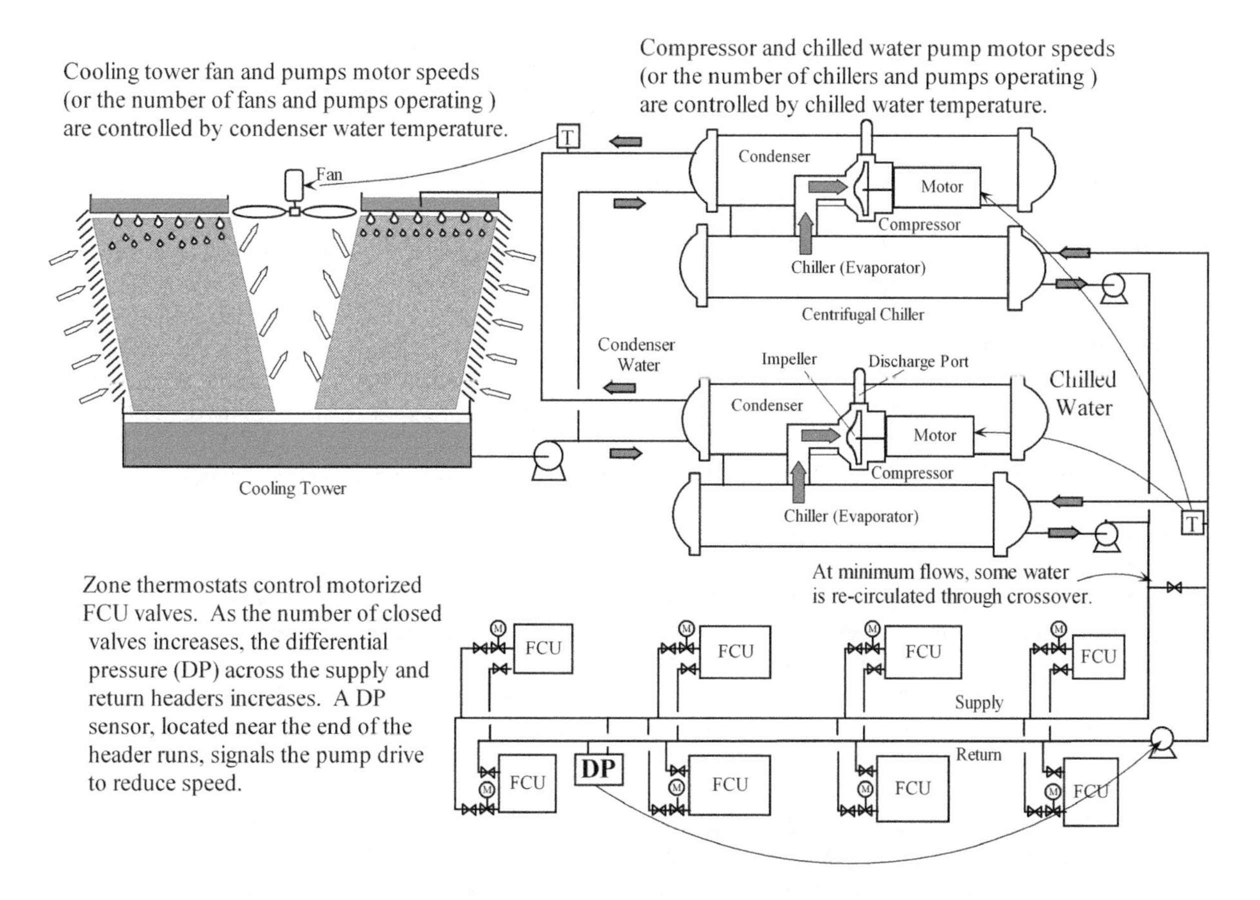

Figure 11.12 Controls for water-cooled variable-volume chilled water systems.

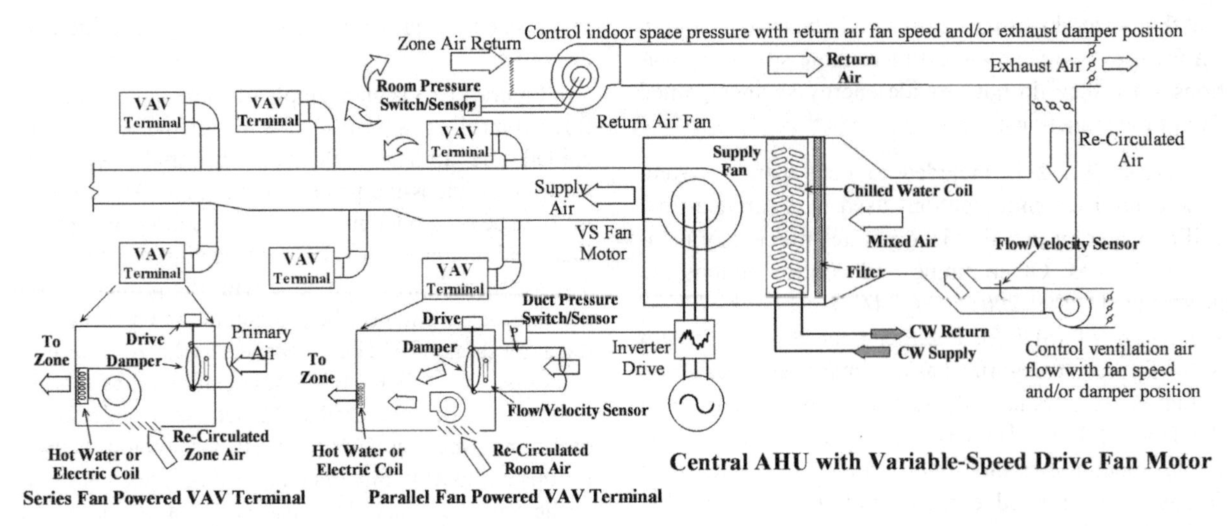

Zone thermostats control the position of VAV terminal air valves (open with increasing room temperature). As valves close, duct static pressure increases and the pressure sensor sends a signal to decrease fan speed. Parallel FPVAV terminal fans are activated when primary air is reduced to provide sufficient air distribution in zone. With series FPVAV, terminal fans run continuously but use higher fractions of zone re-circulated air when primary air is reduced. However, primary air must be maintained at sufficient levels to provide adequate ventilation air. Thus, the heating coils must be activated (even in the cooling mode) to prevent over-cooling or when a zone requires heat.

Figure 11.13 Example of variable air volume system control devices and techniques.

microprocessors to execute software programs that are customized for commercial buildings. Controllers receive input information from sensors that measure values such as temperature and humidity to perform control routines in software programs and to perform control using output signals to actuators, such as valves and electric (or pneumatic) actuators connected to dampers. The operator may enter parameters such as setpoints, minimum ON and OFF times, or high and low limits, but the control algorithms make the control decisions. The computer collects input and positions output devices in a stepwise scheme. Signals are calculated digitally rather than using an analog circuit or mechanical change (ASHRAE 2005, chapter 15). This method of control is referred to as direct digital control (DDC).

Digital controls can be used as stand-alone devices or integrated into building management systems to span multiple HVAC applications. Control routine software can be pre-programmed (firmware) and allow the operator to change setpoints from a terminal. In some cases, the control software can also be modified by the operator. Digital controllers can also have manually adjustable features that are often adjusted through an on-board LED interface or portable hand-held devices.

An additional benefit of DDC control in building automation systems (BAS) is the availability of data generated by the transducers used to provide signals to the controllers. For small incremental costs, these microcomputer-based systems can manage and display information that can be used to optimize system performance or provide required reports. The same BAS can also be used to control and monitor non-HVAC systems (security, safety, etc.).

A discussion of the details of DDC and BAS is beyond the capability of this author, but in addition to the information appearing in the ASHRAE Handbook series (ASHRAE 2005, chapter 15; ASHRAE 2003, chapters 39, 40, 41, and 42), ASHRAE Guideline 13-2000 is suggested. An earlier publication, *Practical Guide to Building Controls* (ASHRAE 1997), which provides an introductory overview of ten important control-related topics, is also recommended.

A Note on Control Protocols

The uniqueness and specificity of software programs used by equipment manufacturers and control developers means that communication between systems is often not seamless. ASHRAE has attempted to moderate the negative impact of the multiple proprietary systems by establishing *BACnet®* (Building Automation and Control Networks) protocol (ASHRAE 2004d). Its purpose is to create interoperability for the variety of computerized building control systems that include computers and controllers. *LonWorks* is another open-protocol system that also provides control of automated devices for HVAC, fire protection processes, lighting, elevators, food service equipment, refrigeration, security, and many other building-related functions (ASHRAE 2003, chapter 39). However, the communication between these and many proprietary systems remains a significant challenge for HVAC engineers.

CHAPTER 11 PROBLEMS

11.1 An air system design requires 6600 cfm with 3.0 in. of water (static pressure). Find the required motor size, drive pulley diameter (if the blower wheel has a 12 in. diameter pulley and the motor is 4-pole), fan efficiency, and motor demand.

11.2 Compute the demand and current of a 6-pole, 230-VAC, 3-phase, 20 hp motor at 100%, 75%, and 50% load.

11.3 Compute the demand and current of a 6-pole, 460-VAC, 3-phase, 15 hp motor at 100%, 75%, and 50% load.

11.4 A 30 DE-25 in. fan is operated at 1195 rpm to deliver 10,000 cfm at 3.5 in. of water. Select an 1800 rpm, 230-VAC, 3-phase motor to drive this fan and specify resulting demand (kW) and current.

11.5 Repeat Problem 11.4 if a motor one size larger than required is specified.

11.6 A pump that operates 3000 hours per year and requires 400 gpm and 55 ft of head is significantly oversized. The curve of the existing pump is shown in Figure 10.15. The impellor diameter is 9.75 in. and the motor is 20 hp. The required flow rate is attained by throttling the pump discharge valve.

a. Estimate the required horsepower and compute the resulting motor efficiency and power demand at the throttled position.

b. Estimate the required horsepower and compute the resulting motor efficiency and power demand for the 20 hp motor if the pump impellor was trimmed to a size shown on the curve that provides the needed flow and head.

c. Estimate the required horsepower and compute the resulting motor efficiency and power demand for the adequately sized motor if the pump impellor was trimmed.

d. Estimate the annual energy consumption for the three above options.

For a cost of $500 to trim the impellor and $1000 to purchase and replace the motor, estimate the simple payback for the options in 11.6b and 11.6c.

11.7 Design a lighting system for a 30 × 40 ft classroom using 2-bulb, 48-in. T-8 fluorescent lighting fixtures with electronic ballasts.

11.8 Compare the demand of the resulting design for problem 11.7 with ASHRAE Standard 90.1-2004 limits for this application.

11.9 A 1000 ft^2 storage area is currently lit with standard 100-watt incandescent bulbs to an illumination of 30 foot-candles for 80 hours per week. Compare the annual operating cost of using existing bulbs and replacing them with equivalent lighting output compact fluorescent bulbs. Include operating cost (at 8¢/kWh), cost of bulbs (at 30¢ each for incandescent and $3.00 each for compact fluorescent), and installation cost (1 hour at $20/hour labor).

REFERENCES

ASHRAE. 1995. *1995 ASHRAE Handbook—HVAC Applications*, chapter 42, Automatic control. Atlanta: American Society of Heating, Refrigerating and Air-Conditioning Engineers, Inc.

ASHRAE. 1997. *Practical Guide to Building Controls.* Supplement to *ASHRAE Journal* 39(9).

ASHRAE. 2000. *ASHRAE Guideline 13-2000, Specifying Direct Digital Control Systems*. Atlanta: American Society of Heating, Refrigerating and Air-Conditioning Engineers, Inc.

ASHRAE. 2003. *2003 ASHRAE Handbook—HVAC Applications*, chapter 39, Computer applications; chapter 40, Building energy monitoring; chapter 41, Supervisory control strategy and optimization; chapter 46, Design and application of controls. Atlanta: American Society of Heating, Refrigerating and Air-Conditioning Engineers, Inc.

ASHRAE. 2004a. *ASHRAE/IESNA Standard 90.1-2004, Energy Standard for Buildings Except Low-Rise Residential Buildings*. Atlanta: American Society of Heating, Refrigerating and Air-Conditioning Engineers, Inc.

ASHRAE. 2004b. *2004 ASHRAE Handbook—HVAC Systems and Equipment,* chapter 40, Motors, motor controls, and variable-speed drives; chapter 42, Valves. Atlanta: American Society of Heating, Refrigerating and Air-Conditioning Engineers, Inc.

ASHRAE. 2004c. *ANSI/ASHRAE Standard 62.1-2004, Ventilation for Acceptable Indoor Air Quality.* Atlanta: American Society of Heating, Refrigerating and Air-Conditioning Engineers, Inc.

ASHRAE. 2004d. *ANSI/ASHRAE Standard 135-2004, BACnet® a Data Communication Protocol for Building Automation and Control Networks.* Atlanta: American Society of Heating, Refrigerating and Air-Conditioning Engineers, Inc.

ASHRAE. 2005. *2005 ASHRAE Handbook—Fundamentals*, chapter 15, Fundamentals of control.

Atlanta: American Society of Heating, Refrigerating and Air-Conditioning Engineers, Inc.

DOE. 1996. Buying an energy-efficient motor. *Motor Challenge Fact Sheet*. US Department of Energy, www.oit.doe.gov/bestpractices/motors/factsheets/mc-0382.pdf.

DOE. 1999. Energy efficiency program for certain commercial and industrial equipment: Test procedures, labeling and certification requirements for electric motors. *Federal Register* 64(192). US Department of Energy, Office of Energy Efficiency and Renewable Energy.

DOE. 2002. Variable speed drive part-load efficiency. *Energy Matters*, Winter. US Department of Energy, Office of Energy Efficiency and Renewable Energy.

GE. 2000. *GE ECM2.3 Series Motors*, GET-8068. GE Industrial Systems, General Electric Company, Ft. Wayne, IN.

Grainger. 2005. Catalog No. 396. Chicago: Grainger.

IESNA. 1993. *IES Lighting Handbook, Application and Reference Volume*. New York: Illuminating Engineering Society of North America.

NEMA. 2003. *Motors and Generators, Standard MG-1*. Rosslyn, VA: National Electrical Manufacturers Association.

Turner, W.C., ed. 1992. *Energy Management Handbook*, 2d ed. New York: John Wiley.

12 Energy, Costs, and Economics

Advances in HVAC equipment efficiency, lighting efficacy, and envelope construction practices continue to be developed and applied. However, data indicate energy consumption in newer buildings is greater than in buildings completed before the energy crisis of the 1970s. Figure 12.1 (EIA/DOE 1999) shows an increase in energy consumption costs for newer buildings despite the implementation of standards and building practices in response to rising energy prices in the 1970s. The only notable decline was for buildings completed in the 1980s, which corresponded to a reduction in the required rates of ventilation air for indoor air quality (ASHRAE Standard 62-1981). Ventilation rates, an important factor in energy use, were returned to higher values in 1989. Equally troubling are the much higher energy consumption costs for buildings with HVAC systems that are generally considered to be energy efficient, such as variable air volume (VAV) and central chilled water systems (CWS) coupled with energy management systems (EMS) and economizer cycles. Figure 12.1 indicates that these systems consume 40% to 50% more energy than commercial buildings with simple unitary equipment.

ASHRAE's energy standard, ANSI/ASHRAE/IESNA Standard 90.1-2004 (ASHRAE 2004a), contains useful mandates for the primary HVAC equipment and building envelope components. If these mandates were universally applied, the values summarized in Figure 12.1 would likely be dramatically different. However, *net* HVAC *system* efficiencies or *net* building requirements are not established. Buildings are assumed to comply if all components fall within the guidelines prescribed in an extensive set of tables. The additive nature of the power requirements of each component is not directly addressed. For example, supply fan power limits (Table 6.3.3.1, Standard 90.1) can exceed electrical demand for chillers (Table 6.2.1J, Standard 90.1) and there are no limits on pump power. Unless care is taken to minimize auxiliary power requirements and optimize primary components, systems with numerous high-efficiency HVAC components can consume more energy than simple systems with much less sophistication. The resulting effect is demonstrated in the energy cost comparisons in Figure 12.1.

Two examples are presented here to demonstrate the value of considering *system* design rather than focus-

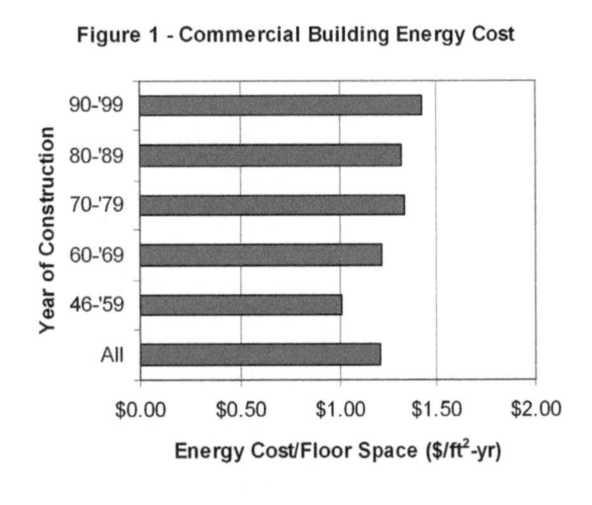

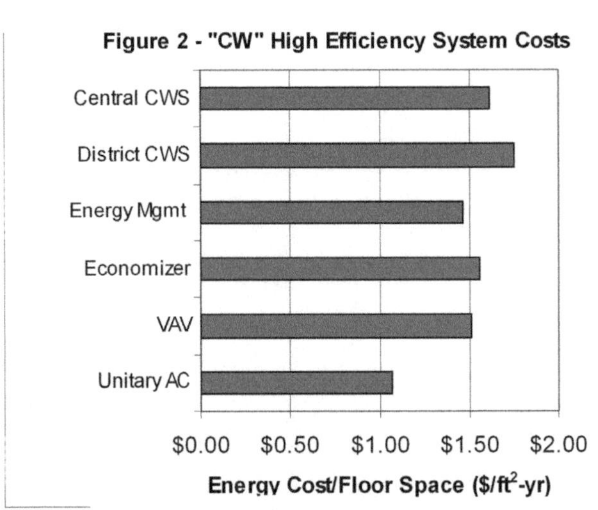

Figure 12.1 Historical and HVAC energy costs for commercial buildings (EIA/DOE 1999).

ing primarily upon component efficiency. The first example demonstrates the value of minimizing auxiliary equipment. Consider a 50 ton AHU that requires an airflow rate of 20,000 cfm and a total static pressure (TSP) of 5.0 in. of water. Data from Table 9.13 of this book indicate a fan motor brake horsepower (bhp) of 26.3 hp is required. The demand can be computed from Equation 11.2 assuming a 95% premium efficiency motor (92.4% is the EPACT minimum for an 1800 rpm motor).

$$W_{motor\ In}\ (\text{kW}) = 0.746\ \text{kW/hp} \times 26.3\ \text{hp}/0.95 = 20.65\ \text{kW}$$

This supply fan demand approaches the amount required for modern centrifugal chillers.

$$\text{kW/ton} = 20.65\ \text{kW} \div 50\ \text{tons} = 0.41\ \text{kW/ton}$$

This electrical power input is converted into heat via inefficiencies in the motor (q_{motor}) if it is located in the AHU, the fan itself (q_{fan}), and the remainder by friction loss in the air distribution system ($q_{air\ friction}$).

$$q_{motor} + q_{fan} + q_{air\ friction} = 20.65\ \text{kW} \times 3412\ \text{Btu/kWh} \div 12{,}000\ \text{Btu/ton·h} = 5.9\ \text{tons}$$

The net capacity of the system is reduced by 12% to 43.1 tons and the specific demand for the supply air fan (based on net capacity) is now equivalent to the demand of a centrifugal chiller.

$$\text{kW/ton (net)} = 20.65\ \text{kW} \div 43.1\ \text{tons} = 0.48\ \text{kW/ton}$$

This relatively high demand illustrates the importance of minimizing air distribution system friction losses when low demand and energy efficiency are priorities.

A second example demonstrates the system savings if water and air are delivered at lower temperatures and flow rates. Note that a Model 060 chiller (Table 5.10) experiences an 8.4% decline in efficiency (EER 15.5 vs. EER 14.2) when the chilled water temperature is reduced from 45°F to 40°F (for 85°F condenser entering water temperature [EWT]). If this chiller serves multiple Model 60-HW-4 chilled water coils, each would have a capacity of 55.2 MBtu/h using 45°F water at 17 gpm with 2000 cfm of supply air (for 75°F/63°F entering air temperature [EAT]). These coils would have equivalent capacities using 40°F water at 13 gpm with 1500 cfm of air (values extrapolated from 42°F to 40°F). Assuming friction losses are equivalent, this adjustment would result in the following:

- A 24% reduction in pump requirement (17 gpm/coil →13 gpm/coil)
- A 25% reduction in fan requirement (2000 cfm/coil →1500 cfm/coil)
- An 8.4% decrease in chiller efficiency (15.5 Btu/Wh →14.2 Btu/Wh)
- A reduction in required pipe size because of the reduced water flow
- A reduction in duct size because of the reduced airflow
- An improved latent capacity because of the lower water temperature and airflow
- A negligible adjustment in insulation cost (increased thickness but reduced area) due to the lower air and water temperatures

The current version of ASHRAE Standard 90.1 (2004) also does little to prevent oversizing of equipment, which will result in increased energy consumption with constant capacity equipment because of excessive start-up losses. Machines of variable capacity that utilize variable-speed motors will not deliver projected energy savings since drive and motor efficiency are substantially reduced at speeds less than 50% of full speed (DOE 2002; Turner 1992). Therefore, the oversized equipment will operate many more hours at lower than necessary speeds where motors and variable-speed drives have very low efficiency.

An alternative path to ASHRAE Standard 90.1 compliance is the Energy Cost Budget Method, which allows "adopting authority approved" simulations to evaluate compliance. General characteristics of simulation programs are provided. The drawback in permitting simulations to substitute for fixed system guidelines or measured performance has been demonstrated (Scofield 2002; Torcellini et al. 2004). A 13,600 ft^2 "green" academic building in Ohio was completed in 2000 at a cost of \$7 million (\$515/ft^2). It won two architectural awards and was heralded as a showcase of "green" construction and a model of sustainable architecture. However, two years of monitored data indicate actual energy consumption was three times higher than projected by the "approved" simulation. The owners paid a substantial premium, and energy costs were equal to comparable conventional buildings in the area.

HVAC SYSTEM DEMAND

ASHRAE Standard 90.1 contains a concise (six pages) and flexible section that sets guidelines for energy-efficient lighting. A summary of this section of the standard is presented in the previous chapter of this book. The format of the standard allows designers the flexibility of improving lighting efficacy, reducing lighting levels, increasing daylighting, and/or providing task lighting in critical areas.

The use of lighting power density guidelines sets the precedent for developing similar power density guidelines for HVAC systems. The need for this is demonstrated by Figure 12.1. Furthermore, the use of maximum allowable power density values provides a single measure of system efficiency that requires both (1) building envelope compliance and (2) HVAC equipment efficiency compliance. Acceptable HVAC power use can be achieved by reducing building load and/or improving HVAC equipment efficiency. The reduction in building HVAC energy consumption would naturally follow if a few simple part-load performance guidelines were applied—guidelines similar to those used in the lighting section of Standard 90.1. Additionally, oversizing (and the associated added capital cost and energy consumption) will be substantially reduced since larger than necessary equipment will contribute to high installed power densities.

The HVAC power density guidelines provide targets that require (1) attention be devoted to minimization of load through energy-efficient envelopes, high-efficacy lighting practices, Energy Star™ equipment, and optimum ventilation air control and (2) selection of high-efficiency heating and cooling equipment with minimal auxiliary equipment. The building area cooled and heated per unit of capacity (ft^2/ton) is maximized and the demand per unit of capacity (kW/ton) is minimized so that the resulting power density is minimized (W_{HVAC}/ft^2).

$$W_{HVAC}/ft^2 = 1000\ W/kW \times kW/ton \div ft^2/ton \quad (12.1)$$

Table 12.1 is a set of guidelines developed from Means's *Mechanical Cost Data* (Means 2004b) for 16 building types. The results shown in the "base" efficiency levels represent standard building practice and

Table 12.1 Building Efficiency Indicators—Floor Area/Unit Cooling Capacity (ft^2/ton)

Building Type	Base	High	Premier
Apartment	450	540	630
Bank	240	290	340
Bar/tavern	90	110	130
Church	330	400	460
Computer room	85	100	120
Dental office	230	280	320
Department store	350	420	490
Drugstore	150	180	210
Hospital	270	320	380
Hotel room	275	330	385
Exterior office	320	380	450
Interior office	360	430	500
New home	600	720	840
Old home	400	480	560
Restaurant	200	240	280
School	260	310	360

are suggested as the base for determining the amount of area that should be conditioned per unit of cooling capacity (1 ton = 12,000 Btu/h = 3.52 kW) in commercial applications. Although Table 12.1 is not climate corrected and applies to cooling load only, it provides average values that reflect energy-efficient practices in commercial buildings in most regions of the US. The values reflect buildings that utilize lighting power densities of ASHRAE Standard 90.1-2004, ventilation air rates compliant with ASHRAE Standard 62.1-2004, and standard envelope insulation levels and window treatments. Office and other interior equipment are reflective of recent advances resulting from programs such as Energy Star™ labeling.

The thrust of this approach is to provide two additional levels of compliance for more advanced designs if building owners desire to invest in higher levels of efficiency. The values in the columns labeled "High" and "Premier" are included as extensions of the base building efficiency. These levels can be achieved by further attention to ventilation air minimization made possible by Section 6.2 of ASHRAE Standard 62.1-2004 (ASHRAE 2004b). In the majority of buildings, rates can be reduced compared to the rates prescribed by 62.1-2004 though careful delivery of the outdoor air to the "breathing zone" and by increased ventilation efficiency. The amount of area cooled per ton can be further increased with daylighting, task lighting in critical areas, enhanced insulation levels, high-performance windows, and in many cases the use of energy recovery units on ventilation air.

The second component in the power density guidelines is to characterize the demand of *all* the HVAC system components. Figure 12.2 illustrates the process and the need for this step. Conventional wisdom assumes that a centrifugal chiller system with variable-speed drive air handlers and series fan-powered variable air volume (VAV) terminals is both advanced and energy efficient. Indeed, all components in the following example comply with ASHRAE 90.1-2004 and practices outlined as energy efficient in the *2005 ASHRAE Handbook—Fundamentals* (ASHRAE 2005). However, note the number of items in Figure 12.2 that require electrical input.

1. Centrifugal chiller
2. Supply air handler
3. Return air fan
4. Chilled water pump
5. Condenser water pump
6. Cooling tower fan
7. Fan-coil units (VAV terminals)

Also note that heat added in the supply air handlers (2), a large portion of the return air handlers

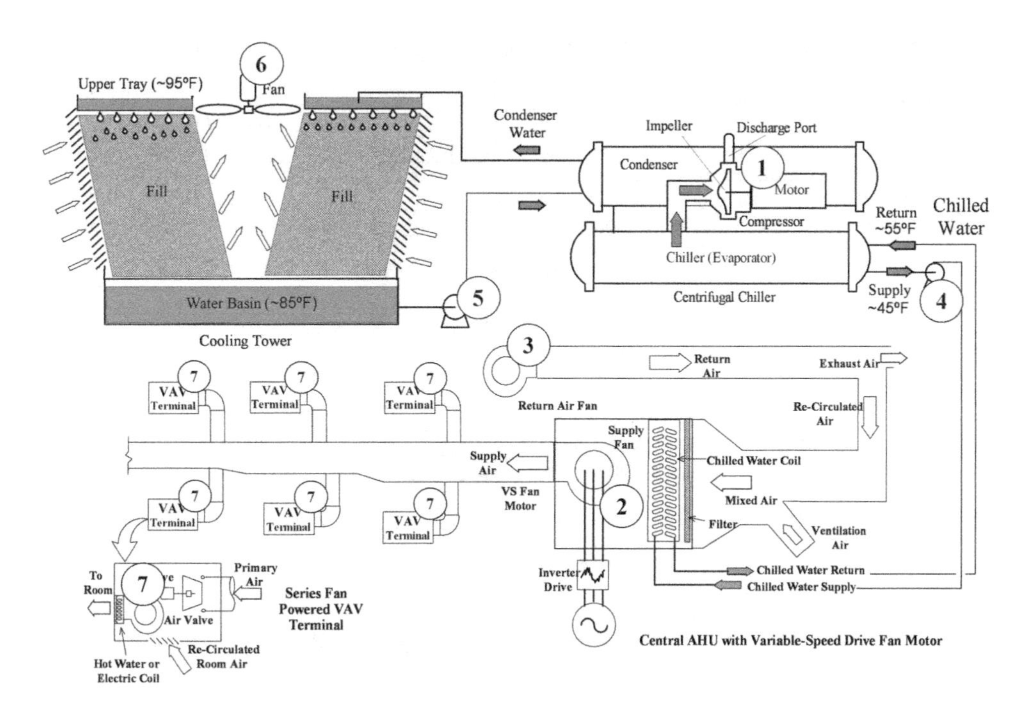

Figure 12.2 Chilled water, fan-powered VAV system with multiple components.

(3), the chilled water pump (5), and the VAV terminal fans (7) negatively impacts the net cooling capacity of the system. Both the power demand and heat input penalties are additive in nature and are not well identified in ASHRAE Standard 90.1 or LEED Green Building guidelines (LEED 2003).

A method of accounting for the combined effects of these individual components has been developed (Kavanaugh 2003). Table 12.2 (from *HVACSysEff.xls* on the CD accompanying this book) is a demonstration of the results of the application of this method to the system shown in Figure 12.2. Note that all of the components used to generate the results are compliant with ASHRAE Standard 90.1. The values in the column labeled "kW" are required inputs. The next column to the right computes the required kW per ton, the next column ("Ton") notes both the chiller input and the penalties for fan or pump heat, the next column provides the power output in hp/ton, and the last column is an accounting of power input per 1000 cfm, the figure of merit for ASHRAE Standard 90.1 supply fan limits.

The efficiency results are disturbing in that the *system* EER and COP are well below values for low-cost unitary equipment (window air conditioners, rooftop units, etc.). However, simple analysis and removal of a large portion of the equipment shown in Figure 12.2 will result in a simple chilled water system that requires no high-pressure supply air handlers, return air fans, or fan-powered VAV terminals. Simple fan-coil units replace the complex VAV system and the large demand requirements. The analysis is conducted on this system using a slightly larger pump to distribute the chilled water. This provides more favorable performance, as shown in Table 12.3. Much of the required fan power and associated heat generation penalties are removed and the efficiency is improved from a dismal EER = 7.5 Btu/W·h (COP = 2.2) to a very good value of EER = 10.5 Btu/W·h (COP = 3.4) at worst-case (design) conditions.

A premier level of performance can be achieved with a ground-source heat pump that has even lower auxiliary equipment demands (Table 12.4). Only three of the seven power requirement categories are required for this application. Worst-case (design) EER is 14.0 Btu/W·h (COP = 4.1).

Table 12.5 is provided as a guideline for three levels of efficient HVAC *system* efficiency in both EER and COP and also in demand limits (kW/ton).

The third step in computing HVAC power density values is to combine the impact of building envelopes and HVAC systems. The two previous indicators (building efficiency, HVAC system efficiency) will alert designers to which components are most likely to need improvement. Table 12.6 results from the application of Equation 12.1 using values in Tables 12.1 and 12.5 to arrive at values for HVAC power density levels for "low-efficiency," "high-efficiency," and "premier efficiency" levels.

This procedure is recommended as a convenient way for the designer to determine which design options are

Table 12.2 "Low-Efficiency" HVAC *System* Demand per Unit Cooling Capacity (kW/ton)

WC Centrifugal Chiller - with FP VAV					Std. 90.1
		kW	Ton	hp/ton	hp/1000 cfm
1. Chiller/Compressor: kW/Ton ----->	0.5	0.5	1	0.6032	
See Next Sheet for Values					
2. Supply Air Handling Unit Total Fan Pressure					
Δh(coil) + Δh(filter) + ESP: Inches of Water ----->	4	0.298	-0.085	0.36	0.90
(For Motors > 1.0 hp) Motor Eff (%) ----->	90				
Fan Eff. (%) ----->	70				
3. Return Fan Pressure: Inches of Water ----->	2	0.149	-0.034	0.18	0.45
(For Motors > 1.0 hp) Motor Eff (%) ----->	90				
Fan Eff. (%) ----->	70				
4. Chilled Water Pump Head - Ft. of Water ------>	70	0.050	-0.014	0.06	
(For Motors > 1.0 hp) Motor Eff (%) ----->	90				
Pump Eff. (%) ----->	70				
5. Condenser Pump Head - Ft. of Water ------>	70	0.063		0.08	
(For Motors > 1.0 hp) Motor Eff (%) ----->	90				
Pump Eff. (%) ----->	70				
6. Cooling Tower or Condenser Fan					
0.065 for Axial - 0.135 for Centrifugal: kW/Ton----->	0.065	0.065		0.08	0.07
7. Zone or VAV Terminal Fans					
0.18 for AC Motors - 0.13 for ECM: kW/ton ----->	0.180	0.180	-0.051		0.18
(For Motors < 1.0 hp)					
	Σ=	1.306	0.82		1.59
	kW/Ton=	1.6			
	EER =	7.498			
	COP =	2.198			

Table 12.3 "High-Efficiency" HVAC *System* Demand per Unit Cooling Capacity (kW/ton)

WC ScrewChiller with Unitary Fan Coils					Std. 90.1
		kW	Ton	hp/ton	hp/1000 cfm
1. Chiller/Compressor: kW/Ton ----->	0.6	0.6	1	0.7239	
See Next Sheet for Values					
2. Supply Air Handling Unit Total Fan Pressure					
Δh(coil) + Δh(filter) + ESP: Inches of Water ----->	0	0.000	0.000	0.00	0.00
(For Motors > 1.0 hp) Motor Eff (%) ----->	90				
Fan Eff. (%) ----->	70				
3. Return Fan Pressure: Inches of Water ----->	0	0.000	0.000	0.00	0.00
(For Motors > 1.0 hp) Motor Eff (%) ----->	90				
Fan Eff. (%) ----->	70				
4. Chilled Water Pump Head - Ft. of Water ------>	90	0.065	-0.018	0.08	
(For Motors > 1.0 hp) Motor Eff (%) ----->	90				
Pump Eff. (%) ----->	70				
5. Condenser Pump Head - Ft. of Water ------>	70	0.063		0.08	
(For Motors > 1.0 hp) Motor Eff (%) ----->	90				
Pump Eff. (%) ----->	70				
6. Cooling Tower or Condenser Fan					
0.065 for Axial - 0.135 for Centrifugal: kW/Ton----->	0.065	0.065		0.08	0.07
7. Zone or VAV Terminal Fans					
0.18 for AC Motors - 0.13 for ECM: kW/ton ----->	0.180	0.180	-0.051		0.18
(For Motors < 1.0 hp)					
	Σ=	0.972	0.93		0.25
	kW/Ton=	1.045			
	EER =	11.48			
	COP =	3.365			

Table 12.4 "Premier Efficiency" HVAC *System* Demand per Unit Cooling Capacity (kW/ton)

Ground Source Heat Pumps with ECM Fans					Std. 90.1
		kW	Ton	hp/ton	hp/1000 cfm
1. Chiller/Compressor: kW/Ton ----->	0.65	0.65	1	0.7842	
See Next Sheet for Values					
2. Supply Air Handling Unit Total Fan Pressure					
Δh(coil) + Δh(filter) + ESP: Inches of Water ----->	0	0.000	0.000	0.00	0.00
(For Motors > 1.0 hp) Motor Eff (%) ----->	90				
Fan Eff. (%) ----->	70				
3. Return Fan Pressure: Inches of Water ----->	0	0.000	0.000	0.00	0.00
(For Motors > 1.0 hp) Motor Eff (%) ----->	90				
Fan Eff. (%) ----->	70				
4. Chilled Water Pump Head - Ft. of Water ------>	0	0.000	0.000	0.00	
(For Motors > 1.0 hp) Motor Eff (%) ----->	90				
Pump Eff. (%) ----->	70				
5. Condenser Pump Head - Ft. of Water ------>	50	0.045		0.05	
(For Motors > 1.0 hp) Motor Eff (%) ----->	90				
Pump Eff. (%) ----->	70				
6. Cooling Tower or Condenser Fan					
0.065 for Axial - 0.135 for Centrifugal: kW/Ton----->	0.000	0.000		0.00	0.00
7. Zone or VAV Terminal Fans					
0.18 for AC Motors - 0.13 for ECM: kW/ton ----->	0.130	0.130	-0.037		0.13
(For Motors < 1.0 hp)					
	Σ=	0.825	0.96		0.13
	kW/Ton=	0.857			
	EER =	14.01			
	COP =	4.106			

Table 12.5 HVAC System Efficiency Guidelines

HVAC System Efficiency Level	EER (Btu/W·h)	COP	Specific Demand (kW/Net Ton)
Base	9–11	2.6–3.2	1.1–1.3
High efficiency	11–13	3.2–3.8	0.9–1.1
Premier efficiency	> 13	> 3.8	< 0.9

more likely to lead to the desired goal of lower demand and efficiency. The application of this procedure highlights the impact of auxiliary equipment on system consumption. Careless specification of primary and auxiliary equipment can result in elaborate and costly systems with high demand and poor efficiencies.

This procedure is not intended to substitute for detailed energy analysis and resulting economic evaluations of heating and cooling alternatives. The procedure does not include several other items that can be connected to building HVAC energy consumption. These include water heating, ventilation air preconditioning, heat recovery, exhaust air fan energy, and refrigeration. However, design load system efficiency is one indicator of seasonal efficiency. If demand at extreme conditions is minimized, other measures to reduce energy consumption (heat recovery systems vs. drives, DDC controls, etc.) will be much more effective. Furthermore, electrical demand is a critical issue in many areas of the US.

Table 12.6 Recommended HVAC Power Density Levels (W/ft^2)

Building Type	Base	High	Premier
Apartment	2.7	1.9	1.3
Bank	5.0	3.5	2.5
Bar/tavern	13.5	9.3	6.7
Church	3.4	2.4	1.7
Computer room	14.0	10.0	7.1
Dental office	5.2	3.6	2.6
Dept. store	3.4	2.4	1.7
Drugstore	8.0	5.6	4.0
Hospital	4.4	3.1	2.2
Hotel room	4.4	3.0	2.2
Exterior office	3.8	2.6	1.9
Interior office	3.3	2.3	1.7
New home	2.0	1.4	1.0
Old home	3.0	2.1	1.5
Restaurant	6.0	4.2	3.0
School	4.6	3.2	2.3

ENERGY CONSUMPTION AND COST—BIN METHOD ENERGY ANALYSIS

The recommended methods for estimating building energy consumption and costs can be as simple as degree-day-based calculations (ASHRAE 2005, chapter 32), which provide an indicator of the severity of local weather conditions, or as elaborate as hour-by-hour energy simulations driven by Typical Meteorological Year (TMY) weather data (ASHRAE 2005, chapter 32). An intermediate-level analysis method is referred to as the bin method (ASHRAE 2005, chapter 32). Various levels of detail are available to enhance the accuracy of this type of analysis, but all are amenable to incorporation into spreadsheets. Therefore, the bin method is presented in this text as a useful compromise for residential and simple commercial buildings.

Accurate modeling of larger commercial buildings requires much more sophisticated building energy simulations that have been field validated and are correctly applied. Readers are encouraged to consider much more in-depth energy simulations and cost analysis models throughout the design process. Simplified models might be used in the initial design phase to narrow options to the most appropriate, and more detailed models are used for analysis in the later stages of design. Table 14-1 from the *ASHRAE GreenGuide* (Grumman 2004) is suggested as one reference to guide the decision as to which level of analysis is optimum at various stages of the design process.

Weather data can be arranged in temperature bins of typically 5°F increments, as shown in Tables 6.6 through 6.10 of this book. The number of hours per year (or month) in which the outdoor temperature falls into a bin is referred to as the *bin hours*. For example, there are typically seven hours per year in Birmingham, Alabama (Table 6.6), where the temperature is between 95°F and 99°F (97°F average). Thus, there are seven bin hours in the 97°F temperature bin. Each location also has a mean coincident wet-bulb temperature (WBT), which is an indicator of the humidity level at the location.

The bin data can be reported in terms of the totals for all periods of the day, or they can be subdivided into four- or eight-hour increments. Of the seven bin hours in the 97°F bin, six occur between noon and 4 p.m. and one occurs between 4 and 8 p.m. This allows a more detailed approach to bin calculations since loads change with time of day and occupancy patterns.

Heating and cooling loads change with temperature and time of day. The bin method permits these changing loads to be computed at the different bin temperatures. For example, Figure 12.3 shows a plot of temperature vs. load for a zone in an office building. At 100°F the cooling load is 30 MBtu/h when the zone is occupied and 10 MBtu/h when it is unoccupied. The occupied cooling load is zero when the outdoor temperature is 35°F, while the unoccupied cooing load is zero when the outdoor temperature is 75°F. In heating, the occupied load is 14 MBtu/h at 12°F while the unoccupied load is 25 MBtu/h. The zero heating load temperatures are 30°F occupied and 50°F unoccupied. The shapes of these load lines are characteristic of an office building with high internal loads and moderate ventilation air requirements during the occupied periods.

The heating and cooling requirements for each bin temperature can be determined from an equation for load lines as a function of outdoor air temperature (t_o) at each bin temperature increment. For example, the occupied cooling load ($q_{c\text{-}occ}$) and heating load ($q_{h\text{-}occ}$) for the two occupied lines in Figure 12.3 are

$$q_{c\text{-}occ}\,(\text{MBtu/h}) = 0.462t_o\,(°\text{F}) - 16.2 \quad (\text{for } 35°\text{F} \geq t_o \geq 100°\text{F}) \tag{12.2}$$

and

$$q_{h\text{-}occ}\,(\text{MBtu/h}) = -0.778t_o\,(°\text{F}) - 23.3 \quad (\text{for } 10°\text{F} \geq t_o \geq 30°\text{F}). \tag{12.3}$$

Figure 12.4 is a similar plot showing the heat pump heating and cooling capacity and electrical demand as a function of outdoor temperature. The values provided represent the performance of the Model 036 unit

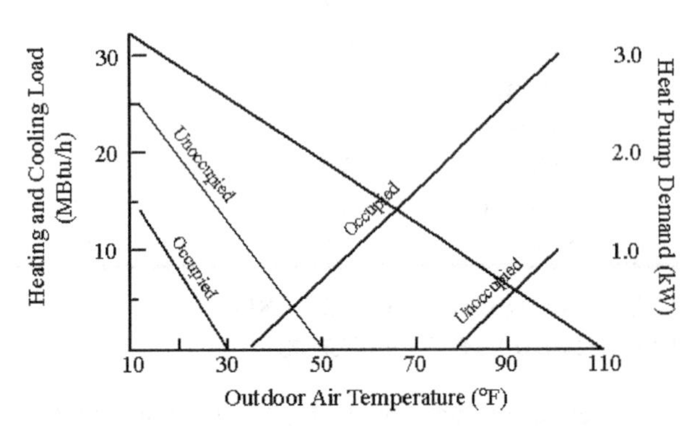

Figure 12.3 Zone occupied and unoccupied heating and cooling loads.

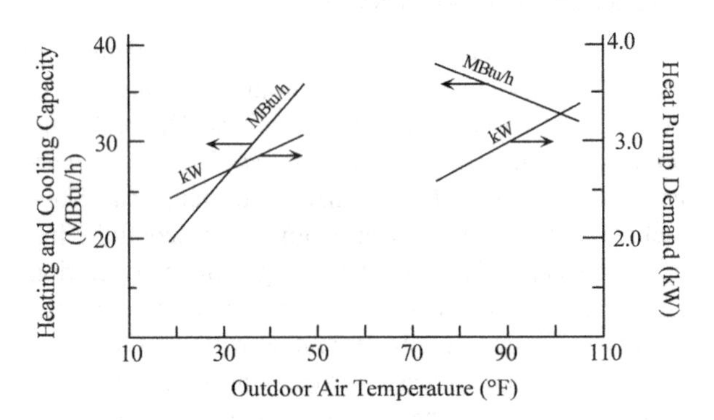

Figure 12.4 Heat pump heat and cooling capacity and demand.

(Table 5.3) for indoor conditions of 75°F/63°F (db/wb) in cooling and 70°F in heating. Useful equations are the capacities and demand as a function of outdoor temperature. For example, equations for total cooling (TC) and cooling demand (kW_c) can be developed as a function of outdoor air temperature (t_o).

$$\text{TC (MBtu/h)} = -0.2t_o\ (°\text{F}) + 53.0 \tag{12.4}$$

$$\text{kW}_c = 0.0265t_o\ (°\text{F}) + 0.5975 \tag{12.5}$$

For each bin the cooling capacity of the unit normally exceeds the load, so the unit will run part of the time. This fraction is referred to as the part-load factor (PLF). For example, using the load at 82°F during the noon to 4 p.m. time period,

$$\text{PLF}_{82} = q_{c\text{-}occ} \text{ at } 82°\text{F (N–4 p.m.)} \div \text{TC at } 82°\text{F} = 21.7 \div 36.6 = 0.59\ . \tag{12.6}$$

Since equipment normally takes some time to reach rated capacity after start-up, a cycling capacity adjustment factor (CCAF) is applied when the unit has to stop and start. For the example at 82°F,

$$\text{CCAF} = 1 - C_d\,(1 - \text{PLF}) = 1 - 0.25\,(1 - 0.59) = 0.90\ , \tag{12.7}$$

where 0.25 is the commonly used default value for the degradation coefficient (C_d).

To find the energy use for each temperature bin ($\text{kWh}_{c\text{-}Bin}$),

$$\text{kWh}_{c\text{-}Bin} = \text{kW}_c \times \text{PLF} \times \text{No. Hours in Bin} \div \text{CCAF}\ . \tag{12.8}$$

To find the seasonal cooling energy use (kWh_c) for the four-hour time period,

$$\text{kWh}_c = \Sigma\text{kWh}_{c\text{-}Bin}\ . \tag{12.9}$$

The energy for auxiliary items (fans, pumps, etc.) can be computed by including their power input with the compressor power if all items cycle on and off with the compressor. Auxiliary energy use for continuous or variable-speed components must be computed independently.

Heating Mode—Furnaces

The procedure is the same for the heating mode. However, the heating system is often more complex since the heater is likely to have both a fossil fuel requirement and an electrical load for the circulation fans. Many utilities use therms (100,000 Btu) as the unit for natural gas measurement and billing.

To find the heating energy use for each temperature bin ($Therms_{h\text{-}Bin\ Temp\text{-}heat}$) based on the furnace heating capacity (HC_{Furn}),

$$Therms_{h\text{-}Bin\ Temp\text{-}heat} = HC_{Furn}\ \text{(MBtu/h)} \times \text{PLF} \times \text{No. Hours in Bin} \div (\text{CCAF} \times \eta_{Furn} \times 100). \tag{12.10}$$

The seasonal heating energy use ($Therms_{heat}$) can be computed by summing the input for all temperature bins for all time periods using the appropriate load (occupied or unoccupied).

$$Therms_{heat} = \Sigma Therms_{Bin\ Temp\text{-}heat} \tag{12.11}$$

To find the fan electrical energy use for each temperature bin ($\text{kWh}_{h\text{-}fan\text{-}Bin}$) (note that CCAF does not apply),

$$\text{kWh}_{h\text{-}fan\text{-}Bin} = \text{kW}_{h\text{-}fan} \times \text{PLF} \times \text{No. Hours in Bin}\ . \tag{12.12}$$

(***Note:*** PLF = 1.0 if the fan is left on continuously.)

To find the seasonal fan electrical energy use ($\text{kWh}_{h\text{-}fan}$),

$$\text{kWh}_{h\text{-}fan} = \Sigma\text{kWh}_{h\text{-}Bin}\ . \tag{12.13}$$

Heating Mode—Heat Pumps

Heat pumps have similar characteristics in heating as they do in cooling. Heating capacity (HC) and power demand (kW_h) are strong functions of outdoor temperature. Again, the heat pump part-load factor (PLF) is computed by dividing the building load at each bin temperature by the unit's capacity at this temperature. To find the heat pump energy use for each temperature bin ($\text{kWh}_{h\text{-}HP\text{-}Bin}$) for both the occupied and unoccupied time periods,

$$\text{kWh}_{h\text{-}HP\text{-}Bin} = \text{kW}_h \times \text{PLF} \times \text{No. Hours in Bin} \div \text{CCAF}\ . \tag{12.14}$$

The seasonal energy use of the heat pump unit ($\text{kWh}_{h\text{-}HP}$) is

$$\text{kWh}_{h\text{-}HP} = \Sigma\text{kWh}_{h\text{-}Bin}\ . \tag{12.15}$$

However, heat pumps (especially air-source) are not able to meet building loads when outdoor air temperatures are low. Normally, this difference is made up with auxiliary electrical resistance furnaces. An indication of this need is when the PLF is greater than 1.0. The amount of auxiliary energy required from each bin is computed by calculating the required heating auxiliary capacity ($q_{Aux.}$) for each bin in which the PLF is greater than 1.0.

$$q_{Aux}\ \text{(MBtu/h)} = q_{Load}\ \text{(MBtu/h)} - q_{HP\text{-}Cap}\ \text{(MBtu/h)} \tag{12.16}$$

$$\text{kWh}_{Aux\text{-}Bin} = q_{Aux}\ \text{(MBtu/h)} \div 3.412\ \text{(MBtu/kWh)} \tag{12.17}$$

To find the seasonal heating energy use of the auxiliary heat energy (kWh_{Aux}),

$$kWh_{Aux} = \Sigma kWh_{Aux\text{-}Bin} \cdot \quad (12.18)$$

The total heating energy (kWh_h) for the system is the sum of the heat pump and auxiliary energy.

$$kWh_h = kWh_{h\text{-}HP} + kWh_{Aux} \quad (12.19)$$

Although the bin method is well suited to use with spreadsheets, one is not included on the accompanying CD. However, Table 12.7 is an example of a typical format for this purpose. The first seven columns are a repetition of the bin data. The next two columns represent the occupied and unoccupied loads (this example is for a continuously occupied building) followed by the unit cooling capacity, PLF, unit power, and finally the cooling energy use. The remaining columns to the right repeat the computation for the heating mode using a natural gas furnace. Note the HC is a constant 55 MBtu/h since capacity is not a function of outdoor temperature. The fuel use is summed in the column labeled "Qh(in)" and the furnace fan electrical use is summed in the "E(h)" column. No auxiliary heat is needed since furnaces are typically sized to meet the load at the most extreme condition. The electric energy use for both heating and cooling is summed in the rightmost column.

The annual energy cost can be determined with the input of unit cost of energy. The unit energy cost is multiplied by the summation values appearing in the lower row of the spreadsheet. Electric utility cost can vary from the heating to cooling seasons, and the spreadsheet can be adjusted to reflect this cost by using different values in the winter and summer. However, in this example, the cost is constant.

A variety of measures are used for fossil fuels. These include ccf (100 cubic feet of gas), Mcf (1000 cubic feet of gas), gallons, pounds, and therm, which is 100,000 Btu. This term evolves from the fact that one ccf of natural gas is approximately equivalent to a therm when combusted. Since the heating value of natural gas varies significantly according to actual components, a therm is a better indicator of the actual heating value of natural gas than a ccf.

The bin method permits the user to define the unit of fossil fuel energy (therms in the example), so both the cost and the number of MBtu per unit are required in this example ($1.05/therm and 100 MBtu/therm). The efficiency of the furnace is also required, and in most cases the annual fuel utilization efficiency (AFUE) (see Tables 5.4 and 5.5) provides sufficient accuracy for constant-capacity furnaces. The efficiency of a multiple- or variable-capacity furnace can be adjusted by inputting the appropriate efficiency via an "if" statement in the spreadsheet when the heating load indicates a shift to a different stage is required.

The example spreadsheet provides a breakdown of the annual energy cost of cooling, heating (fuel), heating (electric), and total. It should be noted that bin weather data are also available in monthly formats (Degelman 1986; InterEnergy 1999) so even more detail is possible.

MAINTENANCE AND REPLACEMENT COSTS

Table 12.8 is included as a brief summary of the repair and replacement costs and required frequency for HVAC equipment. Additional equipment and a more detailed itemization of individual components are presented in Means (2004a). This includes cost of routine maintenance.

Table 12.7 Bin Method Spreadsheet Example

Example Bin Calculation - Electric AC (30 MBtu/h @ 100 F, 80% Efficienct, 55 Mbtu/h Furnace with 500 watt fan)
Birmingham, AL weather data

$/kWh=	0.07
$/Ther=	1.05
MBtu/ther=	100
η(Furn)=	80

oat	mn-4a	4-8a	8a-n	n-4p	4-8p	8p-mn	qc(oc)	qc(uoc)	TC	PLF	W(c)	E(c)	qh(oc)	qh(uoc)	HC	PLF	Qh(in)	W(h)	E(h)	qh(aux)	W(aux)	E(aux)	E(Tot.)
°F	hrs.	hrs.	hrs.	hrs.	hrs.	hrs.	Mbtu/h	Mbtu/h	Mbtu/h		kW	kWh	Mbtu/h	Mbtu/h	Mbtu/h		therms	kW	kWh			kWh	kWh
97	0	0	0	6	1	0	30		30.4	0.99	3.7	26											26
92	0	0	41	139	30	0	27		31.0	0.87	3.6	673											673
87	0	1	158	180	109	1	24		31.6	0.76	3.4	1248											1248
82	2	40	178	202	186	33	21		32.3	0.65	3.3	1519											1519
77	99	143	176	165	165	207	18		32.9	0.55	3.2	1889											1889
72	227	211	138	119	171	208	15		33.5	0.45	3.1	1721											1721
67	206	158	120	150	140	173	12		34.1	0.35	3.0	1179											1179
62	144	153	123	135	125	132	9		34.8	0.26	2.8	735											735
57	144	123	132	118	128	151	6		35.4	0.17	2.7	465											465
52	162	139	159	113	133	160	3		36.0	0.08	2.6	244											244
47	117	114	102	69	108	100	0		36.6	0	2.5	0											0
42	106	133	65	41	84	108							0.0		55	0.00	0	0.5	0				0
37	84	72	32	11	43	80							9.1		55	0.17	46	0.5	34				34
32	88	80	18	8	24	65							18.2		55	0.33	77	0.5	56				56
27	52	58	10	0	7	28							27.3		55	0.50	60	0.5	44				44
22	22	26	8	4	6	12							36.4		55	0.66	39	0.5	28				28
17	7	5	0	0	0	2							45.5		55	0.83	8	0.5	6				6
12	0	4	0	0	0	0							54.5		55	0.99	3	0.5	2				2
												9699					234		170				9869
											$ (c)=	$679						$ (h)=	$12		$ (aux):	$0	
																						Elect =	$691
																						Fuel=	$245

Table 12.8 HVAC Equipment Repair and Replacement Costs (Means 2004a)

Item	Repair Cost Freq. (years)	Repair Cost In-House Cost	Repair Cost Cost w/O&P	Removal and Replacement Cost Freq. (years)	Removal and Replacement Cost In-House Cost	Removal and Replacement Cost Cost w/O&P
Gas boiler—2000 MBtu/h	7	$3,570	$4,160	30	$20,900	$24,700
Gas boiler—10000 MBtu/h	7	$12,470	$14,360	30	$144,100	$169,900
Oil boiler—2000 MBtu/h	7	$1,120	$1,330	30	$22,000	$26,200
Oil boiler—10000 MBtu/h	7	$1,190	$1,420	30	$137,500	$161,900
Cooling tower—300 ton	10	$4,460	$5,300	15	$24,800	$28,800
Cooling tower—1000 ton	10	$10,800	$12,700	15	$66,300	$77,000
AC recip. chiller—50 ton	10	$42,600	$49,500	20	$40,900	$47,700
AC recip. chiller—100 ton	10	$41,900	$50,000	20	$68,500	$79,500
WC recip. chiller—50 ton	10	$40,700	$47,200	20	$34,400	$44,900
WC recip. chiller—100 ton	10	$52,300	$61,900	20	$66,500	$77,400
WC recip. chiller—200 ton	10	$53,200	$63,100	20	$120,200	$139,600
WC centrif. chiller—300 ton	10	$73,000	$85,900	20	$158,300	$184,900
WC centrif. chiller—1000 ton	10	$229,900	$271,000	20	$541,200	$632,900
Absorption chiller—350 ton	10	$13,500	$15,100	20	$229,400	$264,300
Absorption chiller—950 ton	10	$16,300	$18,800	20	$449,000	$514,700
Air-handling unit—CW—1 ton	10	$307	$367	15	$892	$1,040
Fan-coil unit—CW—5 ton	10	$384	$459	15	$1,790	$2,100
Fan-coil unit—CW—20 ton	10	$514	$608	15	$7,430	$8,750
Fan-coil unit—DX—1.5 ton	10	$1,300	$1,550	15	$718	$845
Fan-coil unit—DX—5 ton	10	$1,820	$2,150	15	$1,250	$1,470
Fan-coil unit—CW—20 ton	10	$17,000	$19,800	15	$7,000	$8,330
Heat pump, split—1.5 ton	10	$1,510	$1,800	20	$2,180	$2,630
Heat pump, split—5 ton	10	$2,500	$2,400	20	$4,650	$5,580
Heat pump, split—25 ton	10	$7,300	$8,220	20	$24,500	$32,900
Packaged AC, DX—5 ton	10	$3,280	$3,780	20	$4,930	$5,830
Packaged AC, DX—20 ton	10	$19,800	$22,980	20	$16,300	$18,850
Packaged AC, DX—50 ton	10	$31,100	$36,900	20	$29,000	$33,600
Multi-zone rooftop—25 ton	10	$20,800	$24,200	15	$111,900	$128,300
Multi-zone rooftop—70 ton	10	$29,900	$35,500	15	$154,500	$178,100
Multi-zone rooftop—105 ton	10	$31,200	$37,100	15	$207,800	$240,400
Single-zone rooftop—25 ton	10	$20,700	$24,100	15	$28,300	$33,000
Single-zone rooftop—60 ton	10	$32,500	$38,500	15	$63,400	$74,100
Single-zone rooftop—100 ton	10	$29,800	$35,300	15	$211,400	$243,400
Multi-zone VAV—70 ton	10	$29,900	$35,500	15	$132,900	$154,100
Multi-zone VAV—105 ton	10	$31,300	$37,100	15	$175,800	$203,900
Central AHU—5400 cfm	10	$450	$530	15	$8,450	$9,930
Central AHU—33,500 cfm	10	$2,270	$2,630	15	$35,100	$41,200
Central AHU—63,000 cfm	10	$3,800	$4,410	15	$68,500	$80,900
Gas furnace—25 MBtu/h	10	$690	$820	15	$744	$876
Gas furnace—100 MBtu/h	10	$906	$1,070	15	$1,150	$1,360
Gas furnace—200 MBtu/h	10	$1,880	$2,230	15	$2,380	$2,770

Table 12.8 lists two costs. The "in-house" values represent the costs for the building owner to use in-house staff for the repair or replacement. The costs with overhead and profit ("w/O&P") reflect the values charged to the owner when the tasks were performed by an outside contractor. Table 12.9 presents routine preventive maintenance costs in a similar format. The *ASHRAE Handbook—HVAC Applications* (ASHRAE 2003, chapter 36) discusses maintenance costs and equipment service life. However, the Handbook information is based on data that are more than 20 years old. Therefore, Means data are suggested as a more up-to-date source until the Handbook is updated with more recent information.

Table 12.9 Annual HVAC Preventative Maintenance Costs (Means 2004a)

Item	Size	In-House Cost	Cost w/O&P
Electric boiler	1500 gal.	$244	$299
Gas or oil hot water boiler	120 to 500 MBtu/h	$845	$1,050
Gas or oil hot water boiler	500 to 1000 MBtu/h	$950	$1,175
Gas or oil steam boiler	120 to 500 MBtu/h	$1,100	$1,350
Gas or oil steam boiler	500 to 1000 MBtu/h	$1,200	$1,475
Cooling tower	500 to 1000 tons	$1,375	$1,700
Evaporative cooler	Not listed	$84	$102
AC reciprocating chiller	50 ton	$430	$535
WC centrifugal chiller	Over 100 tons	$1,850	$2,275
WC screw chiller	Up to 100 tons	$910	$1,125
WC screw chiller	Over 100 tons	$1,025	$1,250
Absorption chiller	Up to 500 tons	$370	$460
Absorption chiller	Over 500 tons	$635	$775
Air-cooled condensing unit	3 to 25 tons	$192	$233
Air-cooled condensing unit	26 to 100 tons	$235	$283
Air-cooled condensing unit	over 100 tons	$283	$340
Water-cooled condensing unit	3 to 25 tons	$88	$107
Water-cooled condensing unit	26 to 100 tons	$195	$238
Water-cooled condensing unit	over 100 tons	$228	$275
Compressor, DX refrigeration	to 24 tons	$131	$163
Compressor, DX refrigeration	25 to 100 tons	$172	$214
Air-handling unit	3 to 24 tons	$175	$211
Air-handling unit	25 to 50 tons	$395	$460
Air-handling unit	Over 50 tons	$535	$625
Fan coil unit	No size listed	$187	$231
Air filters, electrostatic	No size listed	$410	$500
VAV boxes	No size listed	$52	$64
Fire campers	No size listed	$63	$78
Fan, axial	to 5000 cfm	$61	$75
Fan, axial	5000 to 10000 cfm	$63	$78
Fan, axial	over 10000 cfm	$68	$83
Fan, centrifugal	to 5000 cfm	$54	$66
Fan, centrifugal	5000 to 10000 cfm	$56	$69
Fan, centrifugal	over 10000 cfm	$59	$72
Hood and blower	No size listed	$165	$198
Pump	Over 1 hp	$70	$86
Forced air heater, gas or oil	to 120 MBtu/h	$385	$475
Forced air heater, gas or oil	over 120 MBtu/h	$495	$605
Packaged unit, air cooled	3 to 24 tons	$173	$211
Packaged unit, air cooled	25 to 50 tons	$237	$288
Packaged unit, water cooled	3 to 24 tons	$173	$211
Packaged unit, water cooled	25 to 50 tons	$237	$288
Split system air conditioner	to 10 tons	$232	$282
Split system air conditioner	over 10 tons	$255	$310
Heat pump, air cooled	to 5 tons	$191	$235
Heat pump, air cooled	over 5 tons	$232	$284
Heat pump, water cooled	to 5 tons	$208	$257
Heat pump, water cooled	over 5 tons	$214	$262

SYSTEM INSTALLATION COSTS

The first cost of HVAC systems is one of the most critical factors in determining the type of equipment that is installed in buildings. In many cases it is the most critical factor. It is important that engineers request and receive information from the owner or project director in writing for the range of funds that are available for the project. At this point it is helpful if the engineer can provide an approximate estimate of the options that are available in that price range. Table 12.10 is a brief summary of the type of information available for this purpose. Six different building types are shown and costs for five HVAC system options are given per unit floor area. The costs are corrected for both building size and cooling load characteristic of the building type. However, no regional corrections are applied, although a detailed listing of factors is available in Means (2004b).

Much more detailed work is necessary to accurately estimate the costs. The mechanical contractor is typically responsible for submitting bid estimates for the system as designed and specified by the engineer. However, the engineer must plan the system with cost in mind. If the design is not well planned, the resulting bid estimate will likely be well off expectations (usually much higher than predicted). The system must then be redesigned and often on a much shorter time schedule. It is in these situations that details are often overlooked and must be adjusted or corrected after the bid has been accepted. These corrections are made via "change orders," which are not competitively bid and will likely be much more expensive than if the correct design had been completed at the time the original bid was submitted.

Therefore, it is essential that design firms utilize convenient but accurate methods to estimate system costs. An example of a spreadsheet tool to estimate ductwork cost based on historical mechanical cost data is given in Table 12.11 (from *DuctCostCalc.xls* on the CD accompanying this book). It has been shown that sheet metal duct costs correlate well to weight. Figure 12.5 is a graphical representation of this relationship (Means 2004b). The cost of duct fittings can be included in this computation in a method similar to the equivalent length method used for friction loss. The spreadsheet tool sums the weights based on the area, thickness (gauge), and number of fittings. This result is multiplied by the cost-to-weight equation shown in Figure 12.5 to arrive at a subtotal cost for the metal duct.

The spreadsheet tool also includes the cost for several duct insulation types and thicknesses. This cost is based on area rather than weight, but note that the area for each section of metal ductwork is computed and displayed. Thus, the total area is available if the entire duct system is wrapped with the same insulation. The area of each different section is also available in cases where various duct insulation options are employed. The spreadsheet tool can also estimate the cost of dampers, return grilles, and supply diffusers. A feature of the tool that is not shown in Table 12.11 is the capability to compute the cost of duct board.

Table 12.12 is a summary of the costs for primary HVAC equipment, including chillers, unitary heating and cooling equipment, and various types of air-han-

Table 12.10 Cooling System Material and Installation Costs in $\$/ft^2$ (Means 2004b)

Average US cost—no regional multipliers applied.

Building Type	Floor Area (ft^2)	Cooling Capacity (tons)	Single-Zone RTU[1]	Multi-Zone RTU[2]	WC Package, Self-Contained[3]	AC Package, Self-Contained[4]	AC Split System[5]
			Material and Installation Cost is US Dollars per Square Foot ($\$/ft^2$)				
Apartment	5000	9	3.92	14.77	3.47	4.39	3.55
	20000	37	n/a	11.22	n/a	n/a	3.93
Dept. Store	5000	15	6.37	22.99	6.65	6.98	5.65
	20000	58	n/a	12.10	n/a	n/a	6.27
Med. Center	5000	12	4.99	22.25	4.41	5.58	4.52
	20000	47	n/a	11.75	n/a	n/a	5.01
Office	5000	16	6.73	24.96	7.23	7.41	6.13
	20000	63	n/a	13.10	n/a	n/a	6.92
Restaurant	5000	25	10.70	34.81	11.43	10.65	10.45
	20000	100	n/a	19.60	n/a	n/a	11.86
School	5000	19	8.21	26.71	8.78	8.96	8.02
	20000	77	n/a	15.90	n/a	n/a	8.35

[1] Single-zone RTU: RTU air conditioner, 1 zone, electric cooling, standard controls, curb, ductwork.
[2] Multi-zone RTU: RTU multi-zone unit, standard controls, curb, ductwork package.
[3] WC package/self-contained: Water-cooled AC, ductwork package, cooling tower, pumps, piping, heating coil.
[4] AC package/self-contained: Air-cooled AC, ductwork package, cooling tower, pumps, piping, heating coil.
[5] AC split system: DX fan coil, condensing unit, ductwork, standard controls, cabinets, refrigeration piping.

dling devices. These costs include the equipment itself and the cost for a contractor to install it (w/O&P). They do not include the costs of the system's ductwork, piping, electrical, controls, etc.

Table 12.13 is a summary of the material and installation costs for three types of piping for sizes ranging from 0.75 to 12 in. diameter. Much more detailed information and computer-based cost-estimating tools are available. The information included in this brief section is intended to be representative of the type of information that is available. Readers are encouraged to invest in more powerful, comprehensive, and flexible tools so that accurate estimates can be provided to building owners and partners in the construction process.

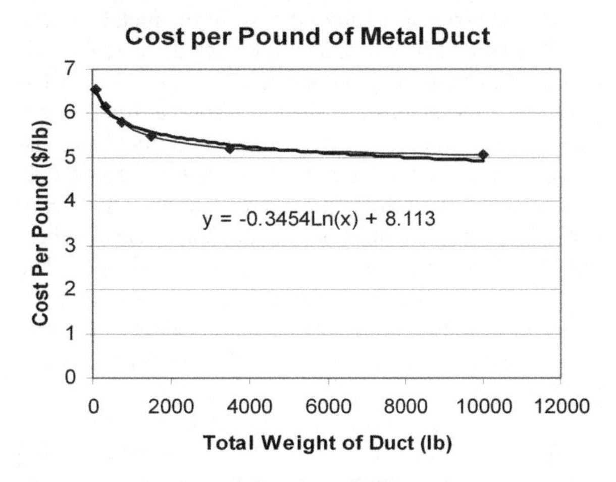

Figure 12.5 Metal duct installation cost.

ENGINEERING ECONOMICS AND HVAC

The economic value of HVAC system options is often the most important constraint in the engineering design process. Unfortunately, the methods for determining economic value vary from client to client. Some owners are constrained on total budget costs, others look at short-term value, and others may consider long-term value. Some owners may dictate the economic evaluation method, while others may consider a variety of measures. In some cases, the engineer may be needed to assist the building owner in determining the wisdom of either replacing an entire system or upgrading and/or repairing an existing system. The engineer must also provide insight into the cost of system maintenance and the economic consequences of failing to adequately cover this expense. In all cases, the engineer must attempt to provide the highest quality system at an affordable cost by estimating the system installation costs, determining the resulting operating costs (energy, maintenance, etc.), and presenting the economic analysis in the most appropriate or in a specified format.

A common analysis option is the simple payback (SPB) format. This method can be useful if the analysis period is brief (< three years) so that the impact of inflation and interest rates is not significant, no money is borrowed for the investment, and no tax savings via depreciation are claimed.

SPB = Installation costs ÷ Net operating (energy, maintenance, tax) cost savings (12.20)

Table 12.11 Example Duct Cost Spreadsheet Calculator

Duct Cost Estimator using *2004 R.S. Means Building Construction Cost Data*

Metal Duct

Length Ft.	W (or D) In.	H (0 if round) In.	Perimeter In.	SM Gauge	90° Els Quanity	45° Els Quanity	Trans. Quanity	Sq? Rnd Quanity	Eqv. L ft	Area ft²	Weight lb	Cost 2004 $
55	20	14	68.0	26	1		2		70	397	298	
25	14	14	56.0	26		2		2	38	177	133	
120	10		31.4	26	8				144	377	283	
			0.0	26					0	0	0	
			0.0	26					0	0	0	
			0.0	26					0	0	0	
			0.0	26					0	0	0	
								Metal Sub-totals		951	713	**$4,168.04**

Insulation

Duct Wrap Area ft²	Type	**Duct Liner** Area ft²	Type	$ Wrap	$ Liner	
951	1.5 in. FR Liner		1.5 in. BlackFR	3299.97	0	$3,299.97
	1 in. FR Liner		1.5 in. BlackFR	0	0	$0.00
	2 in. Unfaced		1.5 in. BlackFR	0	0	$0.00
			Insulation Sub-totals	3299.97	0	**$3,299.97**

Dampers Quanity	Area (ft²)	Cost ($)	**Filter-Grilles** Quanity	Area (ft²)	Cost ($)		
		$0.00	2	4	$300.00	Damper Sub-total	**$0.00**
						Fil.-Grill Sub-total	**$300.00**

Diffusers Quanity	Type	W (or D)	Cost	Quanity	Type	W (or D)			
8	Perf. Lay-In	24	$1,585.14		Perf. Lay-In		$0.00		
	Round		$0.00		Perf. Lay-In		$0.00		
	Rect.-Louver		$0.00		Round		$0.00	Diffuser Subtotal	**$1,585.14**
									$9,353.15

Table 12.12 HVAC Costs—Equipment, Installation, Overhead, and Profit (Means 2004b)

Chillers (tons)	Air-Cooled Reciprocating	Water-Cooled Reciprocating	Air-Cooled Screw	Water-Cooled Screw	Water-Cooled Centrifugal	Water-Cooled Absorp./Duplex
20	24,800	17,700	–	–	–	–
50	37,00	34,500	–	–	–	–
100	64,000	60,00	–	56,000	–	148,000
200	108,000	111,000	113,000	85,000	–	201,000
500	–	–	–	–	186,000	370,000
1000	–	–	–	–	368,000	718,000
Unitary (tons)	**Air-Cooled Comp. Rm.**	**RTU—CAV w/Gas Heat**	**Air Heat Pump**	**Water-Source Heat Pump**	**Packaged w/Elect. Heat**	
3	10,900	4,325	2,400	1,975	–	
5	14,100	5,575	3,600	2,775	6,975	
10	30,000	9,600	9,200	7,525	11,600	
20	40,000	19,300	17,500	15,200	20,900	
50	–	42,400	–	37,300	41,700	
Air Handling (cfm)	**CAV with H & C coils**	**VAV—CW w/coils**	**Makeup Air**	**Evaporative Cooler**		
2,000	4,150	–	7000	570		
5,000	7,975	10,600	8,4000	1,350		
10,000	14,800	13,500	11,300	2,550		
20,000	26,500	24,600	14,700	9,625		
40,000	46,900	–	19,800	14,000		

Table 12.13 Piping Material and Installation Costs (Means 2004b)

Pipe Material	Dia. (in.)	Straight ($/ft)	90°L ($/fitting)	45°L ($/fitting)	Tee ($/fitting)	Red ($/fitting)	Adpt. ($/fitting)
Copper (Type L)	0.75	7.80	25.50	26	41	23	29
	1	9.05	31.50	33.50	51.50	26.50	34.50
	1.5	12.45	41	41	69	34	45
Hangers @ 10 ft centers, Piping 10 ft above floor	2	16.30	51	50.50	82.50	40.50	52
	3	25.50	100	90	167	81	127
	4	39	152	147	279	150	172
	6	89	445	445	680	420	510
	8	133	1350	1250	2150	580	–
Steel—Black (Sch. 40) Threaded ≤ 2 in. Welded > 2 in.	0.75	9.35	36.50	38	57.50	33	–
	1	11.20	39.50	41	63.50	36	
	1.5	13.80	49	53	75.50	55	–
	2	17.40	57	59	91	66	–
	3	27.50	146	144	264	121	–
Hangers @ 10 ft centers, Piping 10 ft above floor	4	36.50	209	206	355	150	–
	6	56.50	330	320	545	268	–
	8	74	470	445	700	325	–
	10	102	640	590	930	420	–
	12	128	805	735	1225	570	–
PVC Plastic (Sch. 80)	3/4	12.15	19.15	20.50	29.50	17.25	14.85
	1	13.55	23	24	34	21.50	17.05
	1.5	17.25	28	30	46	29.50	21.50
	2	19.45	28.50	31.50	51.50	31.50	21.50
Hangers @ 3 ft 4 in. centers, Piping 10 ft above floor	3	24.50	47	55	70.50	–	–
	4	28.50	59	77	91	78	47.50
	6	40.50	148	117	173	120	–
	8	56	315	305	475	223	–
	10	74	–	–	–	–	–
	12	90.50	–	–	–	–	–

The term *net cost savings* is used since an HVAC system that reduces energy costs might result in higher maintenance costs or a greater tax burden if the reduction increases taxable profit.

Longer-term economic analysis generally considers the "time value" of money or inflation rates that SPB does not. Investors must consider the fact that inflation reduces the actual purchasing power of a dollar in the future. Also, some costs inflate at higher rates than others. Prior to 2004, energy costs were inflating at a low rate compared to the consumer price index (CPI). However, future energy costs have the potential for growing faster than general inflation rates, and, therefore, investments in HVAC equipment that lower operating costs may become much more attractive. There are a variety of accounting practices that should be considered in more detail. These include equipment depreciation credits, tax incentives, utility or manufacturer rebates, tax penalties on operating cost savings, investment income opportunities that are lost due to equipment installation costs, benefits of borrowing money at fixed rates (tax incentives on interest payments, fixed payments), and the salvage value of equipment.

Future and Present Worth

The future worth (F) of a present value (P) can be estimated for a constant rate of inflation (i) for a certain number of time increments (n) in years (or months).

$$F = P(1 + i)^n \tag{12.21}$$

The present worth of a future value can be found by rearranging Equation 12.21.

$$P = F(1 + i)^{-n} \tag{12.22}$$

Annual Payments and Costs

Many expenses or savings are computed on an annual basis. The future worth (F) of an annuity (A) is computed by Equation 12.23.

$$F = A\left(\frac{(1+i)^n - 1}{i}\right) \tag{12.23}$$

Conversely, fixed annual values can be converted from the future worth.

$$A = F\left(\frac{i}{1 + i^n - 1}\right) \tag{12.24}$$

Annual values are converted to present worth. The present worth (P) of an annuity (A) is calculated with Equation 12.25. An annuity is defined as a stream of fixed payments or incomes.

$$P = A\left(\frac{(1+i)^n - 1}{i(1+i)^n}\right) \tag{12.25}$$

And present values are converted to annual amounts:

$$A = P\left(\frac{i(1+i)^n}{(1+i)^n - 1}\right) \tag{12.26}$$

A common convention is to generate tables using Equations 12.21 through 12.26 for a variety of interest rates. Table 12.14 is an example of a 20-year table for an interest rate of 5%. Tabulated values are noted in the format of (F/P, i, N) to indicate the value used to convert a present value to a future value (F/P) at a particular interest rate (i) for a number of years (N). For example, Table 12.14 indicates the value of 2.0789 for (F/P, 5%, 15) so that a sum of $1000 invested at 5% interest will be worth $2078.90 15 years in the future. Mathematically this is expressed as

$$P = F \times (F/P, 5\%, 15) = \$1000 \times 2.0789 = \$2078.90.$$

Identical results would be calculated using Equation 12.21.

Many business owners view HVAC economic decisions as part of their overall economic strategy. The benefit of investing capital in a new chiller to generate energy and maintenance cost savings will be competing

Table 12.14 Interest Rate Factor Table for 5% and 20 Years

i =	5	%				
N	(P/F)	(P/A)	(F/P)	(F/A)	(A/P)	(A/F)
1	0.9524	0.9524	1.0500	1.0000	1.0500	1.0000
2	0.9070	1.8594	1.1025	2.0500	0.5378	0.4878
3	0.8638	2.7232	1.1576	3.1525	0.3672	0.3172
4	0.8227	3.5460	1.2155	4.3101	0.2820	0.2320
5	0.7835	4.3295	1.2763	5.5256	0.2310	0.1810
6	0.7462	5.0757	1.3401	6.8019	0.1970	0.1470
7	0.7107	5.7864	1.4071	8.1420	0.1728	0.1228
8	0.6768	6.4632	1.4775	9.5491	0.1547	0.1047
9	0.6446	7.1078	1.5513	11.0266	0.1407	0.0907
10	0.6139	7.7217	1.6289	12.5779	0.1295	0.0795
11	0.5847	8.3064	1.7103	14.2068	0.1204	0.0704
12	0.5568	8.8633	1.7959	15.9171	0.1128	0.0628
13	0.5303	9.3936	1.8856	17.7130	0.1065	0.0565
14	0.5051	9.8986	1.9799	19.5986	0.1010	0.0510
15	0.4810	10.3797	**2.0789**	21.5786	0.0963	0.0463
16	0.4581	10.8378	2.1829	23.6575	0.0923	0.0423
17	0.4363	11.2741	2.2920	25.8404	0.0887	0.0387
18	0.4155	11.6896	2.4066	28.1324	0.0855	0.0355
19	0.3957	12.0853	2.5270	30.5390	0.0827	0.0327
20	0.3769	12.4622	2.6533	33.0660	0.0802	0.0302

Example Problem 12.1

An HVAC contractor decides to defer the purchase of a new service truck and invest a $10,000 bonus on a well-done job in an 8% interest-bearing investment. Compute the future value in this account in seven years.

Solution:

$$F = P(1 + i)^n = \$10{,}000\,(1 + 0.08)^7 = \$17{,}138$$

Example Problem 12.2

Estimate the cost of a 100 ton water-cooled screw chiller in 2014 using an inflation rate of 5%.

Solution:

Table 12.12 estimates the cost of a 100 ton water-cooled screw chiller to be $56,000. Ten years later, assuming a 5% (0.05) inflation rate, the cost will be

$$F = P(1 + i)^n = \$56{,}000\,(1 + 0.05)^{10} = \$91{,}218.$$

Example Problem 12.3

Compute the added annual payment for a $5000 high-efficiency heat pump if the cost can be included in a 15-year fixed rate home loan at a 6% interest rate (a.k.a. annual percentage rate, APR).

Solution:

$$A = P\left(\frac{i(1+i)^n}{(1+i)^n - 1}\right) = \$5000\left(\frac{0.06(1+0.06)^{15}}{(1+0.06)^{15} - 1}\right) = \$514.81/\text{year} = \$42.90/\text{month}$$

with the possibility of investing in mutual funds, the stock market, municipal bonds, and other interest-bearing investments. The money spent on the chiller would not be available for this purpose, so the annual savings or benefit must be discounted at a rate that is equivalent to the interest rate of the lost investment. The money invested in the chiller may also be used to reduce company debt. There are a variety of other impacts, such as depreciation, taxes, mortgage, and variable interest rates, that should be use to more completely evaluate economic decisions. The power and flexibility of common spreadsheets has largely displaced the use of tables for economic evaluations.

Table 12.15 is an example of the flexibility of spreadsheets for economic evaluations. Methods can be used to account for the impact of different inflation rates for energy, labor, investments, and the Consumer Price Index (CPI) (Turner and Kennedy 1984). An investment of $5500 generates an energy savings of $1000 and a reduction in maintenance cost of $125 in the first year. The net cash flow (NCF) is positive and results in a present worth of –$4375 ($1000 + $125 – $5500) at the end of year 1. In year 2, both the energy and maintenance savings are inflated (but at different rates). A discount factor is applied to account for the change in the CPI and a discount rate that detracts from the investment since the original $5500 was not invested at the 4% rate available to the owner. This type of presentation results in a variety of measures that investors might consider, including a ten-year present worth of $4519.91 and a discounted payback of approximately 5.2 years (time for present worth to become positive). The program could also provide the rate of return by adjusting the discount rate (either manually or using the Excel®function Goal Seek) to generate a ten-year present worth of 0.

Taxes and Depreciation

HVAC expenditures that reduce profit or subsequently increase profit by reducing operating costs alter the amount of taxable income. Added profits result in added taxes, which must be included in economic analysis. This puts an added burden on the economics of saving energy or maintenance costs. The inflated net cash flow before taxes must be altered by the effective tax rate of the business.

$$\text{Net Cash Flow before Taxes } (NCF_{BT}) = \text{Energy Savings} - \text{Maintenance Cost} \quad (12.27)$$

$$\text{Net Cash Flow after Taxes } (NCF_{AT}) = NCF_{BT} - \text{Tax Rate} \times NCF_{BT} \quad (12.28)$$

However, businesses can recoup the first cost of investments by depreciating the cost of the HVAC improvement over several years. This is essentially a

Table 12.15 Discounted Economic Analysis with Inflation—Ten Year

	Investment Cost	Discount Rate (%)	Energy Inflation Rate (%)	Main. Inflation Rate (%)	Gen. Inflation Rate —CPI (%)
	5500	4	5	3.5	3.5
	Year 1 Energy Savings	**Maint. Cost***	**Salvage Value in Year 10**		
	1000	–125	0.00		
Year	**Energy Savings**	**Maint. Cost***	**Net Cash Flow**	**Disc. NCF**	**Pres. Worth**
1	1000.00	–125.00	1125.00	1125.00	–4375.00
2	1050.00	–129.38	1179.38	1095.67	–3279.33
3	1102.50	–133.90	1236.40	1067.12	–2212.22
4	1157.63	–138.59	1296.21	1039.34	–1172.88
5	1215.51	–143.44	1358.95	1012.30	–160.58
6	1276.28	–148.46	1424.74	985.98	825.40
7	1340.10	–153.66	1493.75	960.37	1785.76
8	1407.10	–159.03	1566.14	935.43	2721.20
9	1477.46	–164.60	1642.06	911.17	3632.36
10	1551.33	–170.36	1721.69	887.55	4519.91

* If maintenance cost for the alternative is greater than for the conventional system, year one maintenance cost is positive. If maintenance cost is less than for the conventional system, year one maintenance cost is negative.

Table 12.16 Depreciation Percentages (DeGarmo et al. 1997)

Recovery Year	3-Year ACRS/ MACRS	5-Year ACRS/ MACRS	10-year ACRS/ MACRS
1	25/33.3	15/20	8/10
2	38/44.45	22/32	14/18
3	37/14.80	21/19.2	12/14.4
4	0/7.41	211/1.52	10/11.52
5		21/11.52	10/9.22
6			10/7.37
Years 7-10			9/6.55
Year 11			0/3.28

lowering of profit that can offset the impact of the larger tax due to operating cost savings. Table 12.16 lists three-, five-, and ten-year property class depreciation schedules. The Accelerated Cost Recovery System (ACRS) was used beginning in 1981 and was changed in 1986 to the Modified Accelerated Cost Recovery System (MACRS).

The percentage from Table 12.16 is multiplied by the investment cost of the equipment in each year so the yearly net cash flow after taxes and depreciation is

$$NCF_{AT\&D} = NCF_{BT} - \text{Tax Rate} \times NCF_{BT} + \text{Depreciation} \times \text{Investment Cost}\,. \qquad (12.29)$$

The $NCF_{AT\&D}$ replaces the NCF value in non-tax/depreciation economic analysis. It is adjusted for inflation, discounted, and added to the previous year's present worth to calculate the current year's present worth.

Building owners and investors use a variety of measures to evaluate economic decisions, which may be far more complex than the fundamental principles presented in this section. However, the challenge remains for the design engineer to provide systems that give optimum value in the trade-off between investment costs and operating costs while creating an indoor space that is healthy, comfortable, and safe for the occupants.

CHAPTER 12 PROBLEMS

12.1 Compute the HVAC system demand (kW/ton) and efficiency (EER, COP) for a split-system heat pump with a medium-efficiency scroll compressor, an indoor fan with a standard AC motor, and an axial condenser fan.

12.2 Repeat Problem 12.1 using a high-efficiency scroll compressor, an electronically commutated motor (ECM), and an axial condenser fan.

12.3 Compute the HVAC system demand (kW/ton) and efficiency (EER, COP) for a packaged rooftop unit with a medium-efficiency reciprocating compressor, an axial condenser fan, an indoor fan that delivers 1.5 in. of water, and a return fan that delivers 1.0 in. of water. Fan motors are 85% efficient and fans are 65% efficient.

12.4 Compute the HVAC system demand (kW/ton) and efficiency (EER, COP) for a ground-source heat pump using a high-efficiency scroll compressor, a fan with an ECM, and a 50% efficient pump with a 60% electric motor that delivers 25 ft of water head. Assume the entering water temperature (EWT) to the unit is 85°F.

12.5 Compute the HVAC system demand (kW/ton) and efficiency (EER, COP) for a chilled water system (CWS) with a high-efficiency water-cooled centrifugal compressor, 70% efficient chiller pumps with 50 ft of head, 70% efficient loop pumps with 70 ft of head, air-handling units with 75% efficient supply fans that deliver 5.0 in. of water, 75% efficient return air fans that deliver 2.0 in. of water, 70% efficient condenser pumps with 60 ft of head, an axial fan cooling tower, and fan-powered variable air volume (FPVAV) terminals with ECMs. Assume all motors are 92% efficient (except ECMs).

12.6 Repeat Problem 12.5 but replace the VAV system (supply fan, return fans, FPVAVs) with fan-coil units (FCUs) that have a nominal 10 ton/4000 cfm capacity and circulate air with 3 hp fans driven by 85% efficient motors.

12.7 A building in Birmingham, Alabama, is occupied five days per week from 8 a.m. to 8 p.m. During the occupied period, it has a cooling load of 120 MBtu/h at 97°F outside air temperature and a cooling load of 0 MBtu/h at 57°F OAT. During the unoccupied period, it has a cooling load of 40 MBtu/h at 97°F outside air temperature and a cooling load of 0 MBtu/h at 57°F OAT. In heating, the load is 80 MBtu/h at 17°F OAT (occupied), 60 MBtu/h at 17°F OAT (unoccupied), and 0 MBtu/h at 47°F OAT (occupied and unoccupied). It is cooled by a unit with a 125 MBtu/h capacity and 14 kW demand at 97°F and a 141 MBtu/h capacity and 11.4 kW demand at 67°F. It is heated by a unit with a 120 MBtu/h heating capacity with an 80% efficiency with a 1.5 hp 82% efficient fan motor. Compute the annual cost of heating and cooling the building based on 8¢/kWh in the summer and 7¢/kWh in the winter. Natural gas cost is $1.80 per therm (ccf).

12.8 Find the savings for the system described in Problem 12.7 if the efficiency of the cooling unit was improved by 20% (same capacity with 20% lower demand), the efficiency of the furnace is 95%, and the fan is reduced to 1 hp with a 90% efficient motor.

12.9 Repeat Problem 12.7 using a heat pump with the same cooling capacity and a heating capacity of 120 MBtu/h with an input of 11.3 kW at 47°F and 55 MBtu/h with an input of 9.8 kW at 17°F.

12.10 Discuss the economic value of installing a $10,000 energy efficiency package on a $225,000, 30-year, 6.25% APR home mortgage that will lower monthly utility bills by $40. Inflation rates are 8% for energy, 5% for maintenance, and 5% general.

12.11 Repeat Problem 12.10 for an energy inflation rate of 4%.

12.12 Repeat Problem 12.10 for a 15-year, 5.75% APR loan.

12.13 A complete energy retrofit that will cost $200,000 is estimated to provide an annual savings of $30,000. The energy inflation rate is 6%, while the general and maintenance inflation rates are 5%. However, the system will require an additional $3000-per-year service contract. Compute the discounted ten-year present worth of the project.

12.14 Calculate the discounted rate of return on the project described in Problem 12.13 for a ten-year evaluation.

12.15 A ground-source heat pump costs $5000 more than a conventional heating and cooling system. It saves approximately $400 per year in energy costs and $100 per year in maintenance costs. The owner plans to live in this home for 20 years. The energy inflation rate is 7%, the discount rate is 5%, and the general inflation and maintenance rates are 6%. What is the present worth at 20 years and what is the discounted payback?

12.16 Repeat Problem 12.15 for $i_e = 6\%$, $i_g = i_m = 7\%$, and $d = 6\%$ and compare these results with a simple payback analysis.

REFERENCES

ASHRAE. 2003. *2003 ASHRAE Handbook—Applications*, chapter 36, Owning and operating costs. Atlanta: American Society of Heating, Refrigerating and Air-Conditioning Engineers, Inc.

ASHRAE. 2004a. *ANSI/ASHRAE/IESNA Standard 90.1-2004, Energy Standard for Buildings Except Low-Rise Residential Buildings*. Atlanta: American Society of Heating, Refrigerating and Air-Conditioning Engineers, Inc.

ASHRAE. 2004b. *ANSI/ASHRAE Standard 62.1-2004, Ventilation for Acceptable Indoor Air Quality*. Atlanta: American Society of Heating, Refrigerating and Air-Conditioning Engineers, Inc.

ASHRAE. 2005. *2005 ASHRAE Handbook—Fundamentals*, chapter 32, Energy estimating and modeling methods. Atlanta: American Society of Heating, Refrigerating and Air-Conditioning Engineers, Inc.

DeGarmo, E.P., W.G. Sullivan, J.A. Bontadelli, and E.M. Wicks. 1997. *Engineering Economy*, 10th ed. Upper Saddle River, NJ: Prentice-Hall.

Degelman, L.O. 1986. Bin and Degree Hour Weather Data for Simplified Energy Calculations, RP-385. ASHRAE research project report. Atlanta: American Society of Heating, Refrigerating and Air-Conditioning Engineers, Inc.

DOE. 2002. Variable speed drive part-load efficiency. *Energy Matters*. Office of Energy Efficiency and Renewable Energy, US Department of Energy, Winter 2002.

EIA/DOE. 1999. *Commercial Buildings Energy Consumption Survey*, Table C4. www.eia.doe.gov/commercial.html. Washington, DC: Energy Information Administration, US Department of Energy.

Grumman, D.L., ed. 2003. *ASHRAE GreenGuide*. Atlanta: American Society of Heating, Refrigerating and Air-Conditioning Engineers, Inc.

InterEnergy. 1999. *BinMaker™Plus, Weather Data for Engineering*. Chicago: InterEnergy Software, Inc.

Kavanaugh, S.P. 2003. Estimating demand and efficiency. *ASHRAE Journal* 45(7):36–40.

LEED. 2003. *Leadership in Energy and Environmental Design™: Green Building Rating System*, Version 2.1. Washington, DC: US Green Building Council.

Means, R.S. 2004a. *Facilities Maintenance and Repair Cost Data.* Kingston, MA: Reed Construction Data.

Means, R.S. 2004b. *Mechanical Cost Data.* Kingston, MA: Reed Construction Data.

Scofield, J.H. 2002. Early performance of a green academic building. *ASHRAE Transactions* 108(2):1214–1230.

Torcellini, P.A., R. Judkoff, and D.B. Crawley. 2004. High-performance buildings. *ASHRAE Journal* 46(9):S4–S11.

Turner, W.C., ed. 1992. *Energy Management Handbook*, 2d ed. New York: John Wiley.

Turner, W.C., and W.J. Kennedy. 1984. *Energy Management.* Englewood Cliffs, NJ: Prentice-Hall.

Appendix A
Units and Conversions

Table 1 Conversions to I-P and SI Units

(Multiply I-P values by conversion factors to obtain SI; divide SI values by conversion factors to obtain I-P)

Multiply I-P	By	To Obtain SI	Multiply I-P	By	To Obtain SI
acre (43,506 ft^2)	0.4047	ha	in·lb$_f$ (torque or moment)	113	mN·m
	4046.873	m^2	in^2	645.16	mm^2
atmosphere (standard)	*101.325	kPa	in^3 (volume)	16.3874	mL
bar	*100	kPa	in^3/min (SCIM)	0.273117	mL/s
barrel (42 U.S. gal, petroleum)	159.0	L	in^3 (section modulus)	16,387	mm^3
	0.1580987	m^3	in^4 (section moment)	416,231	mm^4
Btu (International Table)	1055.056	J	kWh	*3.60	MJ
Btu (thermochemical)	1054.350	J	kW/1000 cfm	2.118880	kJ/m^3
Btu/ft^2 (International Table)	11,356.53	J/m^2	kilopond (kg force)	9.81	N
Btu/ft^3 (International Table)	37,258.951	J/m^3	kip (1000 lb$_f$)	4.45	kN
Btu/gal	278,717.1765	J/m^3	kip/in^2 (ksi)	6.895	MPa
Btu·ft/h·ft^2·°F	1.730735	W/(m·K)	litre	*0.001	m^3
Btu·in/h·ft^2·°F (thermal conductivity k)	0.1442279	W/(m·K)	met	58.15	W/m^2
Btu/h	0.2930711	W	micron (μm) of mercury (60°F)	133	mPa
Btu/h·ft^2	3.154591	W/m^2	mile	1.609	km
Btu/h·ft^2·°F (overall heat transfer coefficient U)	5.678263	W/(m^2·K)	mile, nautical	*1.852	km
Btu/lb	*2.326	kJ/kg	mile per hour (mph)	1.609344	km/h
Btu/lb·°F (specific heat c_p)	*4.1868	kJ/(kg·K)		0.447	m/s
bushel (dry, U.S.)	0.0352394	m^3	millibar	*0.100	kPa
calorie (thermochemical)	*4.184	J	mm of mercury (60°F)	0.133	kPa
centipoise (dynamic viscosity μ)	*1.00	mPa·s	mm of water (60°F)	9.80	Pa
centistokes (kinematic viscosity ν)	*1.00	mm^2/s	ounce (mass, avoirdupois)	28.35	g
clo	0.155	m^2·K/W	ounce (force or thrust)	0.278	N
dyne	1.0×10^{-5}	N	ounce (liquid, U.S.)	29.6	mL
dyne/cm^2	*0.100	Pa	ounce inch (torque, moment)	7.06	mN·m
EDR hot water (150 Btu/h)	43.9606	W	ounce (avoirdupois) per gallon	7.489152	kg/m^3
EDR steam (240 Btu/h)	70.33706	W	perm (permeance at 32°F)	5.72135×10^{-11}	kg/(Pa·s·m^2)
EER	0.293	COP	perm inch (permeability at 32°F)	1.45362×10^{-12}	kg/(Pa·s·m)
ft	*0.3048	m	pint (liquid, U.S.)	4.73176×10^{-4}	m^3
	*304.8	mm	pound		
ft/min, fpm	*0.00508	m/s	lb (avoirdupois, mass)	0.453592	kg
ft/s, fps	*0.3048	m/s		453.592	g
ft of water	29,989.07	Pa	lb$_f$ (force or thrust)	4.448222	N
ft of water per 100 ft pipe	98.1	Pa/m	lb$_f$/ft (uniform load)	14.59390	N/m
ft^2	0.092903	m^2	lb/ft·h (dynamic viscosity μ)	0.4134	mPa·s
ft^2·h·°F/Btu (thermal resistance R)	0.176110	m^2·K/W	lb/ft·s (dynamic viscosity μ)	1490	mPa·s
ft^2/s (kinematic viscosity ν)	92,900	mm^2/s	lb$_f$·s/ft^2 (dynamic viscosity μ)	47.88026	Pa·s
ft^3	28.316846	L	lb/h	0.000126	kg/s
	0.02832	m^3	lb/min	0.007559	kg/s
ft^3/min, cfm	0.471947	L/s	lb/h [steam at 212°F (100°C)]	0.2843	kW
ft^3/s, cfs	28.316845	L/s	lb$_f$/ft^2	47.9	Pa
ft·lb$_f$ (torque or moment)	1.355818	N·m	lb/ft^2	4.88	kg/m^2
ft·lb$_f$ (work)	1.356	J	lb/ft^3 (density, ρ)	16.0	kg/m^3
ft·lb$_f$/lb (specific energy)	2.99	J/kg	lb/gallon	120	kg/m^3
ft·lb$_f$/min (power)	0.0226	W	ppm (by mass)	*1.00	mg/kg
footcandle	10.76391	lx	psi	6.895	kPa
gallon (U.S., *231 in^3)	3.785412	L	quad (10^{15} Btu)	1.055	EJ
gph	1.05	mL/s	quart (liquid, U.S.)	0.9463	L
gpm	0.0631	L/s	square (100 ft^2)	9.29	m^2
gpm/ft^2	0.6791	L/(s·m^2)	tablespoon (approximately)	15	mL
gpm/ton refrigeration	0.0179	mL/J	teaspoon (approximately)	5	mL
grain (1/7000 lb)	0.0648	g	therm (U.S.)	105.5	MJ
gr/gal	17.1	g/m^3	ton, long (2240 lb)	1.016	Mg
gr/lb	0.143	g/kg	ton, short (2000 lb)	0.907	Mg; t (tonne)
horsepower (boiler) (33,470 Btu/h)	9.81	kW	ton, refrigeration (12,000 Btu/h)	3.517	kW
horsepower (550 ft·lb$_f$/s)	0.7457	kW	torr (1 mm Hg at 0°C)	133	Pa
inch	*25.4	mm	watt per square foot	10.76	W/m^2
in. of mercury (60°F)	3.37	kPa	yd	*0.9144	m
in. of water (60°F)	249	Pa	yd^2	0.8361	m^2
in/100 ft, thermal expansion	0.833	mm/m	yd^3	0.7646	m^3
To Obtain I-P	**By**	**Divide SI**	**To Obtain I-P**	**By**	**Divide SI**

*Conversion factor is exact.

Notes: 1. Units are U.S. values unless noted otherwise.

2. Litre is a special name for the cubic decimetre. 1 L = 1 dm^3 and 1 mL = 1 cm^3.

Figure A.1 Unit conversion factors to/from inch-pound and International units (*2005 ASHRAE Handbook—Fundamentals*, chapter 38).

Table A.1 Unit Conversion Factors

LENGTH	1 m = 3.2808 ft = 39.37 in = 100 cm = 10^6 μm = 10^{10} Angstrom
	1 ft = 0.3048 m = 12 in = 30.48 cm = 0.33333 yd
	1 km = 1000 m = 0.621 mi, 1 in = 2.540 cm = 0.0254 m
	1 mi = 5280 ft = 1760 yd = 1609.4 m
AREA	1 m^2 = 10.76 ft^2 = 10^4 cm^2 1 acre = 43,560 ft^2 = 4047 m^2
	1 ft^2 = 144 in^2 = 0.09291 m^2 = 929.1 cm^2 1 ha = 10^4 m^2 = 2.47 acre
VOLUME	1 gal = 0.13368 ft^3 = 3.785 L = 4 qt = 8 pints = 16 cups = 256 Tbsp
	1 L = 10^{-3} m^3 = 10^3 ml = 10^3 cm^3 = 1.057 qt = 0.03531 ft^3
	1 m^3 = 35.31 ft^3 = 1000 L = 264.1 gal = 1.308 yd^3
TIME	1 hr = 60 min = 3600 s, 1 yr = 52.14 wks = 365 days = 8760 hr
MASS	1 kg = 1000 g = 2.2046 lbm = 35.27 oz. Av. = 0.068521 slugs
	1 lbm = 0.4536 kg = 453.6 g = 16 oz. Av. = 0.031081 slugs
FORCE	1 N = 1 kg·m/s^2 = 0.2248 lbf = 10^5 dyn = 10^5 g·cm/s^2
	1 lbf = 4.448 N = 4.448 × 10^5 dyn
ENERGY	1 J = 1 kg·m^2/s^2 = 107 g·cm^2/s^2 = 1 N·m = 0.7376 ft-lbf
	1 Btu = 778.16 ft·lbf = 1.055 × 10^6 ergs = 252 cal =1.0550 kJ
	1 cal = 4.186 J = 3.088 ft·lbf
	1 kcal = 4186 J = 1000 cal = 3.968 Btu
	1 erg = 1 g·cm^2/s^2 = 10^{-7} J
	1 Q = 10^{18} Btu = 1.055 × 10^{21} J = 10^3 Quad
	1 kJ = 0.94781 Btu = 0.23884 kcal = 1 kPa·m^3 = 6.242 × 10^{21} eV
POWER	1 W = 1 kg·m^2/s^3 = 1 J/s = 1 N·m/s = 3.412 Btu/hr
(Heat Transfer Rate)	1 hp = 550 ft·lbf/s = 33000 ft·lbf/min = 2545 Btu/h = 746 W
	1 kW = 1000 W = 3412 Btu/hr 1 MW = 3.412 × 10^6 Btu/hr
	1 ton (refrig.) = 12000 Btu/h = 3.517 kW
PRESSURE	1 atm = 14.696 lbf/in^2 = 760 torr = 101,325 Pa = 29.92 in Hg
	1 bar = 10^5 Pa = 14.504 lbf/in^2 = 10^6 dyn/cm^2 = 0.98692 atm
	1 Pa = 1 N/m^2 = 1 kg/m·s^2 = 1 J/m^3 = 1.4504 × 10^{-4} lbf/ft^2
	1 lbf/in^2 = 1 psi = 6894.6 Pa = 2.0418 in Hg = 144 lbf/ft^2
	1 in Hg = 3376.8 Pa = 0.4898 lbf/in^2 = 13.57 in H_2O
	1 in H_2O = 248.8 Pa = 0.0361 lbf/in^2=5.2 lbf/ft^2
Volume/Mass Flow Rate (Water)	1 gpm = 3.785 Lpm = 8.33 lb/min = 0.06297 kg/s
Volume Flow Rate	1 cfm = 28.32 Lpm= 0.472 Lps = 0.0283 m^3/min = 1.699 m^3/hr
POWER/AREA	1 W/m^2 = 0.3170 Btu/(h·ft^2) = 0.85984 kcal/(hr·m^2)
ENERGY/MASS	1 kJ/kg = 0.4299 Btu/lbm = 0.23884 kcal/kg
SPECIFIC HEAT	1 kJ/(kg·K) = 0.23884 Btu/(lbm·R) = 0.23884 kcal/(kg·K)
	1 Btu/(lbm·R) = 4.186 kJ/(kg·K)
THERMAL CONDUCTIVITY	1 W/(m·K) = 0.5778 Btu/(h·ft·R) [1 Btu/(h·ft·R) = 1.731 W/(m·K)]
CONVECTION COEFFICIENT	1 W/(m^2·K) = 0.176 Btu/(hr·ft^2·R) [1 Btu/(hr·ft^2·R)=5.678 W/(m^2·K)]
DYNAMIC VISCOSITY	1 kg/(m·s) = 1 N·s/m^2 = 0.6720 lbm/(ft·s) = 10 Poise
TEMPERATURE	1 K = 1.8 R , °C = (°F - 32)/1.8, °F = 1.8° C + 32, R = °F + 459.67
g_c	g_c = 1 = 1 kg·m/N·s^2 = 32.178 ft·lbm/lbf·s^2 = 1 ft·slug/lbf·s^2
Universal Gas Constant, $\bar{R}$	$\bar{R}$ = 8.3144 kJ/kmol-K = 1545 Btu/lbmol-R = 1.9872 kcal/kgmol-K

Source: K. Clark Midkiff, PhD, *Course Notes: ME 215, Thermodynamics I, University of Alabama*

Appendix B
Refrigerant Pressure-Enthalpy Charts and Sea Level Psychrometric Chart

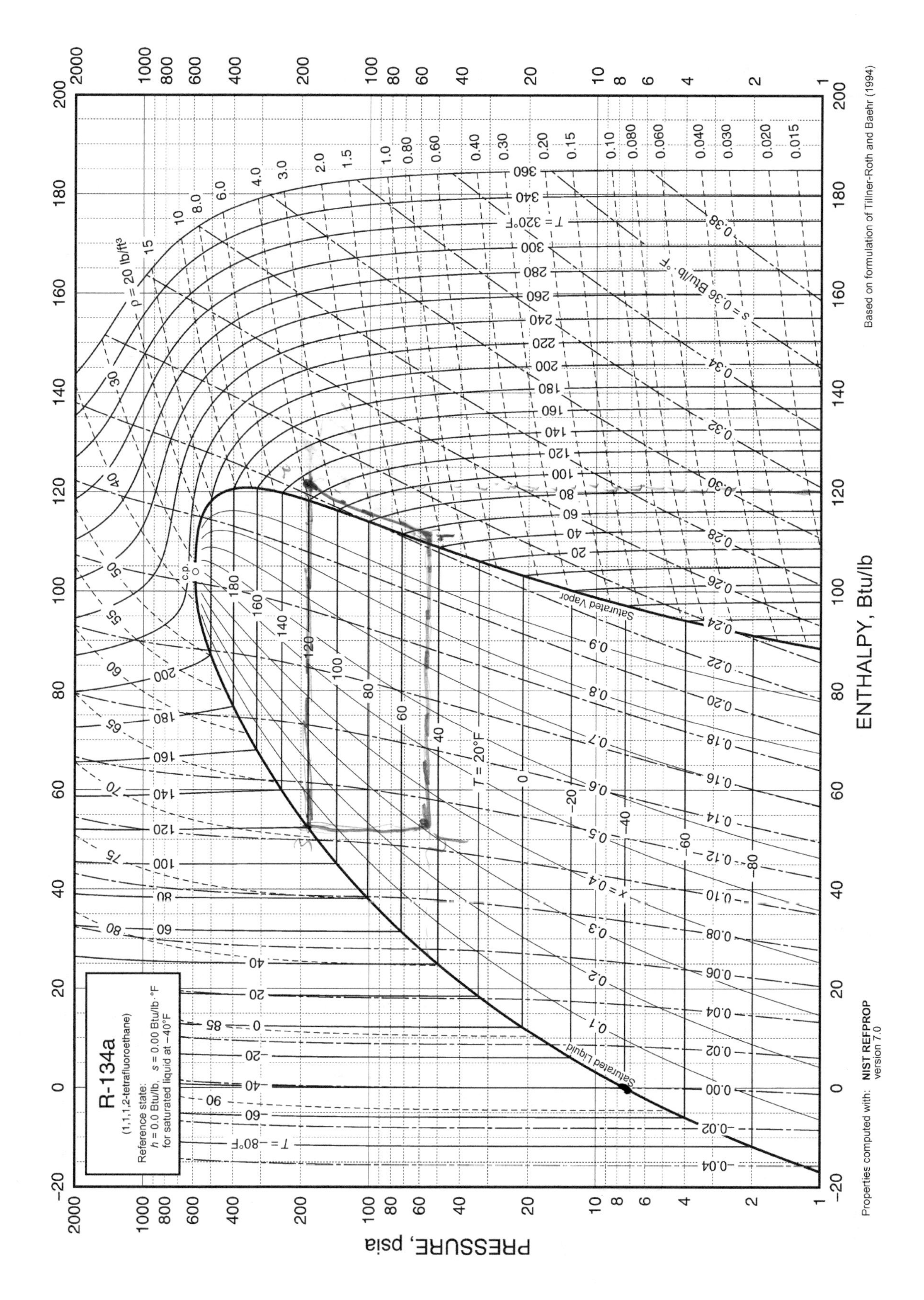

Figure B.1 P-h diagram for R-134a (p. 20.16, *2005 ASHRAE Handbook—Fundamentals*).

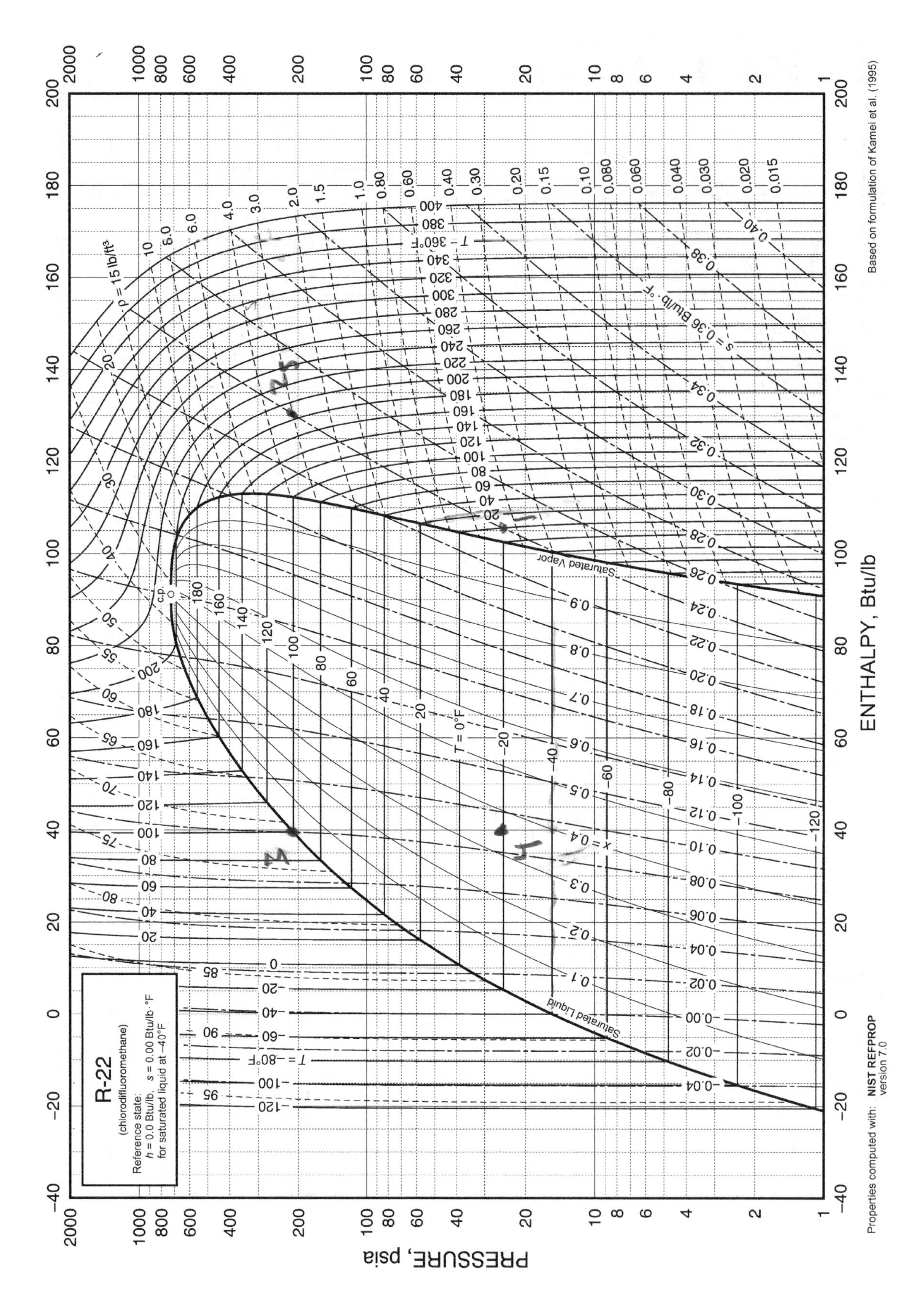

Figure B.2 P-h diagram for R-22 (p. 20.4, *2005 ASHRAE Handbook—Fundamentals*).

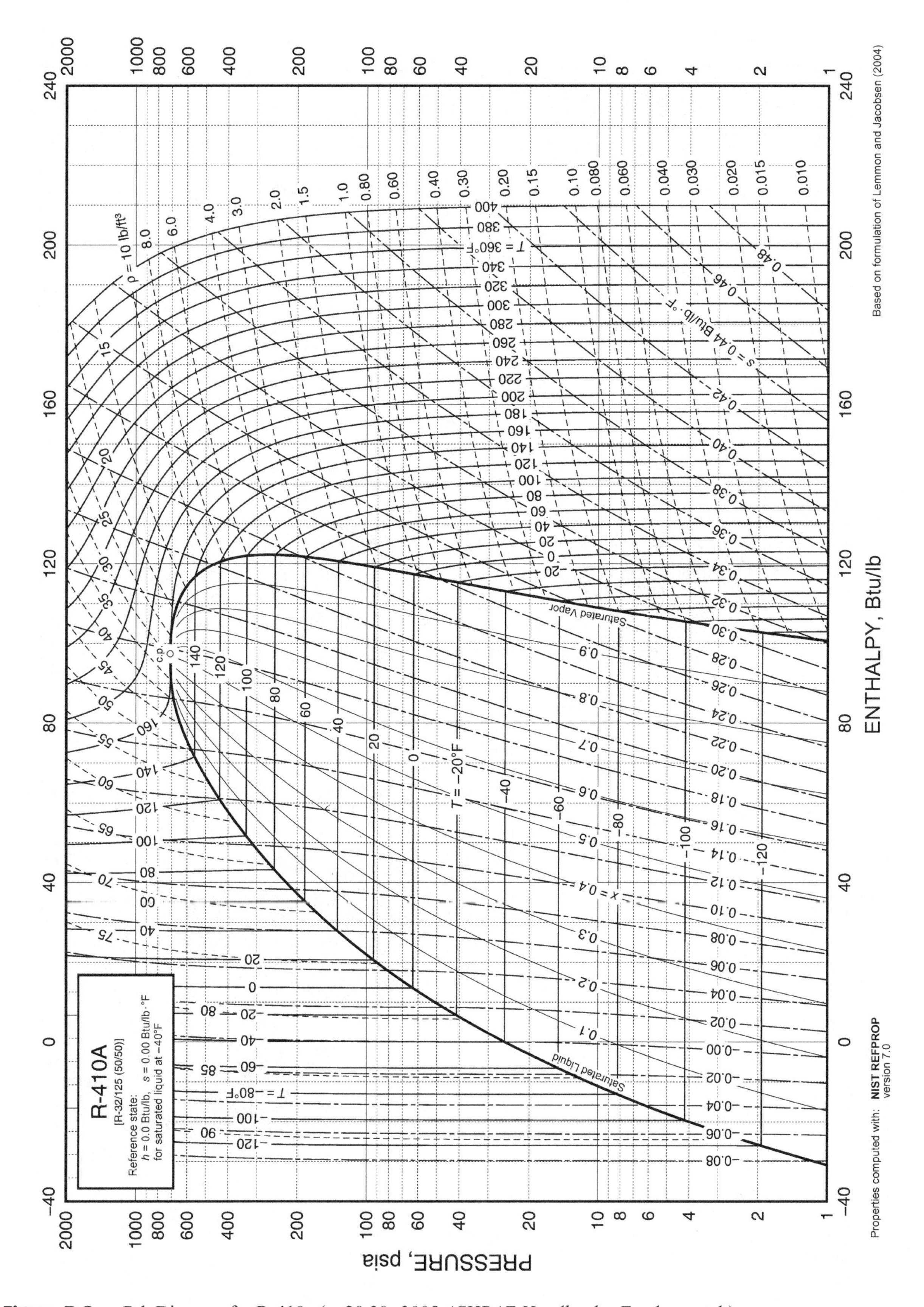

Figure B.3 P-h Diagram for R-410a (p. 20.30, *2005 ASHRAE Handbook—Fundamentals*).

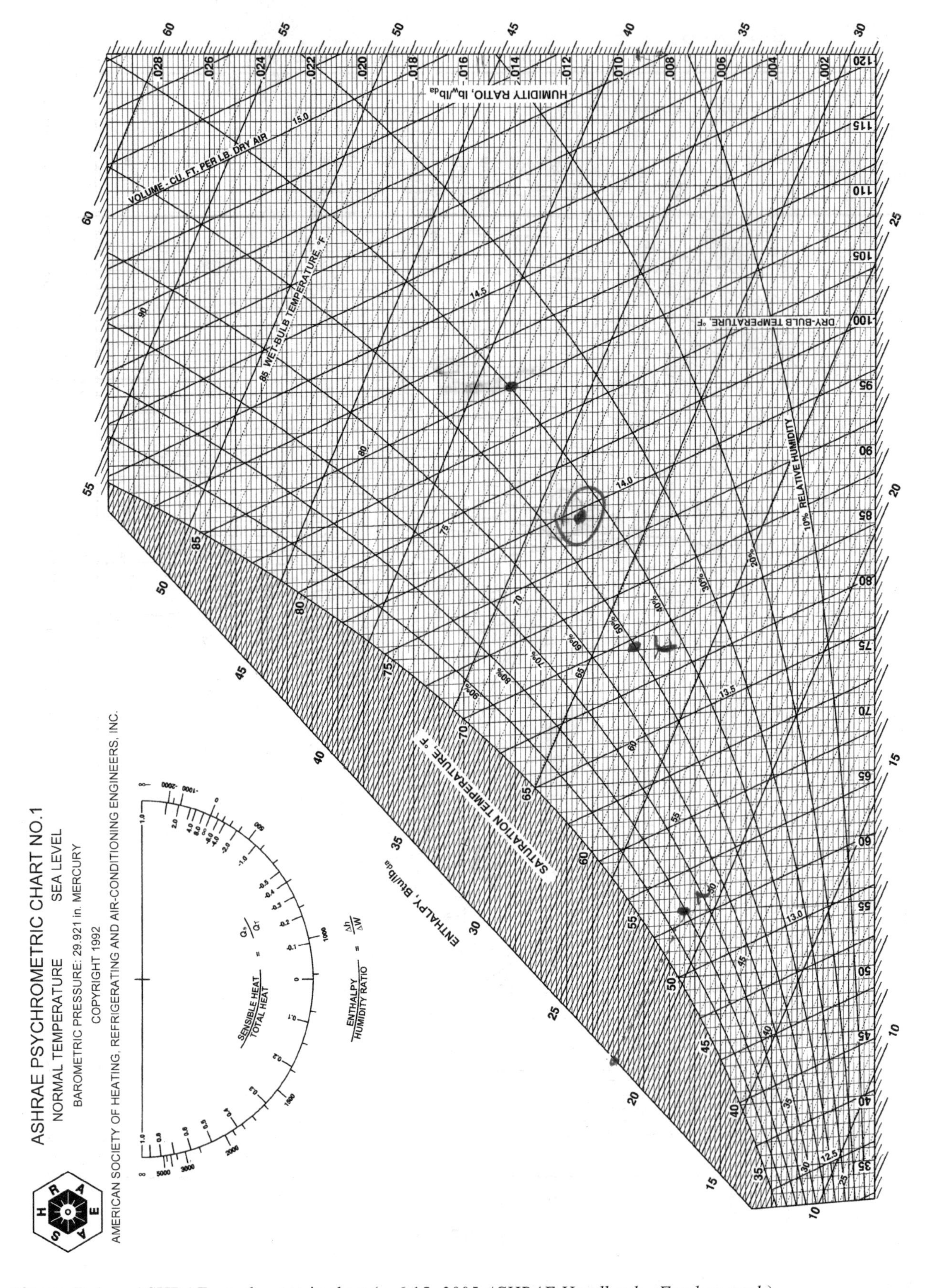

Figure B.4 ASHRAE psychrometric chart (p. 6.15, *2005 ASHRAE Handbook—Fundamentals*).

Appendix C
Floor Plans

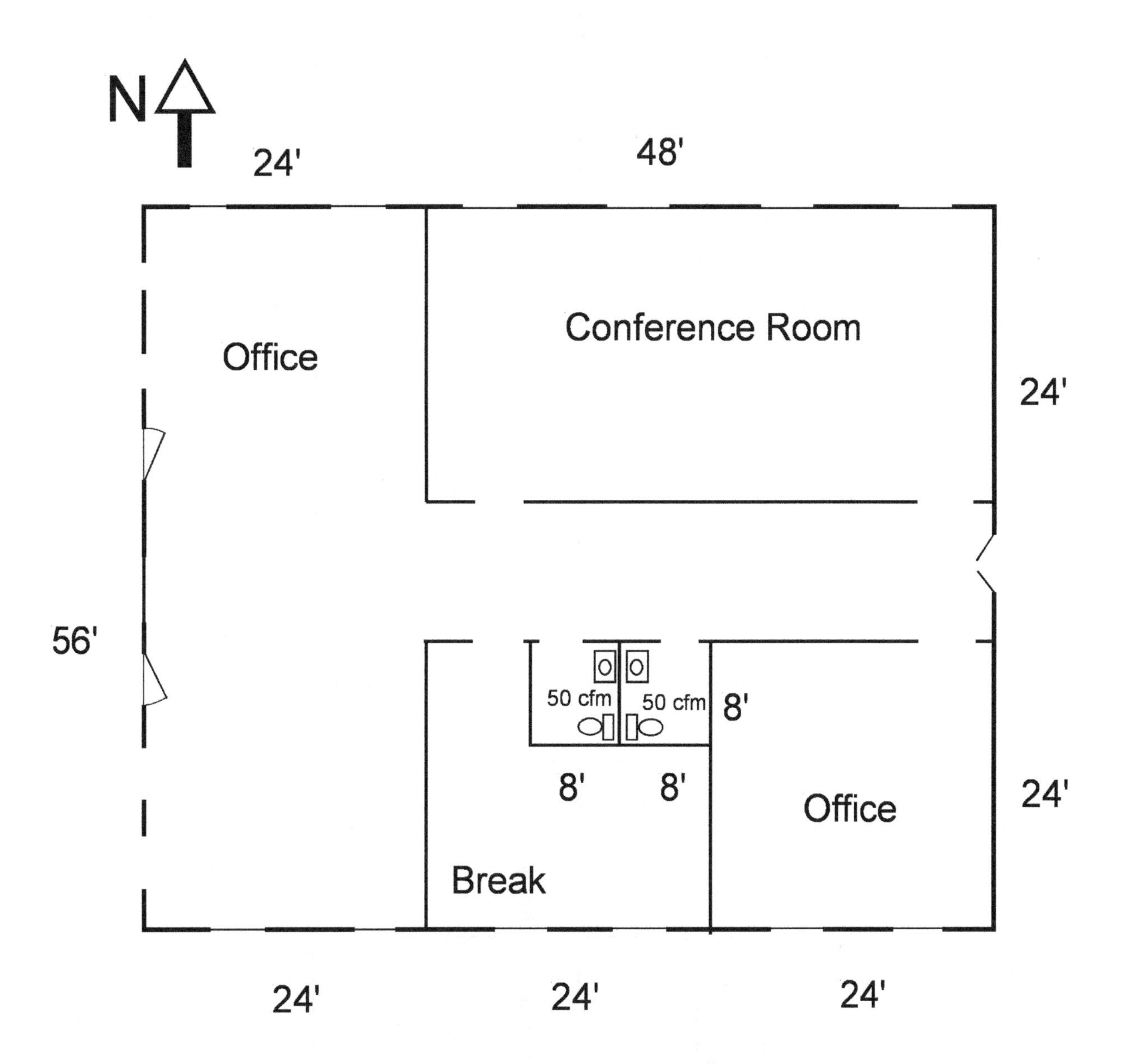

Figure C.1 Small office floor plan.

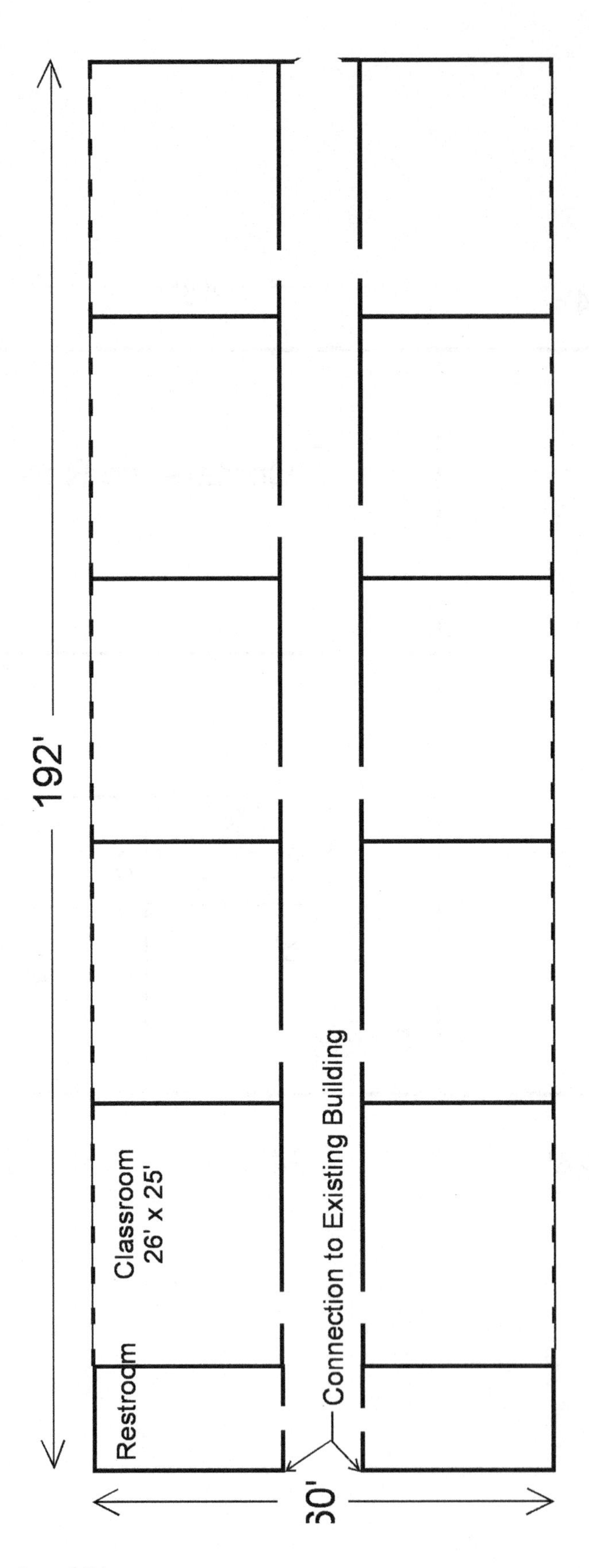

Figure C.2 Classroom wing addition.

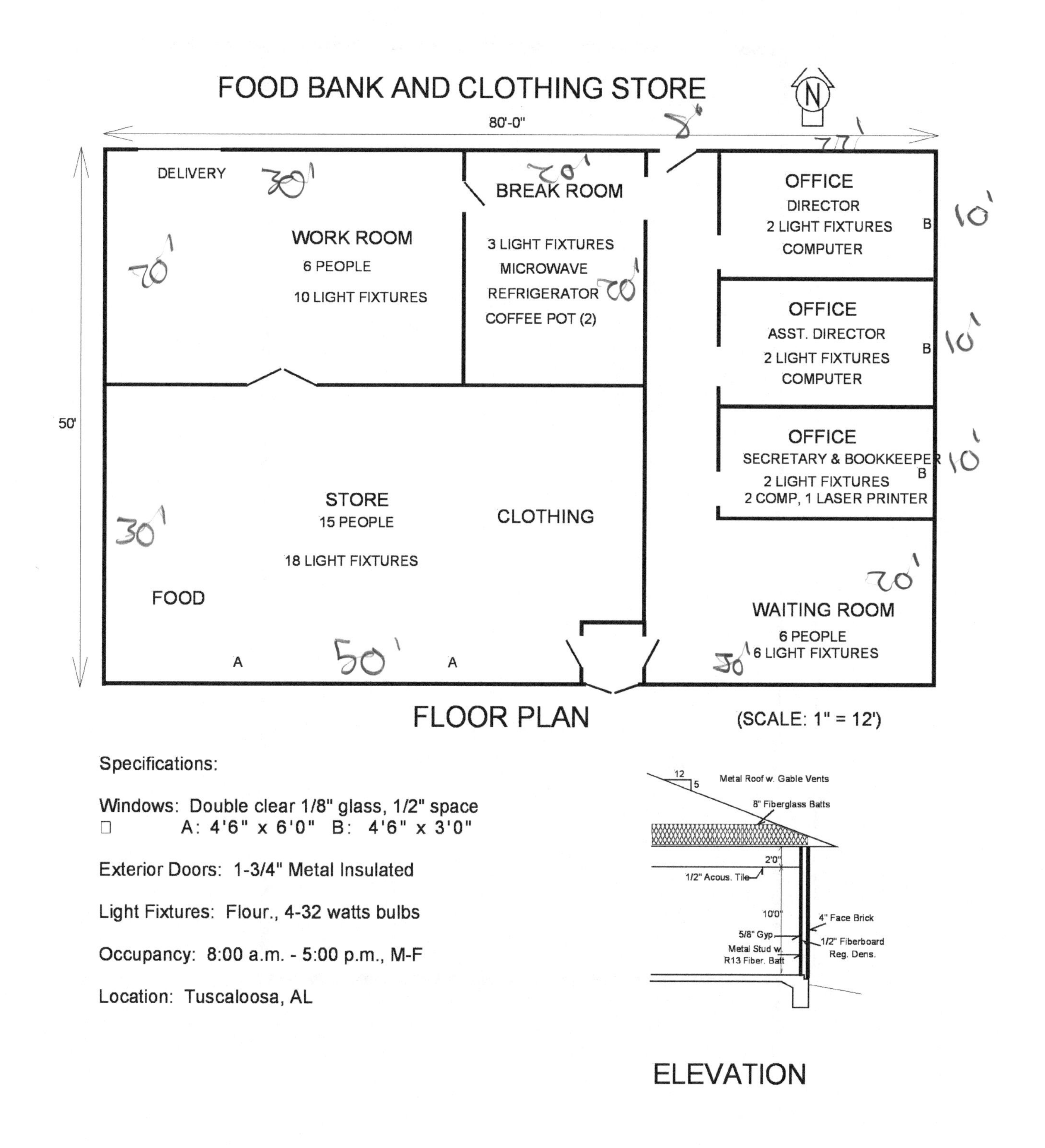

Figure C.3 Food bank and thrift store floor plan.

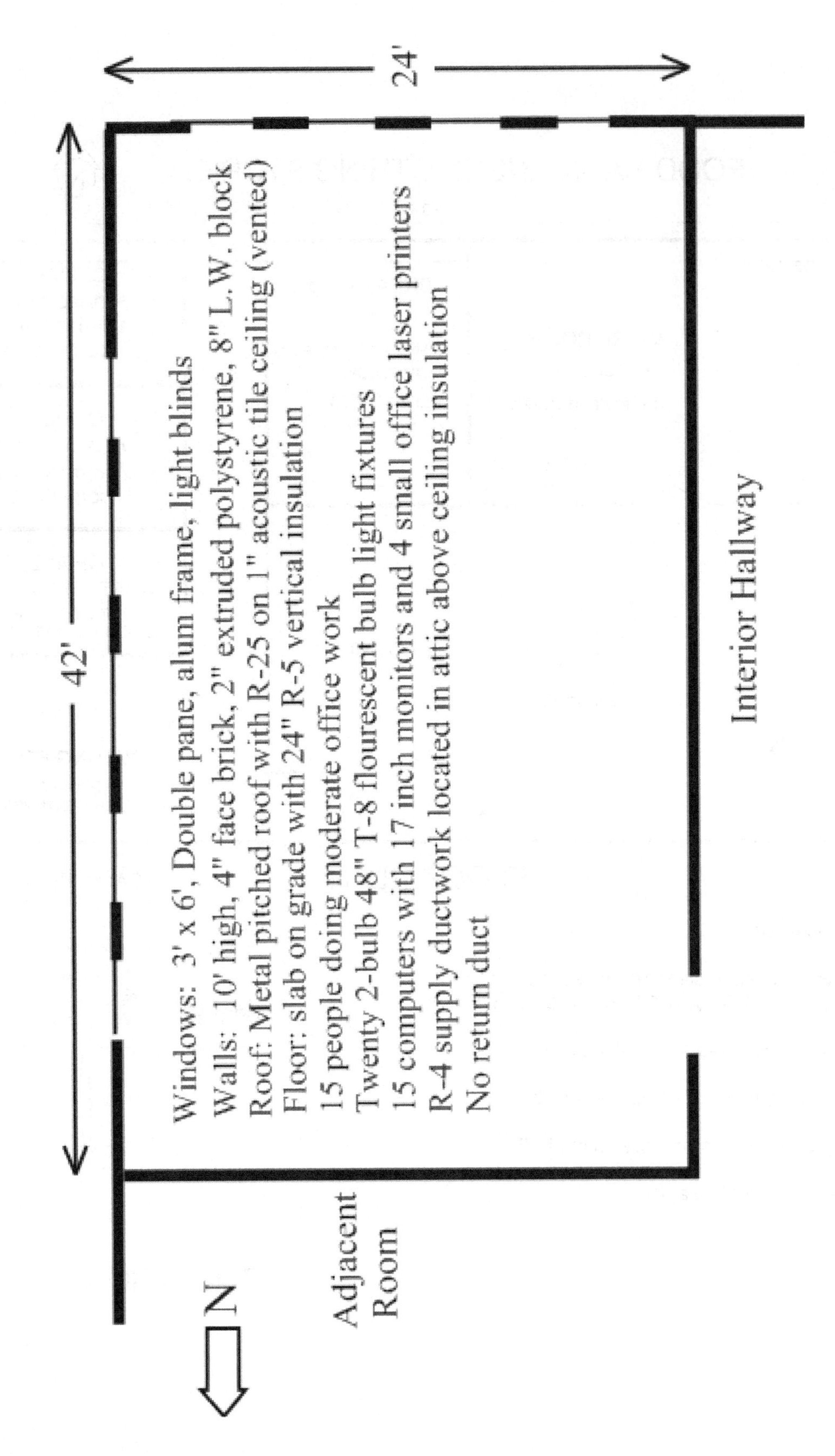

Figure C.4 Single office floor plan.

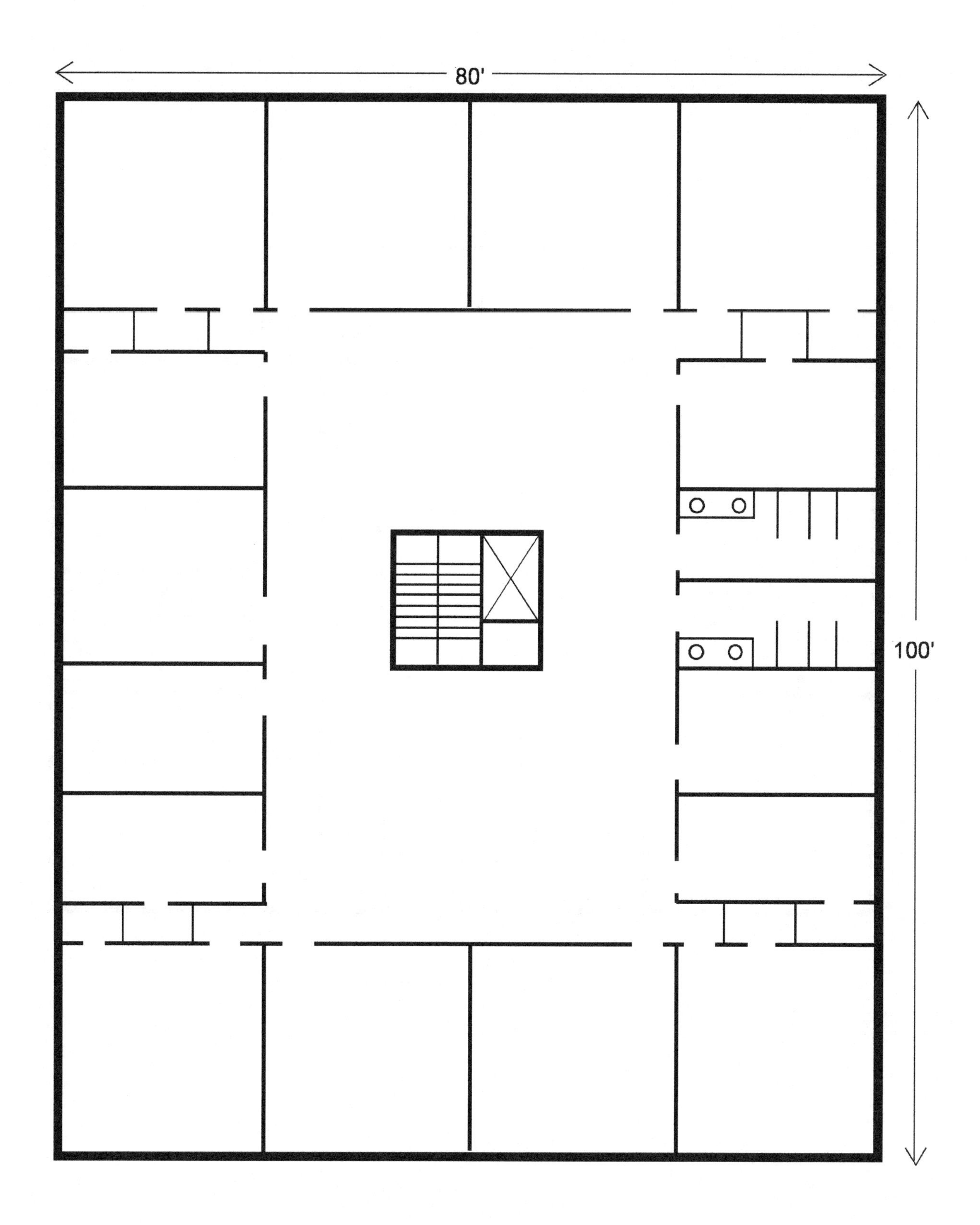

Figure C.5 Four-story office floor plan.

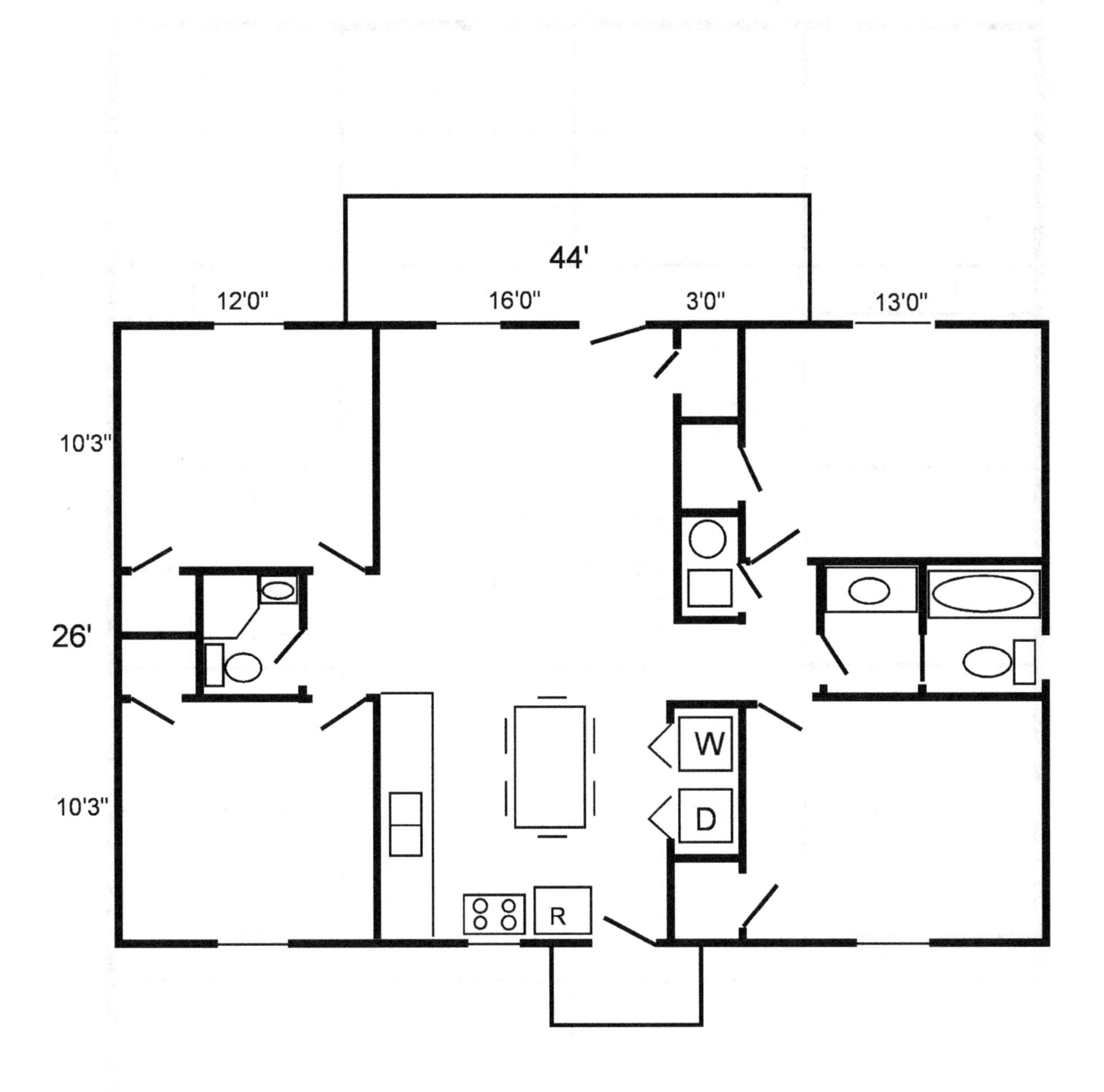

Figure C.6 Affordable housing floor plan.

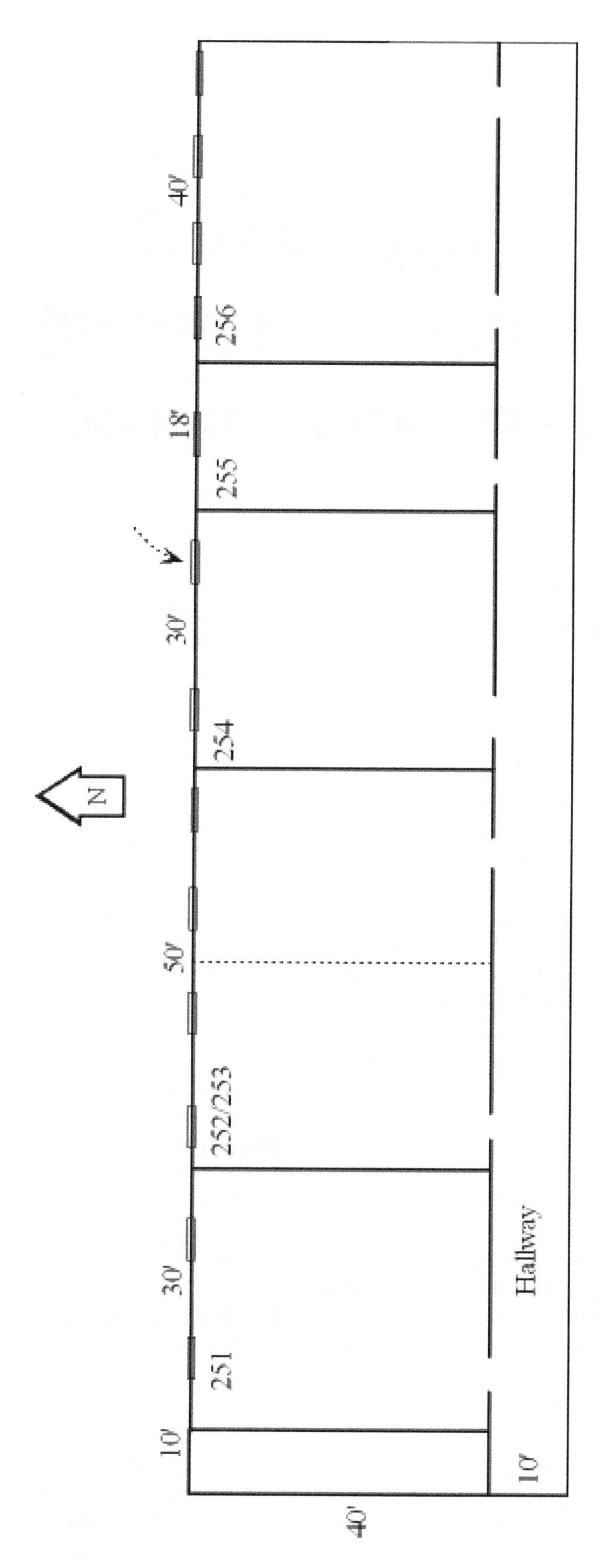

Figure C.7 Classroom wing floor plan.

Appendix D
Developing Engineering Tools with Computer Spreadsheets

VISUAL BASIC MACROS FOR ENGINEERING CALCULATIONS

The use of basic programming language subroutines (a.k.a. macros) can improve the flexibility and function of spreadsheets for engineering calculations. Values from spreadsheet input cells can be passed to subroutines or functions so that complex formulas can be arranged on several lines of code rather than being bunched up on a single line in a spreadsheet cell. For example, the dry-bulb temperature, wet-bulb temperature, and elevation can be passed to a function. The function could use the equations in the *2005 ASHRAE Handbook—Fundamentals* to return relative humidity, dew-point temperature, or humidity ratio to the main spreadsheet program. The spreadsheet could use these values to compute more straightforward quantities (enthalpy, density, specific heat, etc.). A simple example function for pump horsepower in I-P units will be demonstrated.

To activate this tool:

1. Open a "New" Excel®workbook (blank).
2. Click on "Tools" and scroll down to "Macro" and then "Visual Basic Editor."
3. If a long set of instructions appears after this step (or the following step), click on the small "x" in the upper right corner to clear the window.
4. In the project window (that appears on the left side of the screen), click on the Excel®symbol that says "ThisWorkbook." Click on "Insert" and then "Module" and a blank worksheet will appear.
5. Type the word "Function" followed by the variable that is to be calculated and the required input variables in parentheses. For the pump power example: Function PumpHP(HeadinFeet, FlowinGPM, EffinPercent). Then depress the "Enter" key.
6. "End Function" should appear automatically on the worksheet and the name of the variable should appear in the small right window above the worksheet.
7. Type the code to compute the value: PumpHP = (HeadinFeet * FlowinGPM) / (3960 * EffinPercent / 100).
8. Save the file and then close it and return to Excel®
9. On the spreadsheet, define cells (or create names) for the three input variables (HeadinFeet, FlowinGPM, EffinPercent).
10. Go to the cell in which you want to compute the pump horsepower and in the function mode (click on the window after "*fx*" at the top of the worksheet) type:

= PumpHP(HeadinFeet, FlowinGPM, EffinPercent)

The pump horsepower will appear in the cell if these steps have been followed.

DROP-DOWN BOXES

1. Click on the cell that you want to contain a drop-down box.
2. In "Toolbar" on the file menu, go to Data →Validation.
3. When "Setting Tab" appears, click on "Drop-Down Box" and select "List."
4. In the "Source" box, type in the list of words, numbers, or the combination of them that you want to appear in the drop-down box, separated by commas. For example: if you want to select a type of exterior building covering, type: "brick, vinyl siding, 3/4 in. hardboard, 5/8 in. T-111 plywood."
5. You can also put in an optional message for users by selecting the "Input Message" tab.

An example of a typed list is shown in the following example for three types of interior duct surfaces: PVC, Galv Steel, Fibrous Liner.

USING DROP-DOWN BOXES IN MACROS OR "IF" STATEMENTS

1. After completing the "Drop-Down Boxes" procedures, either go to a macro or write a set of nested "if" statements in a spreadsheet cell (maximum of 7 "ifs" allowed in current version of Excel®).
2. In a macro, recall that lists made of text must be set off with quotation marks.
3. Using the drop-down box list in a macro would look something like:

```
Function Roughness (e)
 'Roughness (e) in feet
 If roughness = "PVC" then e = 0.0001
 ElseIf roughness = "Galv Steel" then e = 0.00025
 ElseIf roughness = "Fibrous Liner" then e = 0.01
 End If
End Function
```

4. Call the macro from a cell to display the value selected by the user in the drop-down box. For example, if the drop-down box was written in cell F14:

   = Roughness (F14)

5. Or use the macro to compute a value in a cell. As an example, the relative roughness can be computed if the diameter is also known:

   = Roughness (F14) / Diameter

In this example, the value in the macro was computed from a single variable from a drop-down box. However, values can be computed from several other variables that do not have to come from drop-down boxes. Consider the PumpHP macro in the first example above. The values for HeadinFeet and FlowinGPM could be entered as numerical values that are typed into a cell, while the value for EffinPercent could be provided via a drop-down box.

Nomenclature—HVAC Terms, Abbreviations, and Subscripts

AC	alternating current, air cooled, or air-conditioning
adp	apparatus dew point
AAHP	air-to-air heat pump
ACCA	Air-Conditioning Contractors of America
ADPI	Air Diffusion Performance Index
AHU	air-handling unit
AIA	American Institute of Architects
ANSI	American National Standards Institute
ARI	Air-Conditioning and Refrigeration Institute
ASD	adjustable-speed drive (a.k.a. variable-speed drive, VSD)
ASHRAE	American Society of Heating, Refrigerating and Air-Conditioning Engineers
BACnet®	building automation and control networking (ASHRAE Standard 135-2004, open control protocol)
BAS	building automation system
BEP	best efficiency point
bhp	brake horsepower
Btu/h, Btuh	heat rate unit (British thermal units per hour)
c	cooling
C	loss coefficient (duct fittings)
CAV	constant air volume (flow rate)
C_v	flow coefficient (flow in gpm that results in Δp = 1.0 psi)
CCAF	cycling capacity adjustment factor
CF	correction factor
CFC	chlorofluorocarbon (refrigerants)
cfm, CFM	cubic feet per minute (airflow rate)
CLF	cooling load factor
CLTD	cooling load temperature difference (°F)
COP	coefficient of performance (watts/watt)
CO_2	carbon dioxide (concentration that is frequently used as an indicator of ventilation air level)
CT	(cooling tower or current transformer)
CWS	chilled water system
Δ	delta (difference)
Δh, ΔH	differential head
Δp, ΔP	differential pressure
D	diameter
dB	decibel (sound power or pressure) or dry bulb (temperature)
db	dry bulb (temperature)
D-B	design-build (method of managing building projects)
DDC	direct digital control
DOAS	dedicated outdoor air system
dp	dew point or differential pressure
DX	direct expansion (of refrigerant)
E	energy (electrical ≡ kWh or thermal ≡ Btu)
EER	energy efficiency ratio (Btu/W·h or MBtu/kWh)
EFLH	equivalent full-load hours (c ≡ cooling, h ≡ heating)
EIA	Energy Information Administration (US Department of Energy)
E.I.T.	engineer in training
EMS	energy management system
EPACT	Energy Policy Act (1992)
ESP, esp	external static pressure (in. of water)
f	frequency (Hz, cycles per second)
FC	fluid cooler
FCU	fan-coil unit
FPI	fins per inch (for finned-tube coils)
FPVAV	fan-powered variable air volume
ft	feet (distance or unit of head [ft of water])
GCHP	ground-coupled heat pump (a.k.a. closed-loop GSHP)
GHG	greenhouse gas
GHP	geothermal heat pump (a.k.a. GSHP)
gpm, GPM	gallons per minute
GSHP	ground-source heat pump
GWP	global warming potential
η	efficiency
h	heating (Btu/h, kW), head of liquid (ft), specific enthalpy (Btu/lb), heat transfer coefficient (Btu/h·ft^2·°F)

H	heat (Btu, J); enthalpy (Btu)
HCFC	hydrochlorofluorocarbon (refrigerants)
HDPE	high-density polyethylene (piping material)
HFC	hydrofluorocarbon (refrigerants)
HRU	heat recovery unit
hp, HP	unit of power (horsepower = 0.746 kW) or heat pump
HSPF	heating seasonal performance factor (Btu/W·h)
HVAC	heating, ventilating, air-conditioning
Hz	frequency unit (cycles per second)
IAQ	indoor air quality
IAT (t_i)	indoor air temperature
ID	inside diameter
IESNA	Illuminating Engineering Society of North America
K	turbine flow meter constant (cycles per gallon)
kW	kilowatt (unit of power or heat rate)
kWh	kilowatt-hour (unit of electrical energy)
kW/ton	electrical demand per unit cooling capacity ($kW_{refrig.}/kW_{elect.}$)
L, l	latent heat or liter
L_p	sound pressure level (dB)
L_w	sound power level (dB)
LMTD	log-mean temperature difference (°F)
Lpm	liters per minute
MBtu/h or MBtuh	heat rate unit (British thermal units per hour × 1000)
MERV	minimum efficiency reporting values (for air filters)
MW	molecular weight
μ	fluid viscosity (lb/ft·s)
NC	noise criteria
NEMA	National Electrical Manufacturers Association
NFPA	National Fire Protection Association
NFRC	National Fenestration Rating Council (fenestration windows)
NIST	National Institute of Standards and Technology
NPSH	net positive suction head (ft of liquid)
OA	outside air (a.k.a. ventilation air)
OAT (t_o)	outdoor air temperature
OD	outside diameter
ODP	ozone depletion potential
PE	professional engineer or polyethylene
PLF	part-load factor
ppm	parts per million (typically used to describe CO_2 concentration)
psi	pounds per square inch (unit of pressure)
PTAC	packaged terminal air conditioner
PTHP	packaged terminal heat pump
PVC	polyvinyl chloride (piping material)
q	heat rate (Btu/h or kW)
Q	volumetric flow rate (gpm, cfm, Lps, m^3/s)
R	thermal resistance (a.k.a. R-value ≡ $h \cdot °F \cdot ft^2/Btu$, $°C \cdot m^2/W$)
R_a	gas constant for air ($ft \cdot lb_f/lb_m \cdot °F$)
RAC	room air conditioner
RC	room criteria
Re	Reynolds number ($Re = \rho DV/\mu$)
RH	relative humidity (%)
RTU	rooftop unit
rpm, RPM	revolutions per minute
RMS	root mean square
ρ	density (lb/ft^3)
s	specific entropy (Btu/lb·°F)
S	entropy (Btu/°F)
Sch	schedule (pipe dimension)
SEER	seasonal energy efficiency ratio (cooling) @ 82°F OAT Btu/W·h)
SC	shade coefficient
SCL	solar cooling load factor ($Btu/h \cdot ft^2$)
SHGF	solar heat gain factor
SHR	sensible heat ratio
t, T	temperature (°F,°C)
TC	total cooling (capacity)
TEWI	total equivalent warming impact
TH	total heating (capacity)
TP	total pressure (also p)
TSP	total static pressure (also p_s)
T-stat	thermostat
TXV	thermostatic expansion valve
ton	cooling capacity (12,000 Btu/h, rate required to freeze 2000 lb of water (32°F) in 24 hours)
tsp	total static pressure (inches of water, Pa)
u	specific internal energy (Btu/lb)
U	internal energy (Btu)
V	velocity (fps, fpm, m/s) and in some cases volumetric airflow (ASHRAE Standard 62.1)
VAV	variable air volume (airflow rate)
VFD	variable-frequency drive (a.k.a. VFD)
VP	velocity pressure (also p_v)
VS	variable speed
VSD	variable-speed drive (a.k.a. VSD)
w, W	power (kW)
WAHP	water-to-air heat pumps
wb	wet bulb (temperature)
w.c.	water column (inches of water head)
WC	water cooled
WLHP	water-loop heat pump
WWHP	water-to-water heat pump
x	mole fraction

Index

D

E

F

G

H

I

J

L

M

N

O

P

R

S

T

U

V

W

Z